Noise Control
Manual for
Residential Buildings

Other McGraw-Hill Titles of Interest

Noise Control
Manual for
Residential Buildings

David A. Harris
Building & Acoustic Design Consultant
Port Ludlow, Washington

Boston, Massachusetts   Burr Ridge, Illinois
Dubuque, Iowa   Madison, Wisconsin   New York, New York
San Francisco, California   St. Louis, Missouri

Library of Congress Cataloging in Publication Data

Harris, David A., 1934-
    Noise control manual for residential buildings / David A. Harris
       p.   cm.
    Includes index.
    ISBN 0-07-026942-4
    1. Dwellings—Soundproofing. 2. Dwellings—Environmental
Engineering. 3. Acoustical engineering. 4. Dwellings—Noise.
I. Title.
TH1725.H37   1997
693.8'34—dc21
                                           97-9504
                                           CIP

McGraw-Hill

A Division of The McGraw-Hill Companies

3 4 5 6 7 8 9 0 BKM BKM 9 0 9

ISBN 0-07-026942-4

*The sponsoring editor for this book was Zoe G. Foundotos, the editing supervisor
was Bernard Onken, and the production supervisor was Suzanne W. B.
Rapcavage. It was set in Times by David A. Harris.*

McGraw-Hill books are available at special quantity discounts to use as premiums and sales promotions, or for use in corporate training programs. For more information, please write to the Director of Special Sales, McGraw-Hill, 11 West 19th Street, New York, NY 10011. Or contact your local bookstore.

This book is printed on recycled, acid-free paper containing a
minimum of 50% recycled, de-inked fiber.

TABLE OF CONTENTS

INTRODUCTION

Today's residential noise levels demand new standards of construction to reduce noise. Methods of entertainment, modes of travel, time-saving household conveniences, and sophisticated equipment generate a cacophony of sounds in our residences. Most of these sounds are unwanted, thus the definition of <u>noise</u> is unwanted sound. When the word <u>sound</u> is used, it means the broad sense of the term, something we hear. When the word noise is used it always means the sound is unwanted. Unfortunately, most of us do not make this distinction in our everyday lives. We say noise when we mean sound and vice versa. Furthermore, many sounds may be noise to you and beautiful sound to your neighbor. The definition is in the mind of the listener. For example, you may cherish old love songs while your children may consider this noise. Conversely, your children may love to play hard rock at elevated sound levels, making it noise to others. The sound of dripping water may be noise when you're trying to sleep but a valuable clue to the plumber trying to find a leak.

As the sound levels have increased in our society, more of us call it noise. We live with an increasing level of noise in our work, play, and travel. These noise levels are generally accepted as a part of our advancements. However, the noise level has now reached the limit for many individuals, and we now see increased efforts to quiet airplanes, trucks, the workplace, and even our personal transportation. The belief is very strong that we can quiet them and that we deserve peace and quiet, especially where we conduct our personal lives and where we sleep. There is a significant and increasing demand for quiet in our residences. Public knowledge of noise control has also increased, as evidenced in studies that show people want good sound isolation between rooms and demising walls with their neighbors.

This book is dedicated to the quest for quiet in residences including single-family houses, apartments, condominiums, dormitories, motels, and hotels. It is intended as a guide for the home builder, architect, and designer in designing quiet buildings. Since the professional builder cannot always anticipate user needs, this book is especially designed to assist the homeowner; condominium management/owner; and apartment, hotel or motel facilities manager or owner to communicate their specific needs to the architect, designer, and builder. The "do-it-yourselfer" will also find this book helpful, since solutions are provided in easy-to-understand terms. Formulas and mathematical design concepts have been relegated to notes and references. They have not been eliminated, so that the reader may be able to utilize them when and if needed to communicate with the professionals including architects, builders, acousticians, community leaders, environmental specialists, and related trades persons.

Noise control or noise mitigation efforts for various situations are emphasized. Starting with a simple understanding of sound, we discuss how sound is produced, how it travels, how we measure it, and some common concepts. Huge technical texts, loaded with mathematical models, and lengthy technical treatises have addressed this subject. This discussion is limited to basic concepts needed to understand how sound propagates from one place to another in residences. Sound levels, and how to reduce them within a room with sound-absorbing materials are addressed. Solutions provide an explicit definition of the materials required, how much is required, and how to apply them for noise control and good hearing acuity in your home. Sound barriers, or walls that separate a noise source from a receiver, are explained in simple terms. Once the difference between barriers and absorbers is clear, other concepts in acoustics will be easier to understand. Solutions for mitigating noise from all types of sources including your neighbor's loud party or stereo, airplanes, trucks, or a nearby noisy factory are all quantified in precise acoustical specifications.

This book is dedicated to the person who has a desire to find that elusive environmental attribute called "peace and quiet." Even in this time of environmental awareness, we tend to overlook the impact that noise has on our physical and psychological well-being. Noise is one of those environmental elements that is not considered hazardous to our health. We humans have a tremendous capacity for accommodating noise. However, this capacity is rapidly reaching our limits. We already have caused massive hearing loss in our young with the use of boom boxes and personal earphones, loud workplaces, noisy transportation, and the like. It doesn't need to be this way. With a few sound principals and a bit of dedication we can easily preserve our acoustical environment so that we can hear the sound of the spotted owl as it flutters about in an old-growth forest. Better yet, we can keep the noise levels low enough to be able to wake up to the birds singing, a breeze rustling tall grass, or the cooing of a baby sleeping quietly in the next room. If these sounds are music to your ears but you do not want to give up the technological advances that have made our lives easier and more pleasant, this book will provide you with an opportunity to have the best of both environments. A bit of knowledge, common sense, and a pair of good ears are the basic tools for attainment of these goals.

The author's quest for quiet has encountered many interesting pathways. Born and raised on a farm in Wisconsin with no noisy machinery, I soon learned about noise in the army where an M1 rifle and a 155 howitzer left me with a minor but permanent hearing loss. Trained as a builder, most of my work experience was research and development of building materials and systems of construction. This assignment provided a unique opportunity to build and test many systems of construction for their fire resistance, structural capability, and, significantly, their ability to provide good acoustical attributes.

Sixteen years' of experience in technical market development provided an understanding of the building materials, architectural acoustics, and noise control market. Introducing new materials, concepts, test procedures, and systems of construction was an eye opening experience. Heretofore, the author believed that "If you invented a new mousetrap, the world would beat a path to your door." Market development techniques quickly demonstrate why the inventor is lost in the shuffle. The concept; "Create a need and fill it" takes on great meaning. Without good market analysis, planning, and product introduction, a wonderful invention becomes a waste of time in today's business world. Likewise, marketing programs that become too aggressive, lead to products best described as "super duper sound suckers" with an emphasis on the last word.

As an acoustical consultant for the last 10+ years, I continue to be amazed by how lackadaisical we are as a society about acoustics and noise control. While everyone agrees that noise is a significant environmental issue with physical, political, and economic overtones, we in North America seem unconcerned about noise. Pollutants that make headlines and receive the "big bucks" tend to be those that cause cancer, like asbestos. We fail to recognize that noise will cause equal pain and suffering. As a result we spend billions on environmental hazards but cannot find a few dollars for noise control.

While these experiences provided technical knowledge, it was experience as a homeowner that most qualifies me to write this book. Each new home (we moved 16 times in my career) provided a new challenge in attaining an acceptable level of "peace and quiet". These experiences have proven beyond a shadow of doubt that a quiet solution does not require a height tech solution. The best solutions are found by using the KISS principle: Keep It Simple, Stupid.

After 35 years of experience in this industry, the author is annoyed by the slow progress in the residential construction industry. Unlike the computer technology explosion and several other industries, we still build our residences in a fashion very similar to those constructed by the second wave of settlers. While the first settlers used logs, stone and mud, those that followed had saws to make boards and nails as fasteners. Gypsum wallboard, plywood, and composition roofing materials became common in the early 1900s. The prized home of today is still traditional. Likewise, noise control became a full science in the 1950s with many of the sound transmission loss tests still being referenced today. Yes, we have advanced the electronics used to measure sound and we have a wide range of test procedures. However, the basic concepts developed in the 1950s are still valid. Even the basic units are the same. Moreover, a decibel is the same unit everywhere in the world.

Yes, things have changed. Between 1950 and 1985, building material manufacturers spent a substantial budgets on research, development, testing, and technical marketing. Architects and designers became dependant on manufacturers to provide state-of-the-art expertise and test data. Because of the slump in the economy, corporate takeovers, downsizing, etc. , funding disappeared for research and technical service in the building industry. Architects, designers, and builders must now either do their own testing, development, and technical analysis or rely on an independent expert or consultant. They are ill-equipped to handle the technology on their own. Those who cannot afford to employ an expert simply ignore acoustics all together or rely on antiquated information. As a result, we have experienced a rollback to the 1960s in terms of architectural acoustics. It is this regression that causes the author the most dismay. Hopefully, this book will fill the enormous gap in the knowledge of acoustics by today's builder and designer. More importantly, it is my sincere hope that buildings of tomorrow will provide the "peace and quiet" that everyone deserves.

David A. Harris

Chapter 1 NOISE CONTROL PRINCIPLES

1.1   INTRODUCTION AND BASIC ACOUSTICAL CONCEPTS

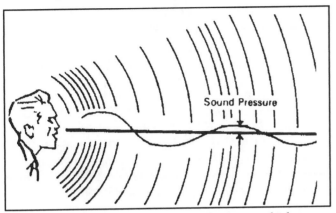

Figure 1-1   Our voice produces sound pressure which increases surrounding air pressure that travels in waves.

Noise is unwanted sound. What may be beautiful music to you may be noise to another and vice versa. Note that noise is always a sound but not all sounds are noise. Sound is produced when an object or surface vibrates rapidly enough to generate a pressure wave or disturbance in the surrounding air (Figure 1-1). The primary sound sources we experience are composed of vibrating surfaces and turbulent air (Figure 1-2). Some examples of a vibrating surface are a tuning fork, drum head, vibrating reed, a resonating metal panel on a dishwasher, and our vocal chords. Examples of turbulent air include a jet engine, compressed air escaping from a nozzle, and the soft sound of a breeze in pine trees.

Many sounds are a combination of both a vibrating surface and turbulent air. For example, a brass instrument makes sound by a combination of the lips vibrating in the mouthpiece and turbulent air both in the mouthpiece and down the tubes of the instrument. In the case of musical instruments, the bass sounds are amplified by one or more devices. A violin uses a wood cavity to amplify and shape the quality of the sound from the vibrating string. A tuba amplifies the sound of the vibrating lips and turbulent air by amplifying and focusing the sound down a long narrow pipe. For those who have experienced the didjiridu, a fascinating instrument of the Australian aborigine, voice sounds are amplified via a long hollow wooden tube to produce the unusual, monolithic, and eerie music.

In buildings, the sound of the furnace or air-conditioning fan is an example of turbulent air. Turbulent air is created in two ways. At the source, the fan blades disturb the air to create air movement down the duct. At the air outlet or register, sound may be generated by the flow of air over the turning vanes or as the duct turns a corner. The duct tends to amplify these sounds unless treated with sound-absorbing materials. This same fan may also be the source of vibrating surfaces if the fan housing is made of loosely fitting parts. Liquids under pressure also produce sounds in a fashion similar to turbulent air.

With the advent of electronics, we now

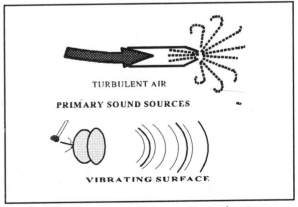

Figure 1-2   Primary sound sources are vibrating surfaces and turbulent air.

1

have the option of electronically amplifying sounds. Examples are the radio, stereo, public address system, and electric guitar. In the case of an electric guitar, the vibrating string is sensed by an electronic amplifier that sends a signal to a speaker, which in most cases is an example of a vibrating surface activated by the electronic signal.

Sound is a form of energy. The intensity of sound could be measured in classical energy units. However, the amount of energy in a sound is a very different scale than we normally encounter. For example, the sound of 100,000 persons yelling at the top of their lungs at a sporting event might generate enough energy to light a 100-watt light bulb in the center of the field of play. Likewise, the human ear hears sounds on a different scale.

Sound levels are usually measured on a scale that corresponds to the way in which the ear responds to loudness. The units we use universally are called decibels. This is an international unit that is the same in English or metric. The abbreviation for decibels is dB. One decibel is one-tenth of a bel. This measure of sound loudness or sound intensity was named after Alexander Graham Bell. The decibel scale expresses ratios of sound as factors of 10 greater or less than a reference sound. For example, a power amplifier with a gain of 10 dB has multiplied the input power by a factor of 10. Likewise, an amplifier with a gain of 40 dB has multiplied the power at the input by a factor of 10,000 (i.e., 10 times 10 times 10 times 10) (Figure 1-3).

APPARENT LOUDNESS	RELATIVE LOUDNESS	SOUND LEVEL (dB_A)	SOUND PRESSURE (psi)	COMMON SOUNDS
Deafening	128	130	0.01	Threshold of pain, jet plane takeoff
Very loud	32	110	0.001	Thunder, artillery, nearby riveter, elevated train, boiler factory
Loud	2	70	0.00001	Noisy office, average street noise, radio/TV
Moderate	½	50	0.000001	Average home/office, quiet conversation, quiet radio/TV
Faint	⅛	30	0.0000001	Quiet home/office, quiet conversation,
Very faint	1/32	10	0.00000001	Rustle of leaves, whisper, THRESHOLD OF AUDIBILITY

Figure 1-3   Relationship of common sounds to various rating scales that represent both objective (dB scale) and subjective ratings for noise.

Human perception of sound is affected by frequency (or pitch) as well as by the loudness measured in decibels. Frequency is based on the number of vibrations that a specific sound makes in a unit of time (i.e. cycles per second). The international unit of measure (both English and metric) is called hertz, abbreviated Hz. The more vibrations, the higher the frequency. For example, the whine of a mosquito's beating wings has a much higher frequency than the slower vibrations of a distant train rumble (Figure 1-4).

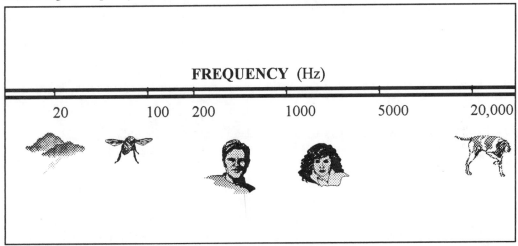

Figure 1-4 Relationship of common sounds to the frequency rating scale. Thunder is approx. 20 Hz, a bee buzz = 150 Hz, normal speech = 250 to 4000 Hz, and a dog whistle = 15,000 Hz.

Generally, sounds below 20 Hz are perceived as shaking. We feel these sounds rather than hear them. The young with good hearing perception may hear sounds as high as 16,000 Hz. Most of us encounter some type of hearing loss as we age. Typically, this loss occurs in the higher frequencies. Frequencies above 10,000 Hz are not heard by a major portion of the population. Speech sounds are prevalent between 500 and 2000 Hz. Most of the sounds that we encounter in buildings occur between 125 and 5000 Hz. Therefore, achitectural acoustics test standards cover test the frequencies between 125 and 5000 Hz (Figure 1-5).

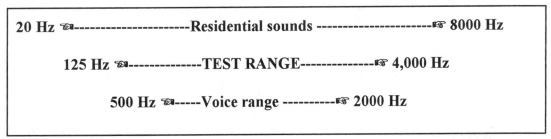

Figure 1-5 Residential Sounds may range from 20 to 10,000 Hz. Tests cover 125 to 5,000 Hz, since these sound frequencies are the most intrusive.

3

We are generally more sensitive to high-frequency sounds than low-frequency sound. To demonstrate this fact, researchers subjected a large group of individuals to a multitude of sounds with frequencies across the entire range of human hearing. They found that two sounds of equal loudness on the decibel scale do not sound the same when they are of a different frequency. For example, if you play a pure tone such as a note on a piano at 250 Hz, it does not sound as loud to the human ear as a sound of equal loudness in decibels at 100 Hz. This perception also changes for louder or quieter sounds. On the basis of this research, acousticians have prepared a set of equal loudness curves (Figure 1-6). Sounds of the various frequencies shown along the curved lines will sound equally loud to the typical human ear. In an effort to make a composite of this complex perception of sounds, acousticians have developed a weighting network for most sound level meters

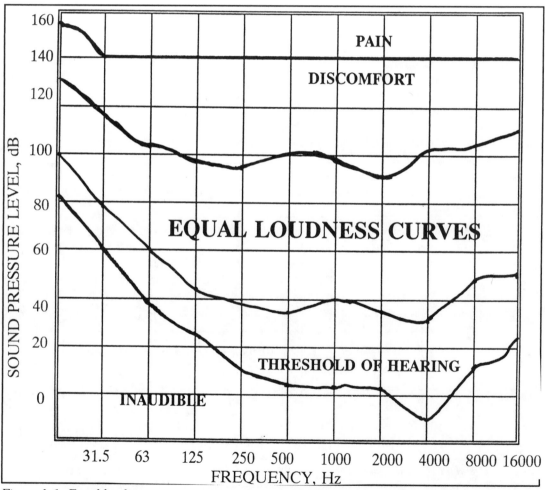

Figure 1-6   Equal loudness contours represent sounds of different frequencies which sound equally loud to the average listener. Actual sound level can be very different.

4

that compensates for the way our ear hears. This scale is known as "A-weighting" and is called the A scale of a sound level meter. If a sound level is given in decibels without any identification of the frequency at which it was measured, most acousticians will assume it means decibels on the A scale.

1.2 GENERAL CONSIDERATIONS

1.2.1 Preliminary Planning

In a typical residential environment, the source of noise may be any number of devices or activities. The noise they generate may be either external or internal relative to the listener's location in the structure. In addition, the noise may be constant, as in the case of an exhaust fan, or intermittent, as are kitchen appliances. **The most practical way to control noise is generally at the source.** Quiet equipment should be selected. If this is not practical, such strategies as placing appliances on a nonvibrating surface, enclosing them, or repositioning noise generators relative to the listener may be employed (Figure 1-7). Source quieting is often a simple matter of turning off the offending noise source. Or it may be a matter of purchasing equipment that is quieter than a competitor's. For example, a well-designed dishwasher that uses quiet-design principles can be dramatically less noisy than a cheaper, poorly designed model. The cost may be only slightly higher for the manufacturer to produce. In general, quieter machines will last longer since operating parts are well-balanced, operate efficiently, and have vibrations reduced to the minimum. If you are purchasing equipment, ask for operating sound levels or better yet listen to an operating machine in a quiet environment. It cannot be overemphasized that purchasing quiet equipment is far cheaper than any measures to quiet the equipment after the fact. The decibels-per-dollar attributes for purchasing quiet equipment are very dramatic. We will dwell on this issue throughout this book.

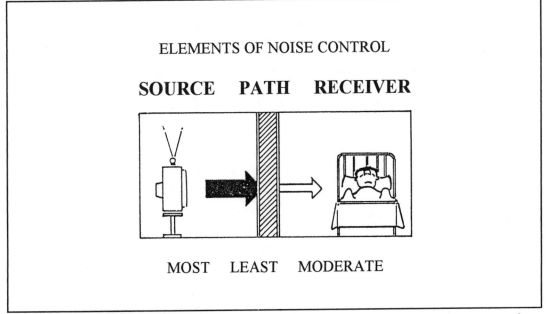

*Figure 1-7 **Source-path-receiver** are the key elements of noise control. Quieting the source is the most practical technique.*

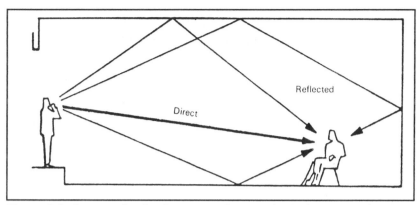

Unfortunately, it is not always possible to control the source of a noise. However, an alternative and effective way to control noise involves the specific paths over which it travels. Paths may be airborne, structureborne, a combination of airborne and structureborne, or reflected.

Figure 1-8 An example of airborne sound is speech. Airborne sound may reach the receiver by direct or reflected paths.

The primary path of sound from a source to a receiver or listener is air. **Airborne sound** is produced by sources which radiate directly into the air, such as voices in conversation (Figure 1-8), street traffic, or the sound of a television or sound center. Builders and designers can control most airborne noise transmission by utilizing floor and wall structures which reduce the level of these noises to the point where they are no longer annoying to the listener (Figure 1-8).

Airborne sound paths may be **direct** or **reflected.** Ordinary sounds may be amplified by reflections in rooms constructed with hard, sound-reflecting materials such as bare wooden floor, gypsum board walls, and uncurtained glass. Along with this annoying amplification comes distortion with echoes. This increased level of noise, also known as reverberant sound, interferes with speech intelligibility. The result is greater sound transmission into other rooms. This problem is treated with sound-absorbing materials such as draperies, upholstery, and acoustical ceiling and wall treatments (Figure 1-9).

The use of sound-absorbing materials for interior room surfaces is not to be confused with construction techniques for reducing transmission between rooms. Acoustical absorption will reduce echoes and sound levels within the source room. Sound barriers contain sound and are covered at length in subsequent chapters (Figure 1-10).

The second path over which sound travels is through the structure itself, by way of structural vibration. This is sometimes called **structureborne sound.** For example, noise can be transmitted by a wall or floor that is set into vibration by direct mechanical contact, as in the case of a heating pipe carrying the vibration from a furnace. Such vibrations may also be transmitted through the structure to adjacent walls and then reradiated as sound in adjoining spaces. Additional examples of activities that can produce structure-borne sound include footsteps, dropped objects, or a slammed door, all typical of impact noise. Hollywood made banging on hot-water pipes a famous example of structureborne sound. The impact of the hammer on a pipe in the basement can easily be heard many floors above because the sound is transmitted over the pipe. Similarly, structureborne noise is also transmitted via structural members such as stiff steel beams (Figure 1-11). Another example is to put your ear to a railroad rail to hear the sound of a train coming many miles away.

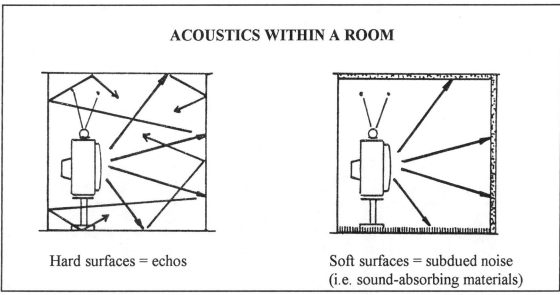

Figure 1-9   Acoustics within a room. Add sound- absorbing material to reduce reverberation (echos) and sound level.

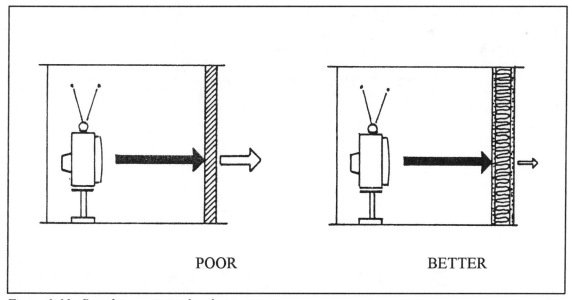

Figure 1-10   Sound transmission loss between rooms.

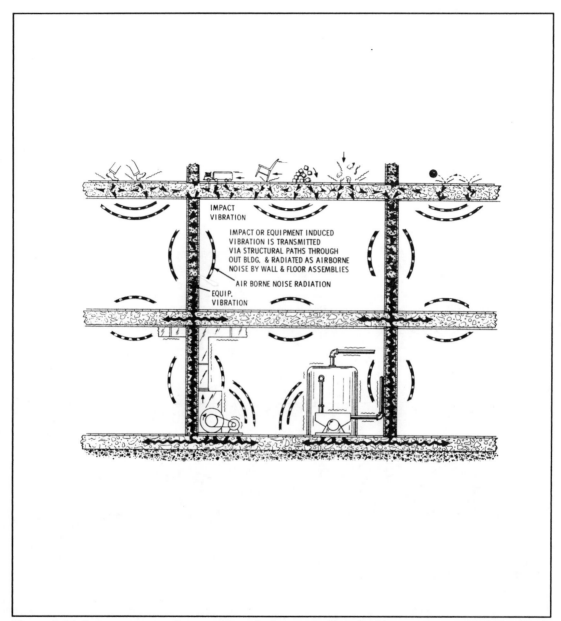

Figure 1-11 Common structureborne noise and vibration paths in buildings. (Source: NBS Handbook 119)

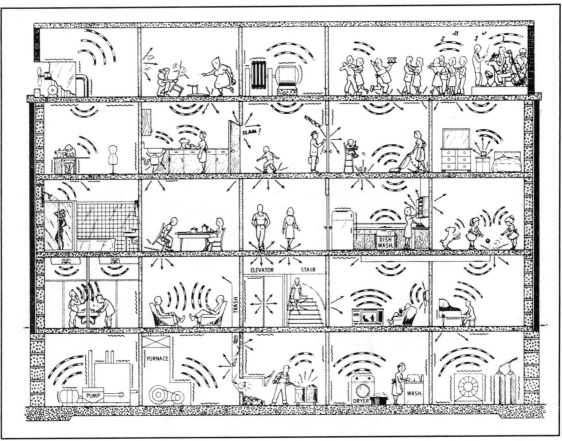

Figure 1-12 Common indoor noise sources may be transmitted via airborne or structureborne paths or a combination of both.

As implied above, sounds may be a combination of airborne and structureborne. For example, a fan may be mounted directly on a surface that resonates or amplifies the sound much like the sounding box of a string instrument. In turn this sound will excite the air surrounding it to transmit an airborne signal to the listener. Likewise, a loud radio may transmit an airborne sound within the source room with such force that it activates the walls of the room. These walls may then transmit the sound via structureborne paths to a far part of a building where the walls of the receiving or listening room activate the air in the receiving room so the sound reaches the ear as airborne sound. It is also possible that a combination of airborne signals and partially structureborne signals may reach the receiver's ears (Figures 1-12 and 1-13).

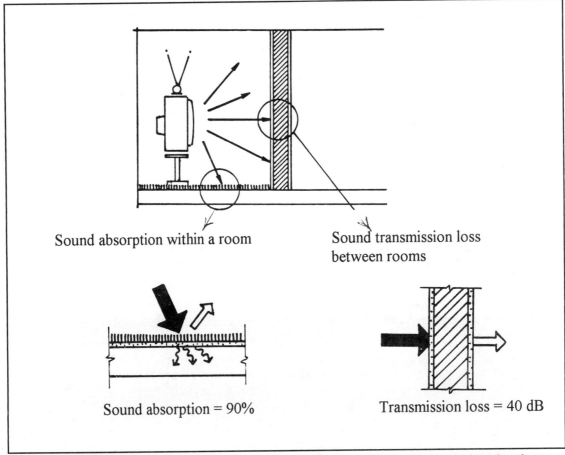

Sound absorption within a room

Sound transmission loss between rooms

Sound absorption = 90%

Transmission loss = 40 dB

Figure 1-13   Sound Absorption vs Sound Transmission Loss. Absorption reduces sounds within the room; transmission loss is improved with a good barrier.

Control of impact noise is treated as a floor problem, since the movement of people and furniture constitutes the major impact noise problem. Floors may be made more acoustically efficient with impact-absorbing carpet & pad. This source treatment is more effective than other basic construction techniques. Note that noise control at the source is the most cost-effective means to reduce noise (Figure 1-14).

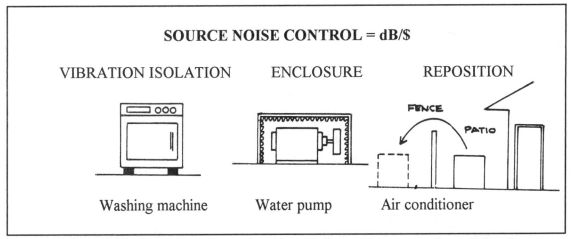

Figure 1-14 Examples of source noise control. Purchase of quiet equipment is more effective than retrofit noise control efforts.

Figure 1-15 A hand held sound level meter (SLM) measures the level of sound in decibels. The A scale is used for most measurements. Many meters can also measure sound levels at individual one-third or one-octave frequencies between 63 to 16,000 Hz. (Photo courtsey Larson Davis Instruments)

An essential ingredient for success in noise control is a **desire** to achieve noise control. Attention to detail and follow-up to see that it is achieved are vital. Effective noise control starts with the inception of the project by working with the site planner and architect, and later by working with the interior designer, general contractor, subcontractors, and construction crews. In many cases the original design concepts may be tempered or changed by local building codes, zoning, local ordinances, and construction requirements. Formal requirements by the many different professionals having input on the design tend to cause changes to the initial design goals.

Building codes, zoning, and local ordinances may include requirements for specific amounts of noise mitigation. In a multifamily building, the demising wall might be required to have a time design fire rating and/or sound transmission class (STC) rating.[1] Design and the technology to achieve a fire rating are quite different than those utilized to achieve a sound rating. Unless the design has been tested for both fire and sound rating, there is a good chance that the STC rating will be compromised and vice versa. Exterior walls may be required to have a particular sound rating, particularly if the residence is near a freeway. The exterior wall may be acoustically acceptable but it may be compromised substantially by poor windows, doors, air vents and the like, also required by code.[2]

While seemingly insignificant to the individual professional, a small change in construction may have a significant impact on the acoustical attributes of the building. For example, the structural engineer might insist that a party wall between two separate living units become a shear wall. If the acoustical design called for resilient attachment of the wallboard or separate, unconnected studs, the revision to a shear wall will likely completely negate the sound transmission loss characteristics of the party wall.

Personal preferences by a contractor can also have a significant negative impact on the acoustical environment of the finished building. A classic example is the heating, ventilation, and air-conditioning contractor (HVAC) who is a member of the Sheet Metal Workers Union. The contractor might install sheet-metal ducts in place of fiberglass ducts. The result will be considerable furnace noise in the living areas. Sound will readily travel down the ducts. And the ducts create a huge flanking path between rooms intended to be acoustically isolated. The ducts act much like a speaking tube used on ships to converse between the engine room and bridge.

The most effective means to combat this situation is to utilize a "systems concept" of design. Instead of specifying particular products using the classic "or equal" phrase, specifications are written in terms of ultimate performance characteristics. Acoustical performance of a wall or floor/ceiling assembly is specified in terms of its structural, fire, and acoustical performance. For example, the HVAC subcontractor is required to certify that the duct system will maintain the acoustical integrity of the barriers. This technique, while limiting to the contractor, will assure that proper attention will be given to the acoustical performance. More importantly, the contractor becomes responsible, morally and financially, to upgrade the final installation to the required acoustical performance criteria. Performance-based specifications allow considerable latitude for substitutions by the contractor while guaranteeing the results meet the design criteria.

1.2.2 Site Planning

Site planners should utilize existing natural sound barriers to locate the structure (Figure 1-16).

---

[1] STC is described in Chapter 2.

[2] Proper acoustical design of exterior apertures is discussed in Chapter 2.

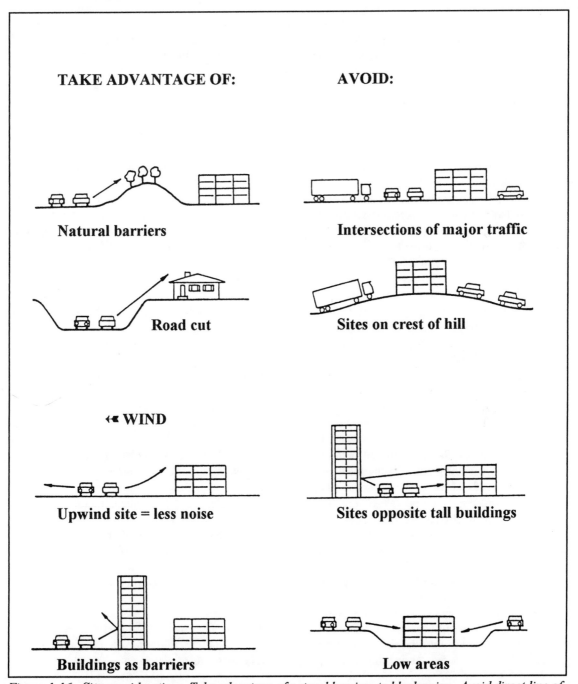

Figure 1-16 Site considerations. Take advantage of natural barriers to block noise. Avoid direct line of sight.

During the site selection process for residential construction, it is essential that the environmental noise conditions at the site be considered. If existing noise data cannot be obtained from regional or local authorities, then an on-site noise survey should be conducted. In California, state law requires every community to have a noise survey. Developers are generally required to conduct a noise survey of their site and delineate noise-mitigating efforts that have been implemented in the site planning to satisfy local noise ordinances. It is important that such a noise survey include measurements made at several locations representative of the conditions at the site, at several times during the day and night, and also several days in the week.

The objective of the measurements is to obtain a record of all the noise intrusions that may impact the site. If a potential noisy neighbor such as a power plant is within sight of the site, it is imperative that measurements be made while the facility is operating. Furthermore, it is the obligation of the person making the sound level measurements to identify whether the site is operating at full, part, or low capacity. It is important that proper procedures be followed in taking noise readings. Therefore, hiring a qualified acoustical consultant who is a member of the National Council of Acoustical Consultants (NCAC) is justified, and may be required, by local authorities. To assist in achieving a comprehensive statement of the site noise, the following sources may be helpful:

- The local airport authority can provide flight count data, aircraft types, and flight patterns for landings and takeoffs under varying weather conditions. Most major airports have a noise contour map showing specific loudness levels.

- Highway commissions and local city planners should be consulted about future expressways, turnpikes, or major traffic arteries. Ask for environmental impact studies and review the "Noise Element" section.

- Ask the local and regional planning authorities to provide information concerning zoning requirements and noise levels. They may have data generated for their jurisdiction regarding present and predicted general and local noise levels. This information may be particularly helpful in identifying a row of shops, stores, office buildings, etc. that may act as a local noise shield from a noisy expressway.

- Building codes in many areas, particularly in the western states such as California, have adopted lot line noise level requirements. This information may lead to the requirement for sound barriers or solid walls of considerable height and length. This information may be located in the local, regional, state, or national building code documents depending on the circumstances governing the site. It may also be contained in the governing municipality noise ordinance.

- Zoning boards may establish minimum lot size, land density, and related information governing the site. This can affect the cost of land in relation to solving sound control problems. As an example, planned unit developments may include considerable public use areas for golf courses, greenbelt areas, playgrounds, lakes and the like.

All this information will assist in establishing the existing and potential noise elements, that presently, or in the future, will impact the site. It is required to design appropriate noise-mitigating measures. Only when the foregoing information is integrated with site noise measurements, terrain, and occupancy, can a reasonable noise-mitigating plan be developed. For a large project, one of several computer analysis programs may be utilized.. Chapter 6 provides details for site noise mitigation.

1.2.3 Outdoor Noise Considerations

Obviously, it is advantageous to locate a building as far away from the noise source as is feasible. However, the rule of thumb, doubling the distance, may be changed by a number of factors. Placing a barrier between the source and receiver is a common procedure. To be effective, noise barriers must be massive and airtight. An example of a poor acoustical barrier is a row of trees. While trees and shrubs are good sound absorbers, they make poor sound barriers. As sound travels through a belt of trees, some attenuation beyond the normal attenuation for doubling of distance does occur. However, the amount of increased attenuation from trees is usually insignificant, particularly if they are deciduous. Only if the trees are densely packed together for a distance of many yards will they be effective sound barriers (Figure 1-17).

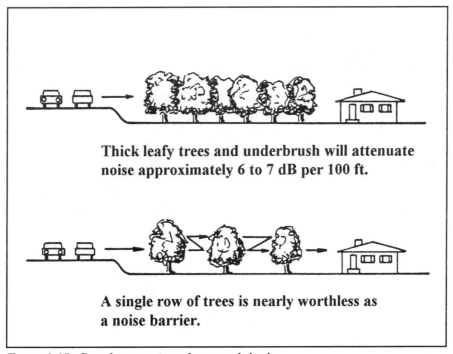

Thick leafy trees and underbrush will attenuate noise approximately 6 to 7 dB per 100 ft.

A single row of trees is nearly worthless as a noise barrier.

Figure 1-17   Sound attenuation of trees and shrubs .

Air is the major carrier of sound waves outdoors. While some sound may be carried through the ground and via structures, nearly all sound transmitted outdoors is airborne. Airborne sound energy dissipates with distance. There is roughly a 6-dB drop in sound level each time the distance is doubled between a source and the receiver. When the sound source is composed of multiple sounds, sound attenuation will be less. Multiple distant sources, such as a highway or railroad, are called a <u>line source</u>. Line source noise is attenuated approximately 3-dB each time the distance is doubled (Figure 1-18)

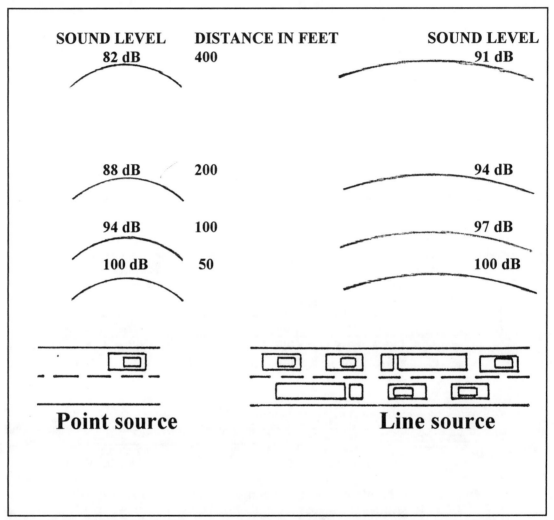

Figure 1-18 Each doubling of distance attenuates sound from a point source by 6 dB. 3 dB attenuation per doubling of distance is typical for line sources.

Note that the hard, smooth surfaces of driveways, parking lots, water such as a pond or pool, and masonry buildings are excellent sound reflectors. Sound will bounce around the barrier, causing overall sound attenuation to be no better than direct sound in the open air.

1.2.4 Walls or Sound Fences

To reduce sound propagation out-of-doors, an obstacle such as a building, wall, or mound of earth can be placed between the noise source and the building. This is a fairly effective way to attenuate high-frequency sound. However, low-frequency sound will be easily diffracted (bent) around the obstacle. Similarly, all sounds may be reflected around the obstacle from another hard-surface material nearby. Because of reflections and diffraction effects, sometimes called "flanking," it is highly unlikely that the barrier will attenuate sounds by more than 15 dB. Except in highly unusual situations, outdoor noise barrier construction can generally be lightweight. Barrier attenuation need be no better than flanking attenuation.

When designing an outdoor sound barrier, the single most important factor is that the barrier must be continuous and have no gaps or holes. The barrier must be high and wide enough to break the line of sight between the source and receiver. Higher barriers are more effective since they create a larger acoustical shadow. Likewise, placing the barrier closer to the source produces a larger acoustical shadow (Figure 1-19). Building orientation to the noise source can be a very effective technique for providing sound attenuation. Examples of poor orientation and better orientation are shown in Figure 1-19.

Of secondary importance, the barrier should be faced with a sound-absorbing material, especially the side that faces the sound source. Sound absorption is particularly vital when the barrier is oriented in such a fashion that sound will be reflected around the end or over the top of the barrier.

When tall buildings are used as a barrier, it is very common to find them not effective because an adjacent structure acts as a flanking path. Sound reverberation is also a factor where buildings are closely aligned. As shown in Figure 1-19, sound will tend to bounce back and forth between parallel surfaces. This effect can actually amplify the noise source in special circumstances. In many instances, reflected sounds can cause occupant noise complaints on the back of a structure in an apparent acoustical shadow.

The addition of sound-absorbing materials to the reflective surfaces will reduce this condition. Unfortunately, weatherproof sound-absorbing surfaces are difficult to achieve. If the buildings are but a story or two high, shrubs and trees will help. The building floor plan can be designed to allow for tall plants adjacent to difficult building locations. A cypress hedge 2 ft thick or an old-growth ivy-covered wall will absorb enough to avoid sound reflections. Other dense foliage and proper selection of evergreens will also provide sound absorption. Note that deciduous plants provide little or no absorption in their dormant period.

1.2.5 Working with the Architect and Acoustical Engineer

An architect and acoustical engineer should be brought into the project **before** a commitment on the site is firm. Working with the planner in the very early stages will provide the architect the opportunity to consider building floor-plan layouts, foundation orientation, the degree of acoustical privacy or performance, and cost. When acoustics and noise control are key elements to satisfying client needs or code restrictions, an acoustical consultant, acoustician or noise control expert who is a member of the National Council of Acoustical Consultants (NCAC) should be a member of the design team.

The decibel-per-dollar impact of developing noise mitigation requirements prior to site selection can be quite dramatic. One of the most devastating events that can happen in a design process is to find, after considerable design effort has been expended, that the site will be next to a proposed airport or under the path of the supersonic transport plane. Site acoustical measurements and careful study of land use plans in the area

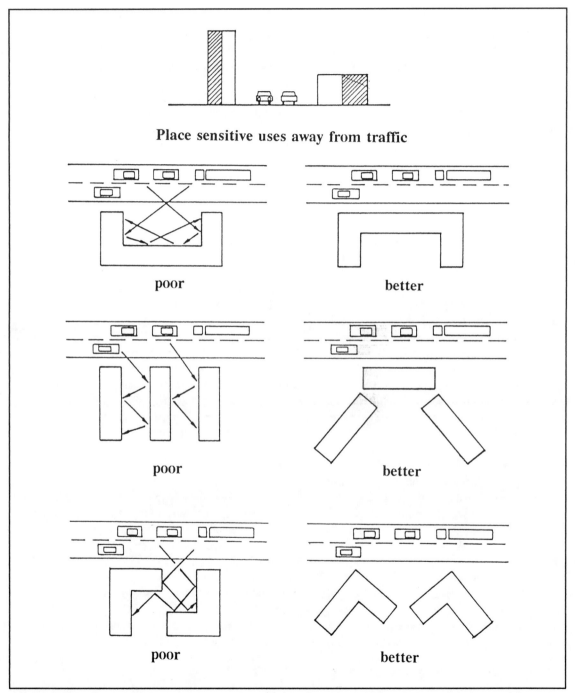

Figure 1-19 Using buildings as barriers. The larger the building the larger the acoustical shadow.

18

may have a significant impact on the ability to meet existing residential building codes.[3]

Acoustical engineers must have direct access and respect of the entire design team. If there is to be an environmental impact analysis, it must contain a "Noise Element" section[4]. Done correctly, a "Noise Element" will discuss all the initial design considerations and provide specific noise mitigation efforts required to make the site compatible with the intended use. While the solutions are complex, the technology is straightforward. With the acoustician as an equal member of the initial design team, it may be possible to incorporate simple design changes up front for little or no additional project cost. Retrofit noise control is notoriously expensive. This subject will be covered in greater detail, along with many examples, in subsequent chapters.

1.2.6 Community Noise Sources

A properly done environmental noise survey should be conducted in conjunction with the site planning process. This survey will identify existing and potential noise sources that will impact the site. It should also identify specific criteria established by code, zoning, or local ordinances. More importantly, proposed sound levels on site are compared with existing criteria for typical occupancy and use. For example, the noise survey report will provide the information required to develop sound attenuation requirements. This information is essential to develop acoustical performance attributes for the noise barriers, building orientation, distances, exterior envelope attenuation, buffer zones, exterior materials, landscaping, and a multitude of site development items.

1.2.7 Interior Noise Sources

Within the building, the noise survey will assist the design team in providing interior acoustical buffers. Corridors, service elevators, equipment rooms, storage facilities, garages, and other facilities where high noise levels will not evoke user complaints can be located on the noisy side or end of the building to isolate more noise sensitive living areas. Within the living unit, kitchens, baths, and utility rooms can be situated on the noisy side. By contrast, bedrooms, dens, conversation areas, offices, reading areas, or media centers might be best located on the quiet side of the building.

Interior courtyards can be noisy due to reverberant sound. Noise is radiated into the courtyard from the surrounding living units. Leaving one end of the court open will reduce the noise level, and, of course, the use of sealed windows, doors, and vents on the court side of the living units will help stop noise in both directions.

---

[3] As a case in point, a single-family residential developer purchased a piece of prime beach property in Los Angeles. After the development was fully constructed, the City of Los Angeles decided to turn the Los Angeles Airport (LAX) into a major international hub. The development was located directly under the immediate takeoff pattern of the then newly introduced jet aircraft. After 20 years of litigation, complaints, and occupant grief, the City of Los Angeles bought the entire development of hundreds of homes, tore them all down and declared it a no use zone because of the noise levels. Billions of taxpayers dollars were expended just to purchase the land at inflated real estate prices. Everyone lost money on this debacle, including the homeowners, landowners, and technical experts. One could easily call it the largest acoustical disaster in modern history. Even more amazingly, 30 years later, another builder purchased property next to the runway, that had been previously condemned as too noisy, and set about to build upscale multifamily dwelling units. Site acoustical measurements, as required by local code criteria, were conducted but generally ignored. Construction commenced with some consideration to upgrading the exterior envelope for control of the aircraft noise. Several years after the project has been completed, this facility is losing occupants because they just cannot stand the noise. The cost of maintaining a large unoccupied structure must be monumental. It is a continuing wonder to the author that noise considerations continue to be ignored.

[4] An example of a "Noise Element" in an Environmental Impact Statement (EIS) is given in Chapter 6.

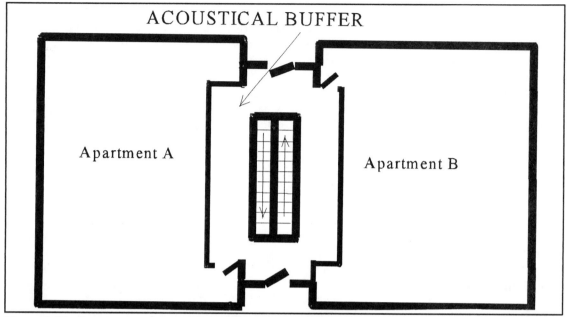

Figure 1-20   Separate noisy areas from quiet zones with buffers such as halls and stairs.

Note that all the noise mitigation elements discussed above require the isolation of those areas that require quiet. This means that the space will now require special ventilation, including air conditioning. Likewise, natural lighting will be limited. As a result, it should become very clear that noise mitigation efforts will have a significant impact on the other environmental elements of the built environment. It cannot be overemphasized that the design team should implement a systems concept to develop performance criteria that are compatible. In other words, the acoustical requirements must complement lighting, structural, HVAC, and other interior environmental attributes.[5]

Significant noise control can be achieved by grouping quiet areas, such as bedrooms, in one location, and the "noisy" areas or activities in another. The den, TV room or "audio listening room" can be adjacent to the bath that separates the loud activities area from the bedrooms. Likewise, the kitchen, utility room, and family rooms should be as far away as possible from the bedroom area (Figure 1-20).

In multifamily units, do not mix floor plans in vertical placement. Instead, stack floor layouts so that kitchens are over kitchens, baths over baths, bedrooms over bedrooms, etc. This is particularly important when the floor covering is not carpet and pad. While proper floor constructions can reduce vertical noise transmission between units, with a vinyl, ceramic, or wood floor it becomes inordinately expensive to provide sufficient airborne and impact noise transmission loss to an adjacent unit. Construction of floor systems with good airborne and impact transmission loss attributes and their ratings is discussed in Appendix 3.

---

[5] A structured planning process and a "systems concept" are recommended where all the building elements are specified by a "Performance Specification." The systems procurement process is discussed at length in Chapter 7 of the book <u>Planning and Designing the Office Environment</u>, D. A. Harris, 1991, Van Nostrand Reinhold (now Chapman & Hall).

Equipment rooms for elevators, air-conditioning units, motors, and fans will be discussed in Chapter 3. In order to ensure that the space allocated for mechanical equipment is adequate, and that the design of the completed building includes the desired noise criteria, close cooperation between the mechanical, electrical, and structural engineers and the acoustician is required. A systems concept should be adopted with all the trades and experts utilizing performance criteria for the interior environment. A planning team approach that utilizes performance specifications is the most effective way to assure that all the acoustical attributes are present in the completed project. While this process may be nothing more than a brief meeting of all the experts, contractor, and subcontractor for a single family residence, the process may become a design team effort with periodic scheduled meetings of the entire team for a large multifamily complex (Figure 1-21).

ACOUSTICAL PERFORMANCE
****

■ **Established by owner/designer**

■ **Compliance by planning team:**

Equipment manufacturer,
installer,
contractor,
subcontractors,
code authority,
engineers,
architect

Figure 1-21   Performance specifications ensure a project is free of noise problems.

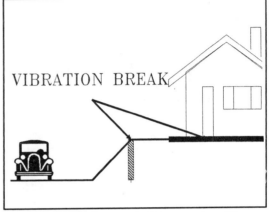

Figure 1-22   Isolate groundborne noise with vibration breaks.

In some locations, vibrations from ground noise can be transmitted through the foundation. The effect in the living unit could be a low-frequency hum or structural vibrations. Noise from expressways, railways, subways, and other sources, such as punch presses, are transmitted through the earth, thus causing foundation vibration. Columns and supporting walls can be isolated at the footer line not only to reduce the interior noise level, but also to prevent structural damage to the building itself (Figure 1-22).

Electrical transformers or substations for the distribution of power can emit a loud hum that is very disturbing. Other annoying sources are heat pumps. Both transformers and heat pumps should be selected for their quiet operation. Unfortunately, sound-dampened or "quiet" transformers and heat pumps are very expensive and are generally made only to special order. In addition, electrical and HVAC engineers like to locate these units in a position that is most practical for hookups. Inevitably, the ideal location for this purpose is on a back patio, just where outside activities such as a barbecue occur. Furthermore, they tend to operate on warm days when the occupants are likely to seek the peace and quiet of the backyard for a nap in the sun. If it is not possible to relocate these units where they will not cause a noise problem, they may be isolated by a sound barrier or sound enclosure.

Sound barriers can be erected close to the units, or better yet they may be totally enclosed. Depending on the particular type of equipment and the enclosure details, it may be necessary to provide ventilation to remove the heat generated within the enclosure. This is generally accomplished by omitting one wall, or using circulating air fans. Note however, that the fan noise may become a source of annoyance. The interior can be constructed with a lead liner or nonporous concrete for a good barrier. It should also be lined on the noise source side with sound-absorbing material such as fiberglass or mineral wool board to absorb the noise at the source.

It is important to remember that, as with a chain, the noise control design is only as good as its weakest link. Some misjudgments can be easily corrected, while you may have to live with others. Attention to detail in the correct design makes construction easier and more effective. For larger projects, a systems approach and performance specifications may be required to assure that all experts and members of the design/construction team maintain the acoustical performance desired.

1.2.8 Working with Construction Crews

When dealing with noise control, how a building is actually constructed is as important as how it is planned. Construction crews must pay attention to details to ensure success. To get the degree of noise control desired, it is necessary to:

- Educate–Crews and their foremen must be taught what is expected of them in terms of their implementation of construction details.

- Train–Artisans must be shown the details of construction which will give good noise control performance.

- Demonstrate–A few demonstrations of the small differences in construction details that mean the difference between success and failure are suggested.

- Follow-up–Supervision in the initial stages will result in the formation of noise-oriented work habits. The first building is the hardest, but they become easier with practice and eventually become habit.

1.2.9 Working with Subcontractors

Working with plumbing, electrical, heating and air-conditioning, and other subcontractors requires attention to detail. Education, training, demonstration, and follow-up are all vital. Rarely are any of these experts knowledgeable in acoustics technology. Even the best in their field are likely not even aware that their normal way of doing things will have a significant negative impact on the ultimate acoustical performance. For example, plumbers will punch through a dividing wall between living units and never give a thought to the potential flanking path.

HVAC contractors who prefer sheet-metal air ducts may replace specified fiberglass ducts with unlined sheet metal or worse yet, put the fiberglass lining on the outside of the duct rather than inside where it will be both a sound absorber and thermal insulator, since this is how they were trained. The result will be a "speaking tube" connection between sensitive rooms.

Electrical contractors like to place outlet boxes back-to-back causing a major flanking path.

Chapter 3 provides more details for each of these subcontractors. It is vital that they be informed of what is required of them and how their work will impact noise mitigation efforts. Let them know what is required.[6] Setting a standard of acoustical performance and insisting on getting it will result in success.

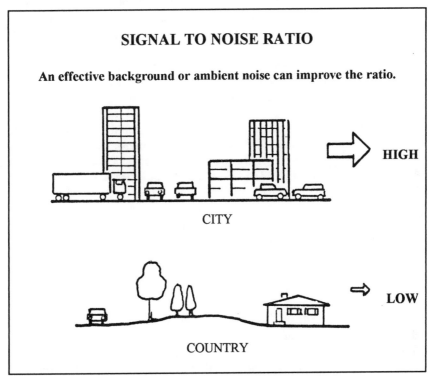

Figure 1-23   Signal to noise ratios are high in the city and low in the country. High background noise levels tend to mask intrusive noise .

1.3 NOISE REGULATIONS FOR OUTDOOR ENVIRONMENT NOISE

Our modern environment has created anomalies in noise control that are difficult to resolve. A rush to acquire the "American Dream," namely a ranch house in the suburbs on a quiet wooded street, coincided with expansions of noisy autos, trains, and aircraft.

The noise problem now is greater for those who moved to the suburbs than it was when they lived in the inner city. The signal to noise ratio became worse rather than better. Where before, general background sound levels masked intrusive noises, the suburbs generally have a lower background; consequently, noise from airplanes, autos, and trains is even more intrusive than ever (Figure 1-23).

We have compensated for this situation in our society by developing noise laws and regulations.

---

[6] Additional details and examples of horror stories are provided in Chapter 5.

These regulations have become very stringent in certain parts of the world. Europe, for example, has very strict noise laws and efficient enforcement. In the United States, the government has developed a series of noise regulations through the Environmental Protection Agency (EPA) and the Department of Housing and Urban Development (HUD). Many states, counties, cities, and local municipalities have their own regulations. Some governmental bodies opened their own noise control agencies when the federal government said it would relegate implementation of the regulations to local agencies. Unfortunately, promised federal funding of local programs was never released.

The Office of Noise Control (ONC) was designed to administer the federal program. While ONC does not exist at this writing because of politics, there are efforts to reestablish the effort. Fortunately, many of the state and local ordinances still exist and there are funds, albeit limited, to enforce regulations. The most significant effort is that of the State of California, which has established minimum levels of performance for most communities and residential construction projects. The model building codes, including the National Building Code (NBC), Southern Building Code (SBC), and the Uniform Building Code (UBC), sponsored by the International Conference of Building Officials (ICBO), all include acoustical criteria for number of circumstances. It behooves the architect, designer, builder, contractor, subcontractors, and owner to be knowledgeable about these regulations. They must know their scope, criteria, and enforcement.[7]

These ordinances can take many forms. A partial list that address noise follows.

1.3.1 Federal Regulations

The Environmental Protection Agency (EPA) has developed guidelines for noise levels. They address specific applications ranging from industrial and commercial environments to residential communities. Their typical requirement is for a developer to submit an environmental impact study on the effect the proposed project will have on noise levels. For example, nearly every airport of significance now has a noise control contour map. It will show noise levels produced under varying weather and flight conditions for the surrounding community.

Similar studies are generally required for any new transportation facilities such as highways and railways, power plants, factories, and racetracks. The Environmental Impact Study (EIS) is usually a requirement for proposed new construction and may be tied to the permit for construction. When noise is considered a potential hazard, the EIS should include a "Noise Element" section. Community noise levels, existing and potential, are identified and compared with appropriate local or governing regulations and accepted practice. Recommendations for noise criteria are made, and noise mitigation options are identified by the following federal agencies:

- In the mid-1970s, EPA championed development of noise labels for noise-producing or noise control products. Unfortunately, no funding was ever made available for this effort and very few products have labels that provide noise control levels or ratings of their effectiveness to quiet.

- The Federal Aviation Agency (FAA) has stringent regulations covering aircraft noise.

---

[7] Additional details on the active programs are discussed in Chapter 4. Specific examples are given including the City of Los Angeles Noise Ordinance, a small municipality noise code, and noise criteria sections of various codes including FHA/VA.

Their studies overlap EPA on noise contour maps. In recent years there has been considerable controversy over which agency controls noise issues. At present, FAA has overridden all other agencies including local and community efforts. The local FAA administrator and/or airport management authority is the one most likely to have copies of the noise contour maps. Every developer, builder, and owner should study these maps carefully before making site commitment. There is no greater disappointment for a residential owner than moving into a new house in a supposedly quiet country environment, only to be awakened the first time a wind shift causes normally unused runways to suddenly become major air corridors. FAA has taken a cavalier attitude about its noise regulations. At present the regulations focus on quieting aircraft, an admirable practice on the surface. However, FAA tends to listen to the aircraft manufacturer and airline problems rather than specific community complaints. Its regulations apply industry-wide, thereby allowing the airlines to avoid costly scheduling of aircraft to specific airports because of local noise regulations. Since the airline industry is in serious financial trouble, FAA has adopted this approach as an effective way to quiet the skies with the least financial burden. The result is considerable disappointment to many local communities in their efforts to quiet their neighborhoods.

• Occupational Safety and Health Administration (OSHA) regulates the degree of noise allowable in industrial plants. Normally this does not impact residential communities. However, if you suspect that a noisy factory, power plant, or the like is near your site, a visit to the local OSHA (both state and federal) office may provide access to noise level data that otherwise may be not available to owners or operators. Typical OSHA criteria will apply to the workers on a construction site. For example, an operator of noisy equipment must be provided an environment that is less than 90 dB$_A$ for an 8-hour. period. For higher noise levels, OSHA requires the equipment to be quieted using engineering controls or management controls. **Temporary measures, such as ear protectors, are not a permanent solution**. OSHA also requires that any worker in an environment that is greater than 85 dB$_A$ must be tested for hearing loss on a regular basis.[8]

• Housing and Urban Development (HUD) provides "Minimum Property Standards" for Federal Housing Authority (FHA) and Veterans Administration (VA) programs (Figure 1-21). In order to qualify for loan financing, the site and living units must comply with the HUD minimum noise standards. Criteria vary depending on location and type of living space. If the site is in the city, with a high background noise level, the criteria are less stringent. A suburban or rural site requires a higher degree of sound isolation between living units (Figure 1-24)[9].

---

[8] A more detailed discussion of industrial noise criteria and control measures may be found in the "Noise Control Manual" by D. A. Harris, 1992, Van Nostrand Reinhold (now Chapman & Hall).

[9] This is a logical extension to the concept of signal to noise ratio discussed in the previous section.

RATING BETWEEN LIVING UNITS

SOUND TRANSMISSION CLASS (STC)

Room separators	Country/city
Living/living	45/40
Kitchen/living	50/45
Bedroom/kitchen	55/50
Bedroom/living	55/50
Bedroom/utility	50/54

IMPACT NOISE RATING (INR)

Bedroom/living	+45/40
Kitchen/bedroom	+45/40
Living/living	+30/25
Bath/bedroom	+45/40

Figure 1-24   Typical acoustical criteria by HUD for FHA/VA-financed housing.

Sound isolation criteria between living units in a multifamily dwelling are the most stringent when adjacent spaces involve bedrooms. The criteria are less stringent if the two adjacent rooms between living units are kitchens or living rooms. Site noise level criteria are relatively new additions. If a site is near an airport, highway transit, or factory, a noise survey may be required, especially if an Environmental Impact Statement (EIS) has indicated noise may be an environmental issue.

Depending on measured and predicted noise levels, an owner/builder may be required to demonstrate that the construction of the exterior envelope will attenuate the exterior noise to a **specified level within the building.** While these regulations may be overlooked by some financiers, the rational is based on extensive studies. For example, in England, noisy sites have been identified as having substantially lower real estate values and greatly reduced resale value.

It behooves a knowledgeable builder, developer, designer, and owner to meet FHA criteria for all projects, be they FHA/VA-financed or not. Any units that do not meet the minimum property standards very likely will be the first units vacated when other housing becomes available. Rent or sales values will obviously be depressed for noncomplying units.

1.3.2 State and Local Codes/Ordinances

Each state and/or local municipality usually adopts or follows one of the model building codes (Figure 1-25). Nearly all of them contain requirements or criteria for noise control. Typical requirements include maximum noise levels at the property line and a minimum sound transmission loss characteristic for party walls between living units (Figure 1-24). Requirements are similar to the HUD Minimum Property Standards. However, the codes vary widely in scope, test requirements, criteria, and compliance. A prudent architect, developer, and builder will obtain details that apply to construction in their jurisdiction from the governing authorities. If in doubt, consult the local building inspectors. They can tell you whether they utilize one of the major codes such as BOCA, SBC, or ICBO, an adaptation, or their own code. Most major cities and counties adopt a basic code and then make modifications to suit their needs. Cities like New York, Chicago, and Los Angeles have adopted codes that are unique to their area. It is important to note that most local authorities, even those who say they have adopted a basic code, have made some changes. More important, local authorities have adopted internal policies on how the code is applied in their jurisdiction. **Do not assume that, because a municipality is in a particular area it uses the predominant model codes.** In fact, if anything is different, it is likely to be the noise regulations and criteria. See Chapter 4 for details.

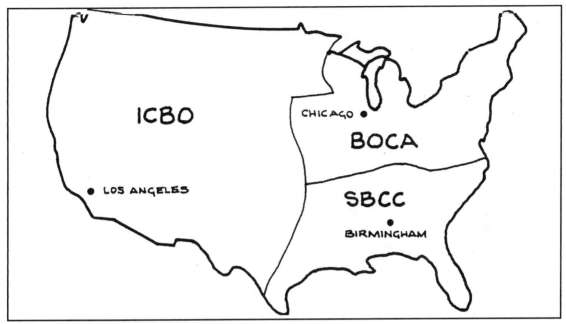

Figure 1-25 Model building code coverage

As an aid to those readers who may do business across local municipal boundaries, a list of the most widely used model codes and where they have been adopted follows:

BOCA-The Building Officials and Code Administrators International Inc. Basic Building Code is used primarily in the midwest and northeast. It contains standard verbiage for sound transmission loss between living units. Actual criteria are usually selected by the local authority.

SBC-Southern Building Code is used extensively in the southern states. Their Southern Standard Housing Code and the One-Two Family Dwelling Code requires minimum sound transmission loss for various elements of the structure such as party walls. Criteria vary depending on the circumstances and local need.

ICBO-International Congress of Building Officials publishes the Uniform Building Code Standards, a Uniform Housing Code, and a One-and Two-Family Dwelling Code. The ICBO (formerly, the Pacific Coast Building Officials Conference) is used extensively by western states and municipalities. Many northern tier states and Midwestern communities also have adopted the ICBO Standards. Their sound control regulations are similar to, but more extensive than, SBC requirements.

NBC-National Building Code was first introduced as a model building code by the National Board of Fire Underwriters now known as the American Insurance Association. Vermont utilizes the code and a number of municipalities have adopted NBC in part or whole. Acoustical criteria are similar but less extensive than ICBO.

Major municipalities, such as New York City, Chicago, and San Francisco, have evolved their own codes over the years. Many have adopted portions of the model building codes. Nearly all contain construction details and/or criteria for noise control. For example, the San Francisco code calls for extensive field confirmation of sound transmission criteria between living units. It is generally more stringent than HUD or model building code criteria.

States and counties have adopted all, part, or combinations of the major codes. Some have totally ignored noise control portions while others have added to the requirements extensively. A case in point is California. Title 24 of the California Code not only requires minimum ratings for demising walls between living units but each project must be evaluated for noise. In most cases this is an extension of the Environmental Impact Analysis. All cities in California are theoretically required to conduct an analysis of their community noise. In addition they are required to adopt a noise ordinance. A typical example is that a developer must submit confirmation from a qualified and licensed acoustician that the project will meet the local ordinances and criteria. This is typically required by the buildings and inspections department prior to acceptance and issuance of permits. If the development is near a freeway or major road, the acoustician will not only take site measurements but will develop a predicted noise model. Construction plans may require special noise walls or barriers, upgraded exterior construction to block noise, submission of computer analysis studies, and funds for acoustical testing at the site for compliance prior to acceptance. Many completed projects were required to have considerable upgrading done to the buildings before issuance of occupancy permits. The cost of compliance can be monumental without proper noise control mitigation.

1.3.3 Noise Complaints

The greatest number of noise complaints come from homeowners and apartment occupants living in residential areas near airports. Development of medium size jet airports into major/international airports, or "Hubs", by individual airlines have become a focus of recent noise complaints by local communities. In addition, major expansion of the jet freighter business has created a large increase in the number of overflights at night. Noisier and larger aircraft have been introduced to make matters even worse. For those who are in the vicinity of a major military air base, noise levels will be significantly higher than commercial facilities because of the types of aircraft employed and their flight conditions. Obviously, an F-16 in a full-alert military takeoff will be noticed by nearly everyone. As a case in point, when Operation Desert Storm was implemented, it was obvious to most citizens that something big was in the offing. As aircraft deployed, these bases become inordinately quiet.

Aircraft noise levels at a home or apartment site will vary, depending chiefly on distance from the airport, the flight path, and the type and altitude of the aircraft. In any event, dwellings near large airports or directly under flight paths may be subjected to rather intense levels of noise. Levels of 90 to 110 dB_A are common. Most airports utilize day/night averages that emphasize night-time noise levels.

The steps for improving sound insulation of houses in moderately noisy neighborhood environments will be discussed first. More difficult problems relative to aircraft noise intrusion are covered at the end of this section and in greater detail in Chapter 6. Generally speaking, most single-family houses are an assemblage of light-frame interior partitions enclosed and supported by a light-frame exterior shell. If this shell were continuous and formed an airtight seal around the enclosed space, it would provide a limited amount of sound insulation. In fact, those elements that improve the energy efficiency of a building are desirable acoustical isolation tactics. Such construction typically has a Sound Transmission Class (STC) rating of 30 to 35. (STC is explained in Chapter 4) However, this shell usually contains several windows and doors. With their customary air and noise leaks, the windows and doors may reduce the STC rating of the exterior shell of the house to a value as low as STC 20. Ways of improving this STC are discussed in Chapter 6. STC ratings for a number of typical construction systems including floors, walls, roof/ceilings, and composite structures including doors, windows, etc. are listed in Appendix 3.

1.4 BASIC DESIGN CONSIDERATIONS

1.4.1 Reducing Neighborhood Noise

If you have a home in a noisy neighborhood and are disturbed by the intrusion of outdoor noise, the following recommendations may alleviate the problem. Fortunately, improvements that yield noise insulation also provide thermal insulation:

1. The first rule of thumb is to install a central heating and air-conditioning system in the house. This will eliminate the need to open windows and doors for ventilation purposes, thereby making it easier to reduce the intrusion of outdoor noise.

2. Existing windows should be caulked or sealed with airtight gaskets.

3. If the noise source is high-pitched (i.e. with predominant frequencies above 2000 Hz), and very directional, install storm windows with caulk or gasket seals on the side of the house facing the source. It is still more beneficial to install storm windows, properly sealed, on all existing windows. Storm windows will also reduce the intrusion of low-pitched noise, noise approaching from other directions, and reflected sounds from nearby structures.

4. Exterior doors should be provided with soft resilient gaskets and threshold seals. This applies particularly to sliding doors which typically have poor seals.

5. Install storm doors equipped with resilient gaskets and threshold seals. Sliding doors also could be provided with storm doors. The use of storm doors, windows, and weather stripping will also conserve energy by reducing air leakage and make it easier to maintain constant indoor temperatures.

6. Seal or caulk all openings or penetrations through walls, particularly around water and gas pipes, electrical cables, and refrigerant lines.

7. Cover mail slots in doors, install hinged cover plates on clothes dryer vents, bathroom and kitchen exhaust ducts, and central vacuum system dust discharge.

8.   Close the fireplace damper, install close-fitting cover plates at clean-out openings and erect a barrier in front of the fireplace opening. Fire doors are acceptable provided they are tightly sealed and have a circuitous path for intake air.

Adoption of the measures listed above will improve the sound insulation of the exterior shell/envelope of the house about 6 dB for low-pitched (i.e. low-frequency) and more than 10 dB for high-pitched (i.e. high-frequency) sounds. These measures will be even more effective when the exterior shell has superior sound transmission loss characteristics. An example of a structure with superior sound transmission loss characteristics is heavy masonry construction. Chapter 7 discusses sound transmission loss in more detail. Appendix 3 provides sound transmission loss data for typical construction systems including walls, roof/ceilings, windows, doors, and composite assemblies.

1.4.2 Reducing Traffic Noise

Homes or apartments of conventional construction that are located near heavily traveled roadways or expressways are particularly vulnerable to the intrusion of excessively high levels of noise. Even if a freeway has a noise barrier, sound levels may be excessive. Since sound mitigation measures[10] may be costly and impractical, the occupants should choose those solutions that provide the most noise reduction at the least cost, that is, the most decibels-per-dollar. In addition to the recommendations described above, the occupant should incorporate, on a step-by-step basis, the following noise control measures, until the noise has been reduced to a tolerable level.

1.   In existing windows, install double-thickness glass panes facing the noise source.

2.   Install storm windows over the same windows. Select storm windows having glass panes mounted or encased in rubber gaskets (Figure 1-27).

3.   Line perimeter surfaces between the existing windows and the storm windows with a highly sound-absorbent liner such as glass or mineral fiber board.

4.   Replace existing hollow-core entrance doors with solid-core sound attenuating doors equipped with perimeter gaskets and threshold seals (Figure 1-28) .

5.   If intruding traffic noise disturbs sleep and the bedroom faces the roadway, choose a bedroom which is located on the opposite side of the house. This will reduce intruding noise by about 10 dB_A. A room that is partially shielded from the noise source should reduce the level of intruding noise about 6 dB_A.

---

[10] Some builders incorrectly utilize the term "soundproofing." In reality a soundproofed space rarely is achievable as some sound inevitably intrudes no matter what measures are employed. Therefore the author has removed the term soundproof from this text.

GENERAL RULE FOR REDUCING NEIGHBORHOOD NOISE.

Think of the structural envelope as a balloon. Any point that will leak air is a potential acoustical leak. Sealing all air leaks is the most effective technique there is in terms of decibels-per-dollar.
When all air leaks are sealed and the sound is still excessive, proceed to upgrading the sound transmission loss characteristics of the exterior envelope. Start first with windows and doors.

Seal sound leaks!

6. Construct a barrier wall or fence between your house and the roadway. To be effective, the fence should have a solid, continuous surface without any openings or holes. The sound barrier or fence must be long enough and tall enough to shield, or hide, the entire roadway when viewed from the nearest side of the building. Barriers are most effective in reducing high-frequency sounds that travel in a beam-like manner. Low-frequency attenuation is limited since low-frequency sounds more easily bend around a barrier. Flanking paths, such as another structure placed in a position to reflect sounds around the barrier, must be eliminated or rendered highly sound-absorbent. When

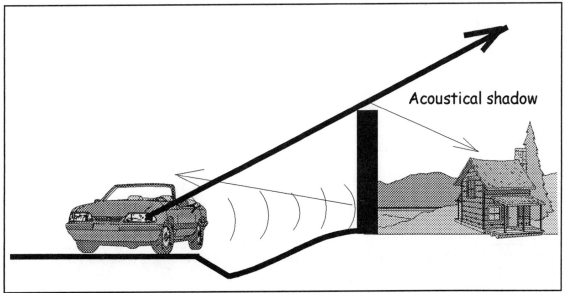

Acoustical shadow

Figure 1-26 Acoustical barriers/fences must be high and long enough to provide an acoustical shadow. They will attenuate highway sounds by 5 to 10 dB_A.

31

these measures have been taken, the barrier or fence may be expected to attenuate the traffic noise by 5 to 10 dB$_A$. The acoustical efficiency of such a barrier may not be improved by massive construction. Therefore, it may be constructed of lightweight materials such as 1- in.- thick boards that are tightly spaced. Note that it must fit tightly to the ground and be tall enough to create a significant acoustical shadow (Figure 1-26).

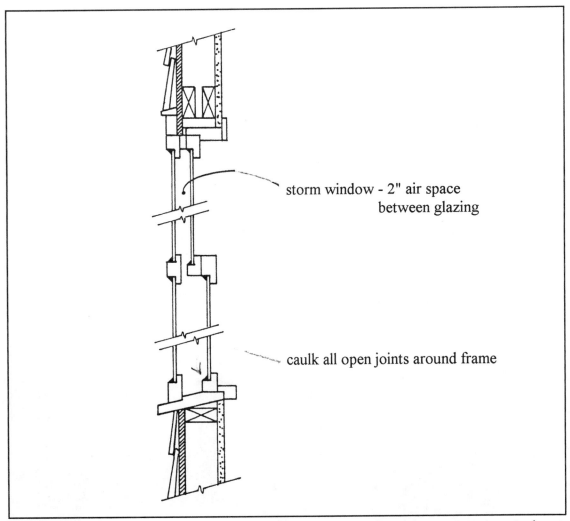

Figure 1-27   Double-hung widow with a storm window fully sealed will attenuate community noise by approximately 27 dB$_A$

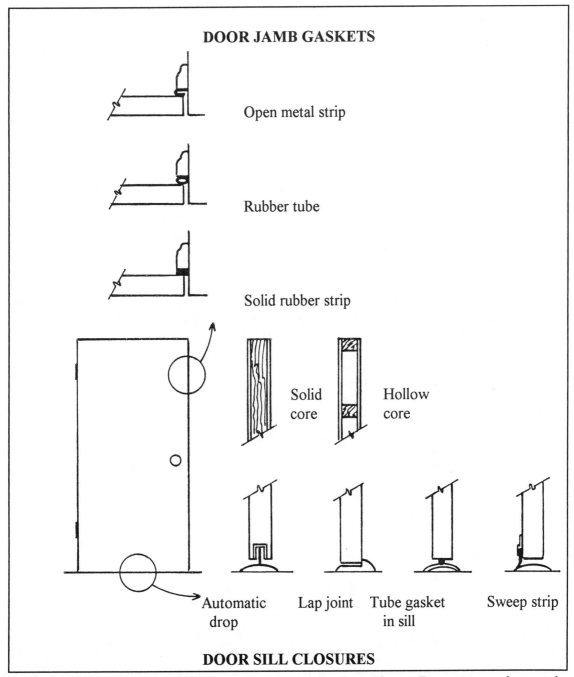

DOOR JAMB GASKETS

Open metal strip

Rubber tube

Solid rubber strip

Solid core

Hollow core

Automatic drop

Lap joint

Tube gasket in sill

Sweep strip

DOOR SILL CLOSURES

Figure 1-28   Door seals are mandatory. Typical occupant seals are shown. For superior seals, special-purpose door frames are required.

1.4.3 Reducing Aircraft Noise

Brief intrusions of high-level sounds are among the most troublesome to remedy. Studies of aircraft flyover noise show that reports of annoyance tend to increase by a factor of only 2 when the flyover rate is increased by a factor of 8. Read optimistically, this means that if you can tolerate one intrusion, you probably can tolerate two. From the standpoint of an acoustical engineer who must quiet an airport area, and the community that must live within the area, the situation must be viewed pessimistically. Reducing the number of landings and departures per hour to one-half their original number may provide only an insignificant dent in the number of complaints from the residents. [11] Moreover, some studies have shown that people do not become accustomed to noise intrusion as time goes on, but remain intolerant. On the basis of these facts, the airline industry has chosen to attack noise at the source. Recent legislation limits the noise level of aircraft. Criteria are being phased in as new aircraft are constructed. Stage 1 quieting is required at present, Stage 2 noise levels are required in some instances, and Stage 3 quieting will soon be required on all new aircraft. Some retrofitting is being done to achieve Stage 2 quieting, but the return in decibels-per-dollar is low. While aircraft will always be a potential noise source that will cause many to relocate far away from the airport, some relief is coming with the quieter aircraft and special takeoff procedures. Unfortunately, most occupants will not be willing to wait for the phase-in, since it will take up to 10 years.

If you are one of the many owners who now find themselves constantly exposed to intense and highly disturbing aircraft noise, consider carefully some basic facts before you decide on any course of action to resolve the problem. The main items to consider before instituting major noise-mitigating measures to a dwelling for aircraft noise are:

1. The cost of mitigating aircraft noise is exceedingly high. While most of the efforts to reduce noise may be justified in terms of thermal comfort and reduced energy consumption, it is recommended that costs be calculated on a life-cycle basis.

2. Although mitigating aircraft noise will reduce annoyance indoors, it does nothing to improve the noise environment outside the structure. If you enjoy gardening, patio parties, or backyard barbecues, you will have to tolerate the aircraft noise as before.

3. The probability is rather low of gaining noticeable relief from the noise in the near future through development of quieter aircraft or changes in aircraft landing or takeoff flight procedures. In fact, the noise environment near a large airport generally tends to grow worse in time, with the continuing expansion in aircraft size and flight operations. Night flights by airfreight operations are a new annoyance.

4. The most effective solution may be to move to a more quiet area, even though it might entail some financial loss in the sale or rental of your home.

---

[11] For a house occupant, this statistic is also important. If you must operate a noisy tool, even though very briefly, it is important to try to stop the noise at the source. Otherwise, you may find yourself the recipient of a number of complaints despite the limited duration of the noise involved.

5.   Be advised of a new threat. Those who moved far away from a major airport in New Jersey, because of aircraft noise, have encountered a new problem. Any low-flying aircraft is more intrusive since a country setting will have very low background sounds. The signal to noise ratio of a low-flying aircraft may be more intrusive in the country than in the suburbs. In addition, air traffic has increased to such an extent that some air corridors are overloaded, causing air traffic controllers to assign air traffic to much lower altitudes than usual. This has become widespread practice.

1.4.4 Major Sound Insulation Modifications

A discussion of the various sound-insulating modifications that may be used to reduce intrusion of intense aircraft noise into a residence follows. (A more detailed discussion may be found in Chapter 7.)

1.   Adopt the noise mitigation and reduction recommendations in the foregoing paragraphs relative to reducing the intrusion of neighborhood and traffic noise.

2.   Treat windows. Windows are typically the weakest link in the acoustical envelope of most structures. Ordinary locked, double-hung windows will generally provide an average noise reduction of about 18 dB. The addition of a storm window will increase the noise reduction to about 24 dB. Caulking both sets of windows improves the sound insulation about 3 dB at the higher frequencies (Figure 1-29). If the aircraft noise is still coming through the windows, despite the fact that they are equipped with storm windows and sealed airtight, then the existing window system will have to be replaced with fixed, well-sealed double-pane windows throughout the house. Note that sealing the windows will require the installation of air conditioning by most codes. A typical upgraded window installation consists of two ¼ in.-thick panes of glass, each encased in a U-shaped, soft, resilient gasket with a 4 in. air space between panes. Perimeter surfaces within the framing are lined with highly sound-absorbent material. This double-window istallation, when properly installed, will increase the sound insulation to approx. 37 dB_A (Figure 1-29).

3.   Seal exterior doors. If the remaining noise now seems to be coming through the exterior doors, remove the molding from the door frame to see if there are any air gaps or noise leaks between the door frame and the wall. The conventional practice of installing prefabricated door kits, in which the doors are premounted in a door-frame assembly, often causes serious sound transmission problems. Builders generally provide an oversized opening to receive the preassembled door and fasten the assembly into place with a few wedges and nails, and completely ignore any resulting air gaps or leaks. These are conveniently covered and hidden from view by the thin finish molding. However, such leaks must be sealed if any improvement in sound insulation is to be expected, and certainly before any existing doors are equipped with gaskets or replaced with upgraded sound isolation doors.

35

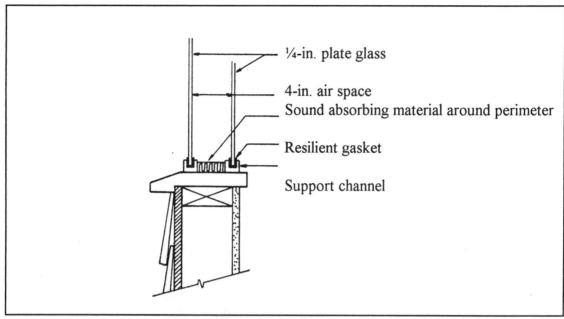

¼-in. plate glass

4-in. air space
Sound absorbing material around perimeter

Resilient gasket

Support channel

Figure 1-29 Double-glazed window detail.

If exterior door sealing efforts are not sufficient, add a tightly sealed storm door. If this is still not sufficient, replace the existing door with a solid-core or acoustically designed metallic door. Remember, the door is only as good as the seals. As the sound transmission loss characteristics of the door design are improved, so must the seals be upgraded. Simple tight seals typical of a weathertight gasket and drop door closure may not be sufficient. Special acoustical gaskets, drop closures, and wedge-type latches may be required.

For a more detailed discussion of sound transmission loss, flanking paths, and leaks, see Chapters 2 and 5. See Appendix 2 to determine the sound transmission loss of a composite partition having a base wall, doors and windows. Appendix 3 contains acoustical data on many typical construction systems and sub-elements such as doors and windows.

Chapter 2 DESIGN CONSIDERATIONS

2.1 INTRODUCTION

The interior acoustical environment of a building is established by two elements: the interior envelope and sound sources. Typically the interior envelope consists of the floor, walls, and ceiling of an identifying space–usually a room. Sound sources found normally in residences include the occupants, be they talking, walking, exercising, or engaging in some other activity, and the mechanical devices such as kitchen appliances, washing machines, vacuum cleaners, stereo, and TV. Each of the sound sources will interact with the acoustical envelope. If the envelope has hard, reflective surfaces, sounds will bounce about or reverberate and possibly even be amplified until they die out. A space with a considerable amount of sound-reflective surfaces will have a long "reverberation time" and sound like a room with echoes. If the interior envelope has sound-absorptive surfaces, the sound sources will be quieted and have a short reverberation time. Rooms with a short reverberation time are preferred for ease in communication. Once room acoustics are established by the amount of sound absorbing materials that occur in the source room, these sounds may then be transmitted to adjacent rooms or living units. The physics of transmitting sounds is called <u>sound transmission</u>. Barriers that reduce the transmission of sound are rated by a term called <u>sound transmission loss</u>.

An interior acoustical envelope usually consists of hard surface materials that readily reflect sounds. Examples of highly reflective materials include gypsum wallboard, plaster, glass, wood, and ceramic surfaces. Examples of materials that have good sound absorption properties include room decorations and furnishings such as carpet and pad, lined draperies, acoustical ceiling materials, acoustical wall coverings, upholstered furniture, and pillows. As you can readily see, materials that are good sound barriers are impervious, hard, and heavy. Materials that are good sound absorbers are soft, fuzzy, and porous. As a comparison, we find that materials that are good sound barriers will also resist water and air passage (glass and concrete for example), while materials that are good sound absorbers are much like a sponge.

This distinction between absorbers and barriers cannot be overemphasized. It is probably the most misunderstood concept in acoustics. This chapter will discuss specific design considerations for those elements that make up the acoustical envelope. First, it will consider **sound absorption within a space** and how to make it better for the occupants in the source room. Second, this chapter will provide specific design considerations for **sound barriers,** including party walls, floor/ceiling systems, and those items that will penetrate the barrier, allowing a flanking path for noise to intrude upon the neighbors.

2.2 ROOM ACOUSTICS AND SOUND ISOLATION

There is an important distinction between room acoustics and sound isolation. Sound isolation is the control of the sound transmission **between** adjacent spaces. Room acoustics has the goals of noise reduction and reduced reverberation (echoes) for good hearing conditions **within** a space. Sound reflections within a room can be controlled with a combination of materials such as an acoustical ceiling, drapes, thick carpeting, upholstered furniture, and sound-absorbing wall treatments. These materials absorb the sound upon impact and thereby reduce the reverberation within the space (see Figure 1-9).

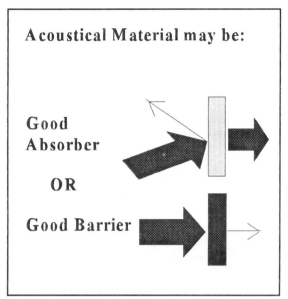

Figure 2-1 Good sound barriers are impervious and dense. Good sound absorbers are soft and porous.

Sound-absorbing materials, when used in a room where noise originates, act indirectly in reducing sound transmission to adjoining rooms by lowering the noise level in the source room. Used in an adjoining room, however, sound-absorbing materials lower the level of background noise, making it easier to perceive transmitted sound. The use of sound-absorbent materials to reduce reverberant sound within a space will provide up to 10 dB_A of noise reduction at best. Accordingly, sound-absorbent surface treatments will supplement, but not take the place of, good sound isolation construction (Figure 2-1).

The principals of sound absorption and sound isolation are dealt with separately since the physical properties of materials are different. **Materials that are good sound absorbers are usually poor barriers. Vice versa, materials that are good barriers are usually poor sound absorbers.** It is a rare material that possesses both of these properties. A good barrier is lead because it is a "limp mass" that has good resistance to air flow. A good absorber is mineral fiber building insulation since it is soft and porous. Closed-cell rigid plastic foam materials are considered sound barriers while open-cell flexible plastic foams are efficient sound absorbers.

Some recently developed products attempt to provide both barrier and absorbent properties by wedding together dissimilar materials. For example, a loaded vinyl sheet has been developed to replace lead. When faced with an open-cell plastic foam material, this flexible sheet will provide reasonable sound barrier and sound-absorbing characteristics. While expensive, this composite material has found considerable use in industrial applications. It is well-suited to quieting noisy equipment in equipment enclosures (Figure 2-2).

The first rule of acoustical design is: Establish the need– better room acoustics or reduced intruding noise.

For better room acoustics ask;

1. Do you want to improve the acoustical environment in the occupied or source space?

2. Is the room too noisy?

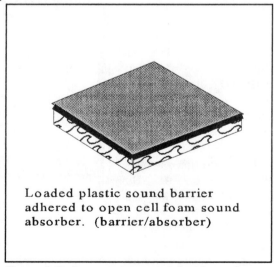

Loaded plastic sound barrier adhered to open cell foam sound absorber. (barrier/absorber)

Figure 2-2 A composite material with good barrier and absorption characteristics.

3. Is the room reverberant? Does it have echoes?

SOURCE ROOM

Use sound ABSORBING material to:
* **Improve speech intelligibility**
* **Lower background noise level**

RECEIVING ROOM

Use a good BARRIER to:
* **Reduce intrusive noise**
Use a good ABSORBER to:
* **Improve speech intelligibility**

Figure 2-3   Design for speech intelligibility within source room. Design for noise reduction with noise barrier between rooms.

If the answer is yes to any of these questions then concentrate on reducing the reverberation time by adding sound- absorbing material (Figure 2-3).

To establish whether noise mitigation is required ask:

Is the major problem intrusion of noise from a neighbor? - or - Do you want quiet in the receiving room?

If the answer is yes, improve the barrier between the source and receiving room.

Figure 2-4 The "cocktail party effect." With a few people conversation is easy. A crowd causes all to raise their voices to be heard.

If you are concerned about hearing speech clearly in the source room, the room acoustics must be designed to absorb sounds.

Sound-absorbing materials do two things: reduce reflected sounds and lower noise levels. Reflected sounds cause a long reverberation time. A room with a long reverberation time will sound hollow or have echoes. Sounds produced in a reverberant space will tend to take a long time to die out, causing difficulty in understanding speech.[1]

Reflected sounds that are reverberating, and not absorbed, will tend to add to new sounds, causing the overall sound level to increase. With a higher noise level, we tend to talk louder causing the background sound level to increase even further. This has been called the "cocktail party effect." At the beginning of the party when only one or two persons are present, we can hear each other with little difficulty even in a reverberant room. As more and more people arrive and begin to talk, the sound level increases to the point where even a closely situated conversants must raise their voices to be heard. Adding sound-absorbing materials to this room will delay the time when "runaway acoustics" occurs (Figure 2-4). A technique and worksheet to calculate the reverberation time of a room is presented in Appendix 2.

By contrast, if you are concerned only with sounds intruding from the source room into the receiving area, the design should focus on reducing the intruding sounds. This can best be handled by building a good sound barrier. The barrier should exhibit good sound transmission loss properties. In addition to a good barrier, adding sound-absorbing materials to the source room will tend to lower the sound source level. In simplistic terms, the noise reduction (NR) and sound transmission loss (TL) of a barrier are derived by determining the sound level in the source room and subtracting the sound level in the receiving room.[2]

$$NR = L_S \text{ minus } L_R \text{ (loudness in source room - loudness in receiving room)}$$

If the noise level has already been lowered by sound-absorbing materials in the source room, that reduced level becomes the source room level, resulting in an equally reduced receiving room level.

2.3 SOUND ABSORPTION

2.3.1 Sound-Absorbing Materials

To be an efficient sound absorber, a material converts impinging acoustic energy to some other form of energy, usually heat. There are three major types of sound absorbers: porous absorptive materials, diaphragmatic absorbers, and resonant or reactive absorbers. Nearly all the sound-absorbing materials designed for use in residences are porous absorptive materials.

Porous absorptive materials are the most well known of the acoustical absorbers. They are usually fuzzy, fibrous materials or soft rubbery foams. Sound-absorbing materials may be covered with an acoustically transparent material for protection. Examples of covering materials are woven fabrics and thin membranes.

---

[1] An example of a reverberant space is an old stone cathedral. The large space, with mostly masonry and glass as the walls, floors, and ceiling, will have a reverberation time of 10 seconds or more at most frequencies. When a person talks from the pulpit, sounds bounce around the space to such an extent that a listener may not be able to identify what is direct sound and what is reflected sound. Since the reflected sound takes longer to reach the listener, it will confuse the listener, making it very difficult to understand speech. Even when the speaker talks very slowly, reflected sounds may interfere with direct sounds. When a large crowd assembles in this space, the people and their clothing provide sound absorption. In medium and small cathedrals, this absorption may be sufficient to reduce the reverberation time to 3 to 4 seconds making it much easier to understand speech. The same events happen in a residential room on a smaller scale.

[2] Noise reduction is normally used for field-type situations. When more precise measurements are required such as may be made in the laboratory or formal testing, the acoustician will determine the sound transmission loss . TL also takes into account the sound absorption characteristics of the rooms being tested. A detailed explanation of TL may be found in Chap. 4 and ASTM E90 and E336.

Sound absorbers may be ceiling tile, wall coverings, carpets, cushions, and heavy lined draperies. Cork was the first material utilized commercially for sound absorption in architectural applications. Soon after, bagasse, a sugar cane by-product, was formed into ceiling tiles. Later, wood fiber tiles and boards appeared. Today, most of the commercially available sound-absorbing materials for ceilings and walls are composed of glass fiber, mineral fiber, or open-cell foam plastic. In all of these materials, sound absorption occurs by causing the sound waves to activate motion of the fibers, membranes, and the air in the spaces surrounding the fibers or voids. Frictional energy losses generate heat, which is dissipated thereby reducing the acoustic energy[3] (Figure 2-5).

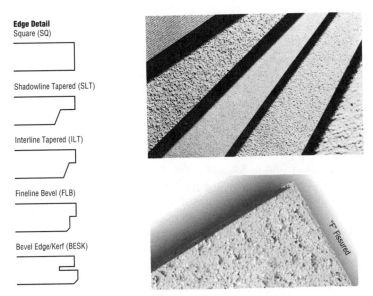

Edge Detail
Square (SQ)

Shadowline Tapered (SLT)

Interline Tapered (ILT)

Fineline Bevel (FLB)

Bevel Edge/Kerf (BESK)

Figure 2-5 Examples of typical sound-absorbing ceiling tiles.

Material properties that affect sound absorption efficiency, in order of importance, are thickness, density, porosity or flow resistance, coefficient of elasticity, and acoustic impedance. In all common sound-absorbing materials, thickness is the key determinant of sound-absorbing performance. For glass and mineral fiber materials and open-cell plastic foam, increasing thickness is nearly 10 times more effective in improving sound-absorption characteristics than any other attribute. The second most important attribute is flow resistance. If sounds are blocked from entering a sound-absorbing material by an impervious coating such as paint or a layer of film over 2 mils thick, sound-absorption values will be reduced dramatically[4] (Fig. 2-6).

A comprehensive list of sound-absorption test results for various materials is given in Appendix 3.

---

[3] A neophyte might think this physical conversion of energy to be a way to conserve energy costs. Unfortunately, the amount of energy converted to heat is so minuscule that it can hardly be measured. As an example, if one were to somehow contain and convert all the acoustic energy from 100,000 fans yelling at a great play in the Super Bowl there might be enough energy to light a 100- watt light bulb. Furthermore, no one has yet found a way to efficiently capture and convert the acoustic energy generated.

[4] A good way to tell if a material has the proper flow resistance to be a good absorber is to blow through it. If there is little resistance to air flow, the material might be a good absorber. A material such as glass will resist any air flow and thus is a poor absorber. This test is also useful for evaluating acoustical material facings. Open-weave cloth will readily allow air flow, while a thick paint film will block sound from entering the sound-absorbing material. Note that a thin film that is 1 mil or thinner may block air flow yet be acoustically transparent.

Note that sound-absorption characteristics will vary depending on the frequency of sound being absorbed and the mounting method. Most mineral ceiling tile materials, such as ⅝ in. mineral or rock wool ceiling panels, are very efficient high-frequency sound absorbers (1000 Hz and higher) but are poor absorbers at low frequencies.

Material	Octave band center frequency, Hz						
(16 in. air space unless noted)	125	250	500	1000	2000	4000	NRC*
⅝ -in. mineral tile	0.25	0.34	0.41	0.56	0.65	0.63	0.50
⅝- in. fiberglass vinyl-faced tile	0.76	0.71	0.60	0.76	0.79	0.74	0.70
1½-in. fiberglass ceiling tile	0.80	0.96	0.88	1.04	1.05	1.0 6	1.00
¼-in. glass	0.05	0.03	0.02	0.02	0.03	0.03	0.05

SOUND ABSORPTION COEFFICIENTS of SOME TYPICAL MATERIALS

* NRC = noise reduction coefficient (average of 250 to 2000 Hz)

Figure 2-6   Examples of sound absorption coefficients.

Efficient absorbers, such as fiberglass ceiling board, are good absorbers across the frequency spectrum. The technique for mounting the material will have a significant effect on sound-absorption coefficients. For example, the same material mounted directly on a hard surface such as gypsum board or plaster may have twice the sound absorbing efficiency if mounted with a 16 in. air space behind the material. For this reason it is vital to know how the material will be installed in order to evaluate the sound-absorbing characteristics. Since most manufacturers make the same material for direct application and for suspension systems, it is important to identify the mounting when selecting a competitive product.

Sound absorption coefficients are obtained in a laboratory by testing in a "reverberation chamber" (Figure 2-7). The American Society for Testing and Materials (ASTM) procedure C423[5] specifies the measurement process. Generally, the specimen, containing 72 square ft. of material, is placed in a large reverberant space designed to cause sounds to impinge on the specimen at all angles of incidence. A sound signal is introduced and then chopped off to allow it to reverberate. The amount of time required for the sound to dissipate to a set level is called the reverberation time. This is then converted by a formula to the sound absorption coefficient (SAC). Measurements are made to determine the SAC at one-third octave intervals between 125 Hz and 5000 Hz.

[5] ASTM Committee E33 on Environmental Acoustics is the standards writing authority for Procedure C423, Test Method for Sound Absorption and Sound Absorption Coefficients by the Reverberation Room Method. It can be found in the current edition of ASTM Standards, Volume 04.06.

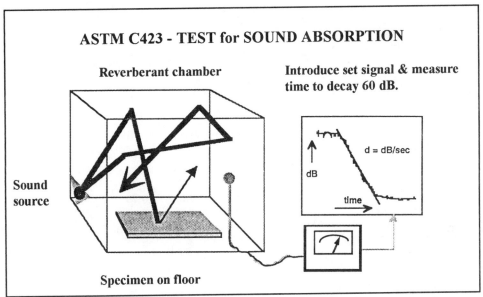

Figure 2-7 Sound absorption coefficients are obtained in a "reverberation chamber" per ASTM C423.

Reported data is usually also given by using a single-number rating system called the noise reduction class (NRC). NRC is determined by averaging the sound absorption coefficients between 500 and 2000 Hz, chosen because this is the frequency range where the information that makes speech intelligible occurs. NRC and sound-absorption coefficients at the individual frequencies generally range from 0.01 to 1.00. NRC can be best described as a rating scheme like a percentage scale. An NRC of 0.01 would be a material like glass that has little or no sound-absorbing capability. An NRC of 1.00 would be akin to open air where all the sound escapes and is not reflected back[6]. Sound absorption coefficients and NRCs of typical materials are given in Appendix. 3. These materials were tested by the reverberation room method known in North America as ASTM C423 (Figure 2-7).

2.3.2 Room Acoustics Design Using Sound Absorption

Listeners hear the combined effects of absorption by the various room surfaces and the attenuation of direct sound due to the distance that sound travels between reflections. If there is no absorbing material on the room surfaces they are considered "acoustically hard" and the room will sound like it is full of echos. If the room surfaces are "acoustically soft" or sound absorbent, the direct sound will dominate and there will

---

[6] Do not confuse NRC with the percentage scale. It is merely an analogy to help the reader understand the process. In actuality sound absorption coefficients and NRCs greater than 1.00 can and do exist. Also note that NRC is an average. It was developed for quick analysis only. We all know the story of the guy who couldn't swim who was told, "you can make it across, the stream is only an average of 1 foot deep." As he proceeded across by foot he drowned in a hole that was 10 ft. deep. Likewise, it would be foolhardy to make a significant material selection based only on NRC. A worksheet on how to select a material with the correct amount of sound absorption for the desired reverberation time in a room is provided in Appendix 2.

be essentially no echos. Speech is easier to understand with no echos or a short reverberation time.

This combined effect is called the room absorption and is given in a metric called sabins of absorption.[7] The total amount of sabins of absorption in a room can be calculated by multiplying the total area of room surface by the sound absorption coefficient of the room surface material and adding the sound-absorption coefficients of individual objects such as seats, furnishings, and persons. The room absorption is usually determined at all frequencies of interest. For speech, the frequencies are 500 to 2000 Hz.

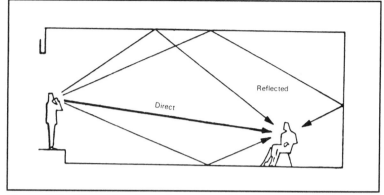

Figure 2-8   Direct & reflected sound combine to impact listener.

The absorption of a given surface is stated in terms of the sound absorption coefficient at individual frequencies or the single-number rating, called the noise reduction coefficient (NRC). Sound absorption for an entire room is given in sabins. One sabin of absorption is equal to one square foot of surface area having a sound absorption coefficient of 1.00. An open window to the sky, which is a perfect absorber, would have 1.00 sabin of absorption. A large surface area of 100 square ft. having an absorption coefficient of 0.80 would produce 80 sabins of absorption.

Absorption of objects is usually stated as a certain number of sabins per object. An upholstered chair can have an absorption of 3.0 sabins. In large rooms and at high frequencies, an appreciable amount of sound energy is absorbed by the air itself. In most living space, size and room air absorption is minimal. The sum of the sabins of absorption in a specific room volume will determine the reverberation time. For most rooms a reverberation time of approximately 2 seconds is desired for clear speech. A worksheet for calculating room absorption and the effect of adding sound-absorbing materials is given in Appendix 2.

2.3.3 Reverberation Time

The amount of reverberation in a room is measured by its reverberation time. This is defined as the number of seconds required for sound energy to die out. For testing purposes, a uniform sound level is produced and cut off precisely. A timer measures the time required for the sound to be reduced by 60 dB.

Reverberation time is a basic acoustical property of a room and depends primarily on the dimensions and sound absorption coefficients of the room surfaces and objects. If the reverberation time is long, say on the order of 10 seconds, the room will be "live" and "reverberant." A room with a 10-second reverberation time may be great for playing chamber music, since the room will mix and blend the music. By contrast, a 10-second reverberation time will make speech communication difficult if not impossible. Words will be garbled and not clear. A room with a 2-second reverberation time will not only make speech easy to understand but will probably reduce the sound level as well. In a typical room the reverberation time is

---

[7] The unit of absorption called the sabin was named after its founder, noted architectural acoustician Wallace Clement Sabine.

for unusual room designs this may not be true.

Room volume is a significant factor in establishing the reverberation time. In a typical residence size room, reverberation time will be rarely longer than 3 to 5 seconds even with no sound absorption. To make a living room acceptable for speech, a reverberation time of 1 to 2 seconds is recommended. This can usually be obtained with carpet and pad and typical overstuffed furniture. However, if you choose hardwood floors and metal furniture, the reverberation time may impede easy conversation (Figure 2-9).

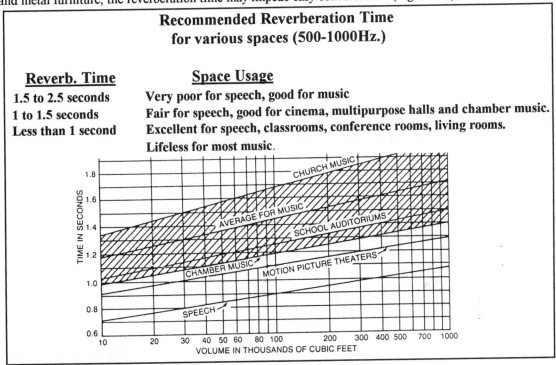

Recommended Reverberation Time
for various spaces (500-1000Hz.)

Reverb. Time	Space Usage
1.5 to 2.5 seconds	Very poor for speech, good for music
1 to 1.5 seconds	Fair for speech, good for cinema, multipurpose halls and chamber music.
Less than 1 second	Excellent for speech, classrooms, conference rooms, living rooms. Lifeless for most music.

Figure 2-9 Recommended optimum reverberation times for various room uses.

In a large room, sound travels farther between room surfaces. Reflections and absorption occur frequently. Other complications occur in real life. Long hallways or large rooms with low ceilings have different reflection patterns from cubic shapes.

Rooms designed for listening pleasure require the shortest reverberation time possible. Modern electronic playback equipment can produce recorded sound that is indiscernible from the original. Older, poor-quality equipment required a reverberant room to mix and meld the sounds for listener enjoyment. Reverberant rooms are no longer needed. **Today's listening environment should take the room out of the acoustical equation.** Electronic equipment controls reverberation. Thus, a room designed for ideal listening pleasure today should have an acoustical envelope (i.e., all interior surfaces) that is highly sound absorbent. Ceilings and walls should have sound absorption coefficients and NRCs at or above 1.00. Floors should be covered with extra-plush carpeting and thick open-cell foam pads. All windows should be covered with acoustically absorbing blinds and thick, heavy lined drapes. Furniture should be overstuffed and soft.

2.4 WALLS/PARTITIONS

2.4.1 Introduction

One of the most common complaints in residences, especially in multifamily dwellings, is "I can hear my neighbor." This usually results when the barrier between units has poor noise reduction characteristics. In a typical field situation, we evaluate the noise reduction of a barrier by determining the sound level in the source room and subtracting the sound level in the receiving room. See section 2.2 above.[8] For example, when the sound in the source room is determined to be 80 dB, and the sound level in the receiving room is 30 dB, we say the noise reduction of the barrier is 50 dB (i.e., 80 minus 30 = 50).[9]

When the situation warrants a detailed analysis in the laboratory or field, including the sound absorption characteristics of the source and receiving room, we utilize the term <u>sound transmission loss</u> (STL). STL is usually tested in accordance with standard procedures in ASTM E90 or E336. A single-number rating called the sound transmission class (STC) is used to make a quick identification of the efficiency of a sound barrier. The terms NR, STL, and STC are used interchangeably by many in the industry. All identify the sound-attenuating capability of a noise barrier. However, each has a precise meaning. For example, NR is used to identify a typical field situation where flanking sound may be predominant. STL and STC eliminate flanking paths and measure only the performance of the barrier in question.

Noise reduction between two adjoining spaces may be caused by the effect of the barrier and/or flanking paths. This section will discuss design attributes of the barrier, assuming there are no flanking paths. Sound flanking paths are significant and probably the most important item in any analysis of noise reduction. Therefore, flanking paths are discussed separately in Section 2.10.

Designs to improve the noise reduction or sound barrier characteristics of walls and partitions may utilize mass, resilient mountings, sound-absorbing core materials and a combination of these principals.

2.4.2 Mass

The effectiveness of a single leaf

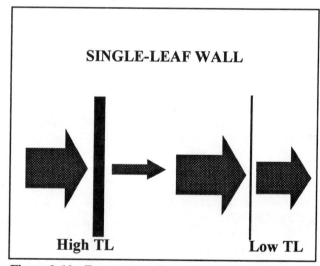

Figure 2-10 - Transmission Loss (TL) for single-leaf walls.

---

[8] Please do not confuse the term noise reduction (NR) with noise reduction coefficient (NRC). NRC identifies the sound-absorbing capabilities of a sound absorbing material. Noise reduction provides a measure of the sound-attenuating characteristics of a barrier. While similar sounding terms, they have quite different meanings.

[9] The unit dB is used without identification here for demonstration purposes. In reality, dB should be identified in a specific frequency band such as 50 dB at the one-third octave center band of 500 Hz, in dB_A, or some fashion that delineates the frequency being considered. If no identifier is mentioned and dB is not used in a generic way, it usually means dB_A. A technique for calculating dB_A from the dB at specific frequencies is given in Appendix 2.

(simple) wall, such as unit masonry (concrete block), concrete, or solid gypsum, depends mainly on its mass (Figure 2-10). Heavy walls of cement, solid concrete, or other masonry can reduce sound transmission. If the surfaces are sealed so they are air impervious, and joints are airtight, then each doubling of the weight can increase transmission loss 6 dB.[10]

Increasing mass is not generally an economically feasible technique for good acoustical design in modern buildings. This is particularly true in North America, where most residential buildings are lightweight wood framing. In Europe, where many multifamily structures are solid masonry, increasing mass for a better noise barrier may be economical. Also, as you increase mass, you may increase stiffness, which could counteract the effect of increased weight. Using mass alone, a noise reduction of 40 dB requires a solid concrete wall nearly 1-ft. thick. Doubling the mass to a 2-ft. thick wall improves the noise reduction capability to 46 dB. For most owners and builders, a wall of this size and mass is not practical. Costs will be more than double and the decibel-per-dollar achievement is clearly not acceptable.

2.4.3 Double Leaf Walls

A prime example of a double-leaf wall is the type used extensively for residential construction in North America. Composed of 2 by 4-in. wood studs 16-in. on center. with gypsum wallboard faces, this wall may be easily upgraded by using resilient mountings and sound-absorbing core materials (Figure 2-11).

Sound striking a double-leaf wall surface is transmitted through the first surface material into the wall cavity. It then strikes the opposite wall surface, causing it to vibrate and transmit the sound into the air of the adjoining room.

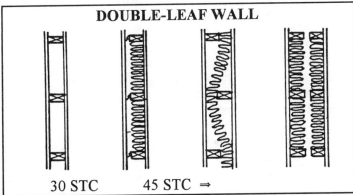

Figure 2-11 Double-leaf wall examples using resilient mounts and a sound-absorbing core .

This is termed <u>airborne</u> <u>sound</u>. When the sound strikes the wall at the stud, sound is transmitted direct through the stud and is termed <u>structureborne</u>[11]. A simple composed of studs and wallboard will provide noise reduction equivalent to a solid mass barrier several times its weight. For example, the STC of a 6 in.-thick concrete wall and the wood-stud and gypsum wallboard wall each have approximately 30 STC.

Upgrading the STC of a lightweight double-leaf wall has been researched by many over the past 30 years. Hundreds of designs have been tested by a multitude of techniques. The most practical methods of

---

[10] Acousticians have dubbed this phenomenon the "mass law." Some test agencies will show the mass effect as a slanted line on reports for comparison purposes.

[11] The terms <u>airborne</u> and <u>structureborne</u> sound are usually reserved for sounds being totally transferred via air or solid structures. They are used here to help describe the noise-transmitting characteristics of a typical wood stud wall system.

upgrading the double-leaf wall include some form of separation or discontinuity between the two wall faces. Effective techniques include double or staggered wood studs, slit studs, resilient clips, resilient furring channels, sound-deadening board, and various damping materials. For additional improvement, it has been determined that adding a sound-absorbing material to the core of these discontinuous structures is quite effective. Note, however, that adding absorptive material to the core is not effective unless the wall faces are decoupled. By adding more mass in the form of additional layers of gypsum wallboard, these lightweight walls can provide up to 60 STC of sound attenuation.

Why does decoupling work? Staggered, double, and split studs minimize the direct mechanical connection between the surfaces. They also tend to reduce the stiffness of the wall. The same effect is gained by attaching one or both faces of gypsum wallboard to the studs with resilient clips or resilient furring channels. The resilient mounts perform much like an automobile spring. Sound transmission is reduced by breaking the sound path. In addition, the air cavity provides vibration isolation between the two sides. Sound in one room striking one side of the wall causes it to vibrate, but because of the mechanical separation and the cushioning effect of the cavity, the vibration of the other side is greatly reduced. Single-stud walls have much higher effectiveness than single-leaf walls of the same overall weight. Discontinuous constructions will improve sound transmission performance by an additional 6 to 10 STC (Figure 2-12).

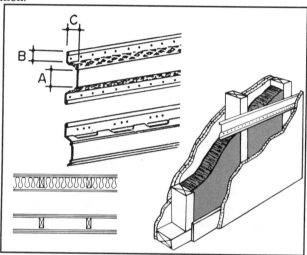

Figure 2-12 *Resilient furring channel examples. Top channel invented by D. A. Harris. A=1¼-in., B=7/16-in., C = ½-in., metal thickness = 25 gauge galvanized.*

What does a sound-absorbing core provide? Sound-absorbing blankets are usually fiberglass building insulation or mineral fiber building insulation[12]. While some are marketed as sound blankets, they are essentially the same materials used for thermal insulation. The blankets further improve the sound isolation performance of the wall by absorbing sound energy in the cavity before the sound can set the opposite wall surface in motion. These blankets will also provide some damping of the vibrating wall surface. When used with a decoupled wall system of staggered studs or resilient channels, sound-absorbing blankets will improve the wall performance by an additional 5 to 10 STC. Caution: Sound-absorbing blankets have not proved effective unless the base wall is discontinuous or decoupled.

Variations of decoupling and absorbing techniques have spawned several interesting products. The

---

[12] Fiberglass insulation is made by spinning molten sand into fibers and a wool-like blanket. Mineral insulation is composed of rock wool. The major differences are uniformity of fibers and melting temperature. Glass fibers melt at approximately 1100° F and rock wool at 2500° F, an important consideration for time-design fire ratings. Glass fibers are uniform in quality while rock wool has a wide variation with lots of "shot" or small pebbles in the blanket. Additional discussion of sound-insulating blankets and the "density myth" is presented in Chapter 5.

first was wood and mineral fiber sound-deadening board. The theory of sound-deadening board is to provide both a decoupler and an absorber in one product. During the 1970s, sound-deadening board was offered by several manufacturers and was quite popular. Unfortunately, the boards did not live up to expectations[13]. Resilient channels and blankets plus drywall steel-stud walls are easier to use and cheaper. To be effective as a decoupler, the outer layer of gypsum wallboard must be attached with thick adhesive midway between the studs, thereby making the board act as a leaf spring. As this technology evolved, it was found that a layer of ¼-in. gypsum wallboard installed in the same fashion was just as effective and cheaper. A second research finding focused on the sound-absorbing blankets. Density became a big issue with considerable controversy about the effectiveness of higher-density blankets. Sprayed-in wood fiber insulation also was used by some builders when it was found that spray installation tended to seal all the cracks and flanking paths[14]. A recent development is the use of damping materials. They are called viscoelastic damping compounds and utilize the principal of "constrained layer damping." Developed for the transportation industry, these materials are the core between two sheets of metal. In 1993, Jeep and several Japanese companies featured these products in national TV advertisements. This same technique has been effectively applied to gypsum wallboard.

2.4.4 Other Interior Walls

Thousands of wall configurations have been invented and tested for their sound transmission loss characteristics by building materials manufacturers. Wall systems are generally broken down into two categories: load-bearing walls and non-load-bearing walls. Most load-bearing systems are masonry or wood studs. Trusses and long-span joists allow non-load-bearing systems to be used in residential buildings, hotels, and singlefamily and multifamily construction (Figure 2-13).

Steel-stud wall systems are popular because they are incombustible, carry 1- and 2-hour time-design fire ratings, and are cheap and easy to construct. Acoustically, 25 gauge drywall steel studs are unique. They are "inherently resilient." The studs act like a wall

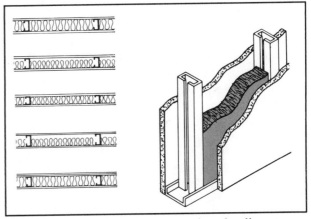

Figure 2-13 *Non-load bearing steel-stud wall systems provide excellent decibels-per-dollar.*

decouplers or resilient attachment. Staggered stud and resilient attachments are not effective, as the studs themselves provide the decoupling action. For example, a simple 24-in. OC. steel stud wall with ½-in. gypsum wallboard faces attached with screws 12-in. OC has a 35 to 39 STC. Adding a sound-insulating blanket to the core boosts the rating to 45 STC. Adding another layer of gypsum wallboard to each face, with the sound-insulating blanket core, yields a 50 to 54 STC, an unsurpassed decibel-per-dollar value.

Selection of the wall system best suited for your project requires consideration of a number of wall

---

[13] Sound-deadening board technology is discussed in Chapter 5.

[14] These techniques are discussed at some length in Chapter 5 - Myths, Misconceptions and Mysteries.

attributes other than their sound transmission loss characteristics. In light frame structures, the demising partition might be load-bearing or non-load-bearing. Many code jurisdictions may require noncombustible construction. Wood studs 16-in. OC are generally used for single-family or two-to four- family dwellings. STC ratings for party walls between living units will range from 40 STC for dwellings in the inner city, where background noise levels are high, to over 55 STC in top-of-the-line condominium projects located in a quiet country setting. There are many other considerations including one of paramount importance, economics.[15]

2.5 FLOOR/CEILING SYSTEMS
2.5.1 System Generalities
Floor/ceiling systems are a major sound barrier between living units in a multifamily dwelling. Like walls, a floor/ceiling system must block airborne sound. In addition, the floor/ceiling system must also reduce impact sounds. This dual function presents a significant challenge to the designer.

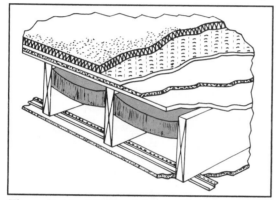

Figure 2-14 *A lightweight wood-joist floor/ceiling assembly must block both airborne and impact-generated noise.*

Airborne-sound isolation may be improved in a floor/ceiling assembly by increasing the mass, providing a resilient mount for the ceiling or floor, and adding sound-absorbing material to the core or joist space. Unlike walls, floors must be strong and stiff to satisfy structural floor load requirements, a requirement that essentially rules out an inherently resilient structural joist. In practice, designers are limited to resiliently mounted ceiling systems, sound-absorbing blankets, and mass for both wood and steel joist systems. Fortunately, most floor/ceiling assemblies inherently have more mass and thickness than walls, making it easier to achieve higher STC ratings. A standard floor/ceiling system with 2 by 10-in. joists, plywood floor, and gypsum wallboard ceiling will have an STC between 35 and 40 (Figure 2-14).

Impact sounds are very difficult to control in North American lightweight wood-joist floor/ceiling systems. As a person walks across the typical 2 by 10-in. wood joist floor with a plywood subfloor and gypsum wallboard ceiling, the impact sounds of footfalls may be intolerable to the occupant below. This stiff lightweight assembly acts like a drumhead under impact by a drumstick. Reducing impact noise from footfalls and falling objects is best handled by treating the source of the sound itself. A carpet and pad works best as it will cushion the impact and substantially reduce the problem. Unfortunately, carpets and pads are not always an acceptable solution. For finish floors that are acoustically hard, such as vinyl floor covering, hardwood floors, and ceramic tile, the designer must employ elaborate means to reduce impact sounds. Many of the ideas used to improve airborne sound also improve impact sounds, such as, mass, resilient mounts, and sound-absorbing core materials. However, none of these solutions are as effective as cushioning the impacts with a carpet and pad.

Impact noise ratings and test procedures are quite different from those for airborne sound. Several

---

[15] See Chapter 4 for more details on acoustical rating requirements by code and governing bodies.

methods are employed in the industry. The most effective one, a live walker test, has not yet been adopted by the consensus-standards writing groups. While not universally accepted by everyone, the tapping machine procedure is widely utilized. In this instance, the floor/ceiling system in question is evaluated by activating a special machine that drops hammers of a precise size, weight, and height on the floor surface. Sound levels are measured in the receiving room below. Results are plotted in one-third octave intervals at frequencies between 100 and 5000 Hz. A single-number rating called the impact isolation class (IIC) is established. A 2 by 10-in. wood joist with plywood floor and gypsum wallboard ceiling has IIC 32. Adding a carpet and pad improves the rating to IIC 66.[16]

2.5.2 Impact and Airborne Design Techniques
To reduce impact and airborne sound transmission through floor/ceiling systems:

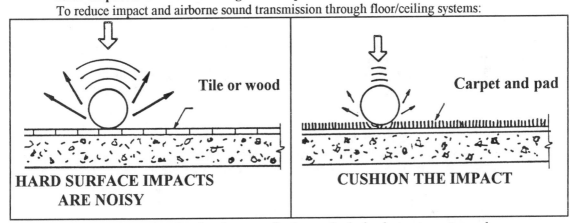

Figure 2-15 *Use of carpet and pad is the most effective method of reducing impact sounds.*

Cushion Impacts- Since direct impacts set a structure in motion, cushioning the impact can significantly reduce the problem. Carpet, carpet padding, and other methods of surface treatment are the most effective way to reduce impact noise (Figure 2-15). Caution: For floors that cannot utilize carpet and pad, the designer must realize that considerable expense is required to provide the degree of impact sound mitigation that can be achieved by damping the impact at the source.

Floating Floors- Transmission of both impact and airborne sounds can be reduced by isolating the flooring from the joist or slab structure with resilient materials. Impact noise isolation increases with the degree of resilience. Like an automobile spring, a floor must deflect to provide isolation. The effectiveness of floating floors is limited, especially at low frequencies, since deflections are severely limited by the structural needs. Since a floor is primarily a support surface in a structure, all the efforts to provide a springy floor will be in direct conflict with the structural requirements. Examples of floating floors include those with resilient materials like fiberglass sound-deadening board, cork, and springs under the sub-floor. Exotic systems have been developed by several manufacturers but are expensive.

Resilient Ceilings- A very common and widely utilized technique for improving both the airborne

---

[16] Airborne and impact ratings for floor/ceiling assemblies are listed in Appendix 3. Impact noise ratings and procedures are discussed in Chapter 4. Examples of designs are given in Chapter 5 and 7.

and impact noise is installing resilient furring channels (Figure 2-12). Attached to the bottom of wood or steel floor joists, spaced 24-in. OC, these channels provide a handy resilient mounting for gypsum wallboard ceiling systems. Resilient channels decouple the structural floor from the ceiling and will improve both the STC and IIC by approximately 5 points. Many other clips and resilient mountings have been invented. However, none has proved as practical as resilient furring channels. Note that the channels require the interior surface of the walls to be resiliently mounted or separated from the framing. This method works more efficiently in big rooms than in small rooms.

Absorb Noise between Floor and Ceiling-Installation of sound-absorbing blankets composed of fiberglass or mineral building insulation between the joists of either a wood or steel joist system will improve the STC and IIC by approximately 5 to 7 points. When sound-absorbing blankets are used with resilient furring channels, both the IIC and STC will be 10 or more points higher than the base assembly. Spray-in cellulose insulation has also proved effective. It has the added attribute of tending to seal cracks.[17] Caution: Adding sound-absorbing materials in the floor/ceiling core will not be effective unless the structural floor assembly is decoupled from the ceiling system.

Mass- Adding mass to the floor system is a very effective way to upgrade both the STC and IIC of a floor/ceiling system. On wood-joist systems, a common practice is to lay down a plywood or gypsum board container over the joists and pour 2 in. or more of lightweight concrete or "gypsum concrete." This technique, utilized with carpet-and-pad- finished floor with a resilient channel suspended gypsum wallboard ceiling and sound blanket core, is one of the best systems available in lightweight wood structures. Note that this system requires upgraded structural members to accommodate the added weight.

For commercial-type structures, where steel joists and concrete floors are the norm, mass may become an integral part of the acoustical design. In this case, the IIC is still greatly improved with a carpet and pad. Likewise, decoupling the structural joists by using a resiliently mounted ceiling and a core of sound-absorbing blankets is very effective. If the ceiling assembly is suspended by 16 in., as is typical in many commercial-type structures, resilient furring channels may not be needed.

Absorptive Ceilings-Sound-absorbing ceilings add a feeling of quiet and well-being to any room. By absorbing sounds created within the room, they prevent the buildup of reverberant sound due to reflections, thereby reducing annoying distortion and resultant sound levels. Note that adding sound-absorbing ceilings will not appreciably improve either the STC or IIC rating of an assembly. Also note that most sound-absorbing ceiling materials are poor sound barriers and should not replace gypsum wallboard, plywood, etc.

Damped Floors- A new and very effective material has been developed for lightweight floor systems. Using constrained-layer damping techniques and a viscoelastic core, floor/ceiling systems using a damped plywood structural floor improve both the IIC and STC. STC 57 and IIC 40 have been achieved with wood joists, 1-in. damped plywood, resilient channels, sound-absorbing core, and gypsum wallboard ceilings. With a carpet and pad the IIC is 64, equivalent to concrete floor systems 2 to 3 times their weight cost.[18]

---

[17] Caution is advised on using cellulose materials since they are combustible, will "punk", or burn slowly, for a prolonged period before catching fire, are difficult to monitor during installation, and may tend to short-circuit resilient attachment devices. The author has also been suspicious that sound ratings on systems using cellulose spray-in materials have been achieved on constructions that were not adequately dry for testing. These materials contain considerable water that not only adds mass but will change the stiffness of the gypsum wallboard ceiling.

[18] A case study of quiet floor systems may be found in Chapter 7.5. Test data is listed in Appendix 3.

2.6 WALL AND FLOOR INTERSECTIONS

Like a chain and its weakest link, walls and floor systems designed to stop noise transmission are only as good as their joint or intersection. If the wall and floor are not joined together properly, their value may be lost. Many of the joint defects are actually flanking paths, where the sound travels via an air gap or via the structure.

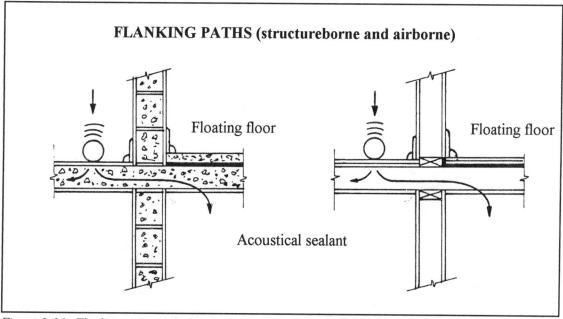

FLANKING PATHS (structureborne and airborne)

Floating floor   Floating floor

Acoustical sealant

Figure 2-16   Flanking noise paths through floors.

2.6.1 Air Leaks.
Acoustical sealants are the best means to provide a permanent air seal. They are made from materials that are permanently elastic to allow the floor or wall materials to move, as they are prone to do from expansion, contraction or outside forces such as earth movement, earthquakes and the like. An air gap of ⅛ -in. at the juncture of an STC 45 partition and a 45 STC floor system will reduce the STC by over 15 points. A permanent tight seal is the most effective way to maintain the acoustical integrity of the wall and floor system[19] (Figure 2-16).

Structural flanking may occur in many ways. The following paragraphs present recommended construction steps in joining walls and floor systems that will permit the floor/ceiling and wall constructions to perform as intended in actual buildings.

---

[19] Acoustical sealants and their use are discussed in Section 2.10.

2.6.2 Floor Joists
When the floor joists are perpendicular to sound barrier walls, two precautions are needed at the floor/wall juncture:
- Block the space between joists to stop airborne sound.
- Reduce transmission of impact and structureborne noise.

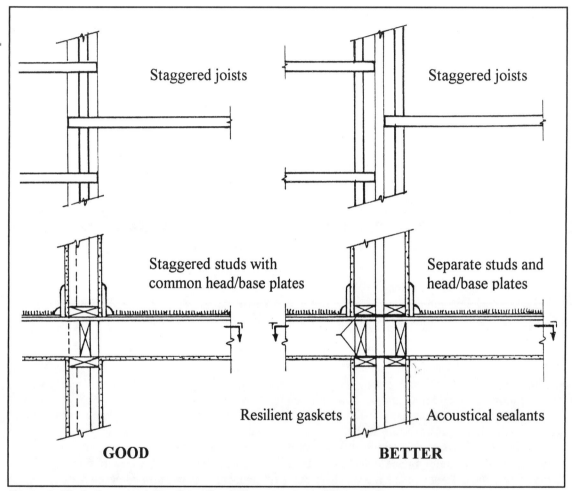

Figure 2-17 Architectural detailing of wall/floor junctions with walls
(Walls perpendicular to floor joists)

The two constructions shown in figure 2-17 both perform well by blocking interjoist spaces and breaking the structural path. The diagram labeled "better" not only breaks the path for sound vibrations but, because of its double blocking, allows plumbing and electrical service to pass through plates, without creating a flanking path. Note that the subfloor should have breaks at the walls to reduce structureborne sound transmission.

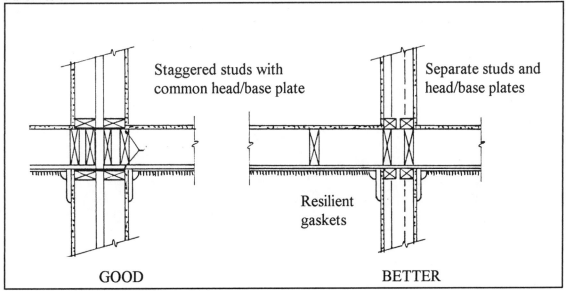

Staggered studs with common head/base plate

Separate studs and head/base plates

Resilient gaskets

GOOD

BETTER

Figure 2-18 Wall/floor detail with walls parallel to floor joists.

When the floor joists run parallel to the sound-blocking walls, the methods shown in figure 2-18 will carry the vertical load and minimize the transmission of sound energy. Regardless of which system is used, all openings, cracks, and material joints should be made airtight with a permanently elastic acoustical sealant.

2.6.3 Stairways

Interior stairways are a source of structureborne and impact noise, particularly in wood frame construction. This problem is prevalent in all multistory construction, even in side-by-side units with the bedrooms on the second floor and the living area on the first floor.

The key concern is impact noise. Impact noise may be mitigated by installing soft floor coverings on the stair treads. A thick carpet over a rubber pad works best. Lightweight or indoor/outdoor carpets are barely better than vinyl stair treads. Plain wood or ceramic stair treads pose substantial impact noise problems. If stair noise is a potential problem, there is no better decibel-per-dollar impact than utilization of a top quality, plush, thick wool carpet installed over a ½" open cell foam rubber pad. No other technique to reduce impact noise on stairs will come close to providing the sound insulation required for the price.

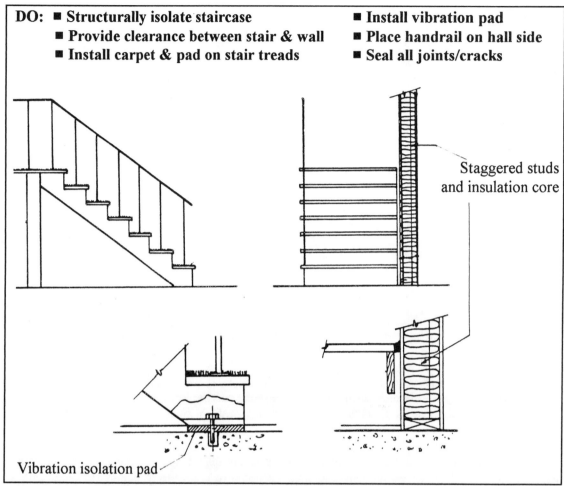

DO: ■ **Structurally isolate staircase**   ■ **Install vibration pad**
      ■ **Provide clearance between stair & wall**   ■ **Place handrail on hall side**
      ■ **Install carpet & pad on stair treads**   ■ **Seal all joints/cracks**

Staggered studs
and insulation core

Vibration isolation pad

Figure 2-19 Noise isolation design elements for stairways.

In wood frame construction, nailing stairway runners directly to adjacent studs will transmit impact and structureborne noise. Isolate these potential construction connections with soft resilient pads. A number of materials designed for sound isolation such as neoprene pads, fiberglass sound isolation board, and spring mounts should be utilized to isolate the stair supports from other structural elements (Figure 2-19). Techniques to help correct stairway problems are:

■   Install thick, heavy, good-quality carpet and pad.
■   When stairs are on both sides of the party wall in wood frame construction, use staggered stud walls with sound isolation blankets in the core. The wall-side stair runner should be held ½ or ⅝-in. (depending on gypsum wallboard thickness) away from the studs to permit installing wallboard behind the runners at a later date. If plaster is to be used, it is

recommended that techniques be employed that will assure complete plastering and eliminate any breaks in the party wall.

■ If the stairway is on one side of the wall only, the stairway side should be completely finished before the stairs are installed. Install sound-isolation blanket or spray-in insulation in the stud spaces. Construct an additional wall of 2 by 2-in. studs and gypsum wallboard on the living wall side or add resilient furring channels and two layers of gypsum wallboard.

■ Stairway landings, either intermediate or top, should receive careful floor treatment with joist spaces completely blocked off against the party wall. Use the header joist to support floors instead of nailing floor joists to studs.

■ To eliminate potential squeaks, utilize adhesive and nails or screws to connect all materials.

■ All stair supports should be isolated from the structural support with a vibration isolation pad. This vibration break will curtail the transmission of structureborne noise to other parts of the building.

2.6.4 Wall Support of Floating Floors

Resilient Underlayments. Resilient floor underlayments are not designed to support walls or pillars. They are designed to carry only the relatively light loading of floors, people, and furniture. It has been the author's experience that any material resilient enough to work acoustically is not likely to work structurally. This is not to say that no resilient underlayment exists. In fact there are a number of exotic or very expensive systems that can be used to isolate structurally transmitted noise. They are typically designed for special circumstances such as supporting buildings near subways, protecting against earthquake and supporting acoustical chambers. In these circumstances, acousticians and structural experts work together to design a system that provides a mass that will move at designed frequencies that do not adversely affect the structural system.[20] If such a solution is necessary, the design should be handled by a qualified expert.[21]

Proper Wood Floor Construction. Support load-bearing walls on a floor plate that rests on subflooring so that the weight is uniformly distributed and structurally adequate. The floating floor should be terminated at least ½-in. from the plate. Fill the resulting crack with a pliable material such as a 5-lb. density fiberglass insulation board as a sill sealer. Seal all air cracks with a permanently elastic acoustical sealant and cover the joint with molding. Note that the molding must be nailed to only one of the surfaces creating the floor/wall juncture to allow movement (Figure 2-20a). Note that this detail is also preferred when a damped sub-floor material such as dB-Ply [22] is used.

---

[20] Materials and systems designed for industrial noise control and vibration isolation are discussed in Chapter 3. A more detailed discussion may be found in *Noise Control Manual* by D. A. Harris, VNR, 1992.

[21] Qualified experts are members of the National Council of Acoustical Consultants (NCAC) - See Section 4.7 for address.

22 dB-Ply is a registered trademark of Greenwood Forest Products, Lake Oswego, Oregon. The product is composed of outer faces of plywood or other subfloor material bonded at the centroid with a damping material. The result is a constrained-layer damped floor panel that reduces impact, and structureborne and airborne noise transmission. See case study in Section 7.5.

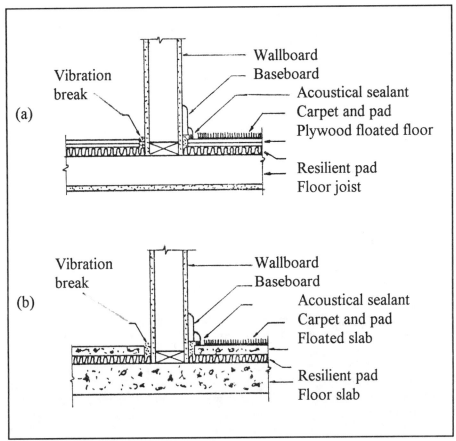

Figure 2-20 Wall/floor juncture detail for floating floor or a damped sub-floor panel.

Proper Slab Construction. Support structural wall on floor plate as in standard wood frame construction. Apply to the slab, a resilient pad composed of at least 5-lb. density fiberglass board covered with a water-resistant film such as polyethylen. Provide an expansion joint of fiberglass board and a permanently elastic acoustical sealant around the entire room. Pour lightweight concrete or gypsum concrete over the resilient pad making sure the pour does not come in direct contact or tie to the wall in any way. Use molding to cover the expansion joint (Figure 2-20). Note that this detail is preferred for any floating floor material.

Preventing Room-to-Room Transmission along Concrete Floors. Floated concrete floors should be confined to a single room to prevent transmission of structureborne or impact noise along the slab under the wall into the adjacent room. Expansion joints around the perimeter of the room and across doorways are vital.

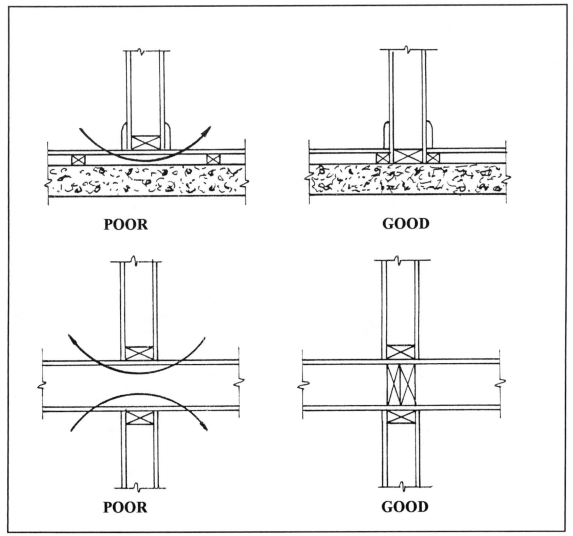

Figure 2-21 - Examples of typical mixed-type construction interfaces.

Preventing Flanking Noises in Mixed Design. Where construction types are mixed, provision to break structural paths between spaces is vital to good noise control (Figure 2-21). Expansion joints or open breaks must intervene between spaces. Caution: Be sure these breaks are airtight and sealed with a permanently elastic acoustical sealant to prevent airborne sound leaks.

Bedroom Hallways. It is standard practice by many builders to use the plenum area created by furring down the hallway ceiling for a hot-air supply duct or conditioned air-ducts. The ducts are often installed before the interior finish contractor arrives on the job and no ceiling is installed above the ducts. This permits impact noise from the hallway to come through the ceiling and down the open stud spaces to the walls of the hallway and adjoining rooms.

The solution to this noise transmission problem is to install gypsum wallboard or plaster surface materials on the base walls and ceilings of the hallway prior to installing the ducts, register outlets, and furred or false ceiling. Defer duct installation until the interior finish contractor finishes the base hallway walls and ceilings. At that time, install the ducts and the furred-down ceiling. This will assure that noise flanking paths are eliminated.

Another solution is to have the interior finishing contractor complete this work before the ducts are run, omitting the furred ceiling section. After the duct installation is complete, install a suspended acoustical tile ceiling system. This will block the transmission paths, provide some sound absorption for the hallway, and allow access to the duct elements for servicing.

2.7 DOORS

Doors nearly always lower the composite wall sound transmission characteristics. For example, a partition having 45 STC is very likely to be reduced to 30 STC with the installation of a door. The door becomes the weakest link in the composite STC rating. Thus the composite rating is essentially governed by the sound transmission loss capability of the door system including the door, door frame, and the door seal.

The single most important factor in improving the acoustical performance of a door is to make it airtight. An air leak is also an acoustical leak, or as acousticians term it, an airborne flanking path. To be effective, the door juncture with the frame, door stop, and threshold must be airtight in the closed position. This is difficult to achieve since the hinges and single latch are prone to a loose fit. To make them air-tight also makes the door hard to operate. With this obvious clash of purpose, door design and installation details must be carefully considered and implemented to maintain acoustical integrity and ease of operation.

The most important factors in good acoustical door installation are:

- Location. Place door where it will cause least problem for noise transmission.
- Construction. Select a door with good sound transmission loss characteristics and tight perimeter seals including head, side, and sill.
- Installation. Be sure all seals are airtight yet easily operable.

For example, if the wall system was designed to have 45 STC, then the door and door assembly must also have the capability of maintaining the same degree of acoustical integrity, namely 45 STC. To do otherwise is a waste of money. Since normal interior doors have 15 to 20 STC, it becomes readily apparent that "acoustical doors" or "sound reduction doors" will be needed for those locations that require sound isolation.

Some typical examples where a sound reduction door should be specified include:

- Access to mechanical equipment or furnace rooms.
- Entry to a common entry hall.
- Kitchens, bathrooms, and listening centers for privacy.
- Between motel rooms that open into a suite and/or any wall that separates occupancies.
- Outside entry doors when the outside noise is excessive.

2.7.1 Door Location

In multifamily dwellings, with internal hallways, doors should be located to avoid a direct sound path. Locate doors so they are staggered across hallways (Figure 2-22). Staggering the doors will provide the longest distance for sound to travel, and thereby provide the opportunity for sound to be diffused and absorbed in the hallway ceiling, wall, and floor materials (provided they are sound absorbent) before reaching adjacent occupancies.

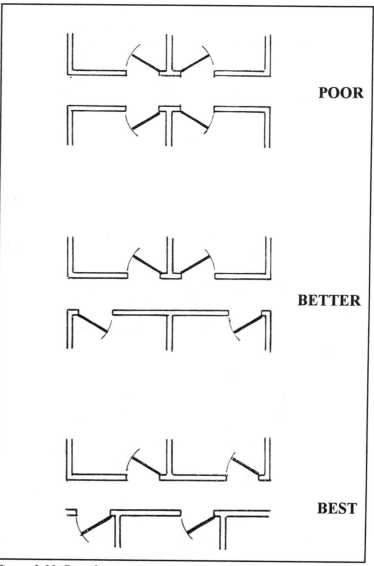

POOR

BETTER

BEST

Figure 2-22 Door locations in common hallway.

Exterior doors should be located away from noisy streets or obvious outside noise sources. Exterior sliding doors are notoriously poor sound barriers. Their inherent design inhibits their ability to provide an airtight seal. Unless you are willing to pay for exterior doors that have an STC that is equivalent to the exterior wall, it is best to locate these doors away from the potential noise source. If the door must be located on the noisy side of the building, consider designing a sound lock. Common in cold climates for conservation of thermal energy, a sound lock is like a small air lock on a space station. The outer door may be closed before the inner door is opened. A typical design is a small portico with a double set of doors. Place sound-absorbing materials in the small room between the doors to further attenuate transmitted noise.

Interior pocket sliding doors have the same inherent problem as sliding exterior doors. With no thermal gasketing of any kind, pocket doors offer minimal sound attenuation. Locate sliding doors only where a visual barrier is necessary. Bottom, top, and pocket sides are impractical to seal for sound conditioning. These doors should not be used to separate kitchens, bathrooms, and mechanical equipment from other living areas even within the same occupancy. For the same reason, pocket doors are not recommended for access doors to mechanical equipment rooms that open into common hallways.

2.7.2 Door Construction

Most lightweight interior doors have very poor sound barrier characteristics. The STC for wood panel, hollow-core, and other interior doors is approximately 15 STC. Interior doors are often undercut to allow for carpet clearance and mounted loosely for easy opening, making them nearly useless as a sound barrier. Moreover, they tend to warp over time, making edge and floor sealing difficult. Doors within a living unit can be made acoustically acceptable for most family situations by providing a weathertight seal. Install a bottom closure and weather stripping at the door jamb. If this action is insufficient, more dramatic and costly measures may be needed. One interim step would be to install a double door or back-to-back doors on the same frame.

Where greater sound privacy is needed, as in motels, hotels, multifamily dwellings, meeting rooms, equipment rooms, and the like, solid wood-core doors may be acceptable. STC ratings of 25 to 30 are common for solid-core doors provided they are outfitted with tightly fitting drop closures, weathertight seals, and latches that hold the door tightly closed. Interior doors for barriers having 40 STC or higher require special manufacture. These doors are usually sold as a door system including the frame, hinges, and lock set. Their performance is nearly always governed by the seal. Installation is critical and should be handled by a manufacturer's representative. It is recommended that specifications for these doors require acoustical field testing to ensure they perform as specified.[23]

Doors for furnace rooms, mechanical equipment and service areas must be fire doors as well as good acoustical barriers. Fire doors are made of steel or other incombustible materials. Their design is complex and highly regulated. While the fire rating is usually assured by considerable testing, an independent inspection program, and a certified product label by the testing authority, there is no such assurance of sound ratings. In fact, sound ratings may have been achieved on an assembly different than the assembly that is fire-rated and labeled. It is recommended that the specifying authority require field testing and/or a manufacturer's guarantee of the pertinent sound rating when the door is installed according to manufacturer's recommendations. It has been the author's experience that advertised sound ratings rarely are achieved in typical field situations. Installation by a person familiar with acoustical isolation practices and the quirks of the particular unit is mandatory and even then suspect. Recent American Society for Testing and Materials (ASTM) procedures for testing the sound transmission loss

---

[23] See Appendix 3 for a listing of door and door system sound ratings. See Chapter 4 for a discussion of sound test procedures. See Chapter 6 for a discussion of the problems of acoustical door design and utilization.

properties of doors and door systems require the door to be operable.[24]

2.7.3 Door Installation

For good acoustical performance, doors must be sealed around the entire door perimeter. The seal must be airtight to be effective (Figure 2-23). Effective acoustical seals can usually be achieved by applying resilient weather stripping on both side jambs and the head jamb It is usually effective on light interior doors and marginally acceptable on solid-core doors. An automatic threshold closer is a must for the bottom to be effectively sealed.

If the door is out of line only ⅛-in., tremendous pressure is created in some places in order to make contact at other places. It is imperative that doors and frames be matched carefully to minimize this situation. In wood doors and frames, normal tolerances are insufficient. Furthermore, differences in temperature and humidity, typical in any residential construction, are sufficient to cause warping over time. Door seals that are designed to be easily adjusted are recommended. Adjustments will be required on site several times a year, especially in cold/warm climates, to accommodate expansion and contraction or other movement. Even a slight air gap will render the door acoustically ineffective.[25] Closures to prevent slamming of doors and to keep them closed are obviously desirable.

When speech privacy is required, solid-core or double doors and an STC greater than 35 is usually required. The door frame is equally as important as the door. Fortunately, most doors are now supplied by manufacturers as door systems or prehung doors. This evolution is clearly in the right direction since tolerances are much better. However, the motivation for purchasing prehung doors is generally to save costs. With pressure to meet the low-cost price war, prehung units are made of the lightest materials possible and with little attempt at making a tight seal. Sound leaks are now very prevalent at the juncture between door jambs and the wall. It is a desirable practice to require all prehung door units to be set in a bed of acoustical sealant.

Drop closures installed on the bottom of the door can be very effective seals if designed and installed properly. It is imperative that the drop closure be adjustable at the job site. The preferred threshold closure will have a permanent threshold installed on the floor with a matching and interlocking shape on the door. (This is referred to as a lap joint in Figure 2-23.) An automatic drop closure is typically activated with a push rod that exerts pressure toward the floor as the door is closed. While effective when first installed, the automatic drop closure tends to get out of adjustment with use. Tube gaskets and sweep strips are useful only on lightweight hollow core or panel doors where minimal sound attenuation is required.

Door latches become a key element to creating a tight seal. A typical passage latch set is designed only to keep the door from swinging open. It provides little, if any, pressure on the door jamb gasket. While this latch set may be adjusted to provide a tight seal, the occupant will soon object to the

---

[24] See ASTM E 1408, Standard Test for Laboratory Measurement of the Sound Transmission Loss of Door Panels and Door Systems. Details are discussed in Chapter 4.

[25] This problem has monumental impact on the ability of the door to block sound. It has been shown that any door that will allow even the slightest seepage of air will reduce the sound transmission loss by a significant amount. A good test is to set off a smoke bomb in a slightly pressurized room and watch for smoke seepage. See Section 2.10 for the effect of an air gap on STC. It is also possible to search for leakage with a common stethoscope.

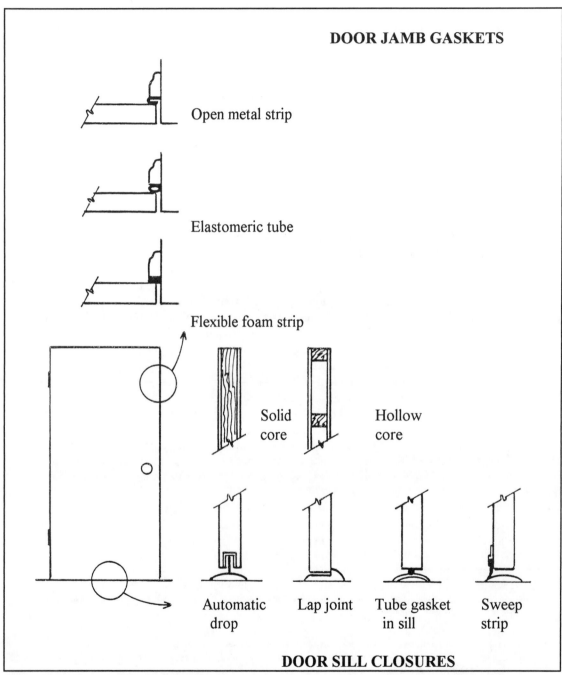

DOOR JAMB GASKETS

Open metal strip

Elastomeric tube

Flexible foam strip

Solid core

Hollow core

Automatic drop

Lap joint

Tube gasket in sill

Sweep strip

DOOR SILL CLOSURES

Figure 2-23 Door jamb and sill details.

difficulty in operating .the door. For a latch set to be effective in providing a seal, some form of cam action is required. This type of latch set is available only from door manufacturers marketing acoustical doors. To aid the seal, some acoustical door manufacturers have designed hinges that will provide a mechanical tightening action as the door is closed. This design is a must if the door system is to achieve a 40 STC field rating. For designs achieving STC 45 and above, the latch set may require a cam action latch with top, bottom, and middle tighteners such as found on commercial freezers.

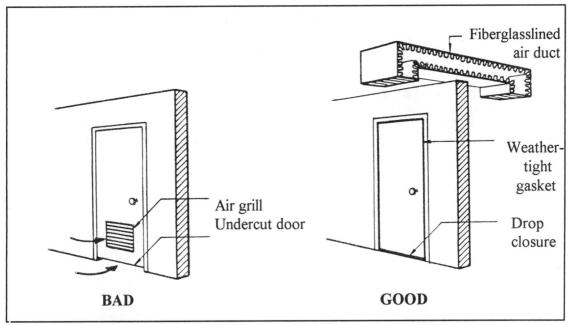

Figure 2-24 Door flanking paths caused by air return.

2.7.4 Door Flanking Paths

Doors and door systems with a good acoustical design may be completely negated by thoughtless designs by other trades. The most typical violations are caused by the heating, ventilation, and air-conditioning (HVAC) system designers who will undercut doors, and install louvered air vents or similar devices. All are designed for unrestricted and easy air flow for return air. Each of these designs will render the efforts to make the door system a better sound barrier totally worthless. These obvious flanking paths must be made airtight to be acoustically acceptable (Figure 2-24).

Designs for air returns that allow easy air flow and provide a degree of sound isolation typically utilize a long and circuitous path that is lined with a sound-absorbing material. A simple version is the fiberglass-lined duct. Note that the fiberglass must be located on the inside of the duct and exposed to the sound to be an effective sound absorber. Film facings tend to reduce the effectiveness. Ducts with the insulation material on the outside of the duct are acoustically worthless. More sophisticated ducts, known as <u>sound traps</u> or <u>sound attenuators</u> are available for difficult situations.[26]

Exterior doors designed to be thermally efficient tend to be acoustically efficient. Requirements for a good weather seal usually provide an airtight seal and reduce flanking paths. Thermally efficient door designs also tend to block sound better than interior doors. Double-pane glazing is also better for blocking sounds. When mounted with a storm door, exterior doors may have an STC of 30. To block sounds from a noisy street or airplane flyover noise, special acoustically rated doors may be needed. Again, field sound testing is recommended to assure the door rating is as intended. Acoustically rated exterior door systems have become more readily available with the advent of recent airport noise abatement programs.[27]

2.8 WINDOWS

In residential construction, windows generally perform three functions: "natural" lighting in daytime, a visual link with the outside, and in the case of operable windows, the potential introduction of fresh air into the building. Windows are generally recognized as the weakest part of the acoustical exterior envelope. Design criteria for the selection of window units for acoustical performance are based on existing or predicted future exterior noise levels at the building site, acoustical properties of the living space, sound absorption characteristics of the room, sound transmission loss characteristics of the exterior walls, and the design noise criteria within the room.

Selection of a land site and placement of the building on that site will establish the level of existing exterior noises. Acousticians term this the community noise level (CNL). Expressways/freeways, airports, and industrial sites are the most common noise concerns, particularly where operable windows are necessary for cooling and fresh air. Installation of year around air conditioning allows the windows to be closed and sealed and greatly reduces window noise problems.

2.8.1 Location of Windows

If possible, reduce the number and size of windows on the noisy side of the building. Since windows are the weak link of the acoustical envelope, the most economical way to maintain the acoustical integrity of an exterior wall is to remove the acoustical holes. Large window areas not only extend the size of the acoustical hole, they make the weak link weaker. In addition, larger-size panes of glass are more likely to vibrate at the same frequency of the sound source and have less noise barrier capabilities. Do not place windows that serve adjacent living spaces close together. Noise will be transmitted out one window into the other. Arrange casement windows so that they open in the same direction. Direct sound

---

[26] Lined ducts and HVAC quieting are discussed in Chapter 3. Sound attenuation values of lined ducts are given in Appendix 3.

[27] Exterior noise attenuation is covered in Chapter 6. Problems directly associated with aircraft noise are being addressed by many local airports. Some have programs that will fund improvements to residences to reduce noise intrusion.

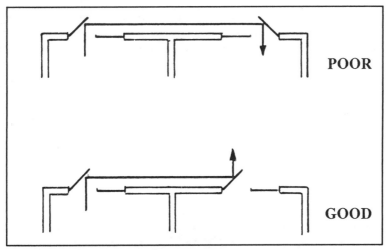

Figure 2-25 Apartment window orientation.

will be blocked by this arrangement instead of being reflected into the adjacent room (Figure 2-25).   Not adhering to this planning premise is certain to produce noise complaints from neighbors. Use sealed window construction in courtyards.

2..8.2 Selection of Glazing
Considerableble acoustical testing has been conducted on homogeneous glass and laminated safety or acoustical glass. Sound transmission loss ratings for glass alone are quite uniform and predictable. For example, a single pane of ¼-in. laminated glass has 35 STC while ⅛-in. plate glass has 24 STC. Some double-pane sealed window units have been tested as high as 45 STC (Figure 2-26).
        The author has found that 80% of the acoustical problems with construction and 20% of the problems are with the selection of the glass itself.[28]

2.8.3 Operable Windows
Windows fail as a sound barrier when they become operable. As with doors, it is nearly impossible to provide an adequate seal if the unit is to be opened and closed with ease.
        Windows with single glazing can at best have an STC of 31 when locked and sealed. Typical wood frame casement windows when locked will perform the best. Wood frame double-hung windows are less likely to be fully sealed even when locked. Plastic-coated units tend to be as good a barrier as their wood counterpart primarily because they are constructed with tighter tolerances. Metal frame windows, both aluminum and steel, tend to be poor sound barriers unless the manufacturer has taken special measures to design them to be noise barriers (Figure 2-27).
        As with doors, windows that have been designed to high thermal efficiency standards tend to be

---

[28] Sound transmission loss and STC data on windows is given in Appendix 3.

better sound barriers. The latch and lock mechanisms are also critical. Casement windows tend to perform better because they generally have two or more locking locations and they utilize a cam-operated closure. Casement units designed with a stepped and overlapped match with the frame and a compressible weather seal will perform acoustically better than double-hung windows.

Double glazing is a practical way to upgrade the sound barrier characteristics of a window unit.

	__STC__
<u>Sheet glass</u>	
⅛-in. thick (plate)	30
¼-in. thick (laminated)	34
½-in. thick (laminated)	32
¾-in. thick (laminated)	40
<u>1-in. Thermal insulating glass</u>	
(½-in. air space)	
2 ply- ¼-in. plate	32
<u>¼-in. Double glazed</u>	
(4¾-in. air space)	46

Figure 2-26 Sound transmission class of various glazing materials

Most modern double-glazed and even triple-glazed windows are fully sealed to prevent intrusion of moisture. This attribute also assures fewer sound leaks. Double glazing provided by using storm windows is also an effective way to upgrade the STC of a window unit. However, storm windows must be fully sealed in place to equate to sealed glazing. Wider spacings between glazing sheets will generally improve the sound barrier characteristics. The added air space also provides room for internal perimeter sound-absorption. Different glazing thickness and materials encourage frequency mismatchs that also may improve sound transmission loss characteristics.

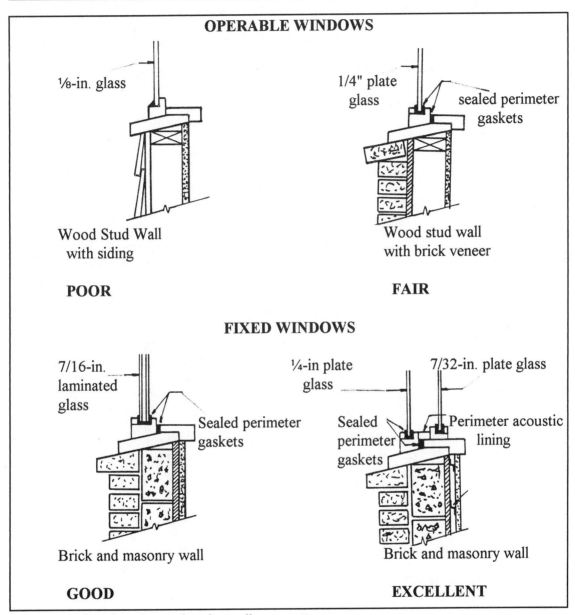

OPERABLE WINDOWS

⅛-in. glass

Wood Stud Wall
with siding

POOR

1/4" plate glass

sealed perimeter gaskets

Wood stud wall
with brick veneer

FAIR

FIXED WINDOWS

7/16-in. laminated glass

Sealed perimeter gaskets

Brick and masonry wall

GOOD

¼-in plate glass

7/32-in. plate glass

Sealed perimeter gaskets

Perimeter acoustic lining

Brick and masonry wall

EXCELLENT

Figure 2-27 Sound insulation of window-wall construction.

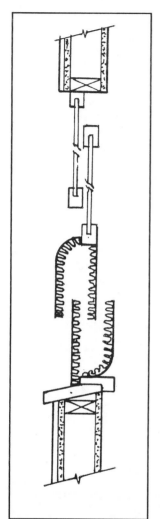

Figure 2-28 Window sound trap.

2.8.4 Operable Windows with Sound Traps

When windows must be left open to provide fresh air or for interior cooling, a sound trap can be utilized to reduce the intrusion of high-frequency noise. Low-frequency noise is difficult to absorb, and may come directly through the assembly and be perceived as a rumble. High-frequency sound, perceived as whistle or screech, is disturbing and easily absorbed.

A simple sound trap design utilizes an air duct with a sound-absorbing lining along a circuitous path. Although cumbersome, the unit will do a fair job of attenuating noise (Figure 2-28).

2.8.5 Fixed Windows

For high-noise areas it is best to install fixed windows. Note that when fixed windows are installed, most code bodies will also require the living units to be equipped with permanent air-handling and air-conditioning equipment. Plate glass is itself a good sound barrier material. Select the thickness of glass to meet the requirements for sound transmission loss (Figure 2-26). Thermal insulating glazing will improve both the thermal and acoustical efficiency. The added air space will provide a better STC than a solid equal thickness. In other words, two layers of ¼-in. glass has essentially the same STC as one layer of ½-in. glass. Separating the ¼-in. glass with an air space will make a better barrier than solid glass. Reduce the possibility of sound flanking by installing the glazing with a permanently flexible mounting and an airtight seal.

The most effective double glazing incorporates two panes of glass separated by at least 8 in. of enclosed air space. To avoid moisture buildup and flanking sound, be sure the seal is airtight. For insurance against moisture, install a jar of silica gel to absorb any stray moisture that may occur in the air space. Design the frame so that one pane can be removed for cleaning purposes.

Design the unit according to the following guidelines:

- Select glazing according to the required sound transmission loss. Use different weights per square foot of glass on each panel to avoid frequency/vibration matching.
- Select the widest spacing between glass that is practical.
- Install glass on flexible (resilient) mounts.
- Utilize a firm frame and provide two structurally separated mounting frames when installing a large window in lightweight construction.
- Add sound absorbent material, such as 1-in. glass-cloth-faced fiberglass board around the reveal edge of the window for cavity sound absorption.
- Safety glass has been shown to be a better acoustical barrier than solid glass. The laminate material acts as an internal damper, making the glass perform much like a constrained-layer-damped material.

Windows that overlook airports and freeways may require considerably more sound attenuation than normal. In this case, the design, construction, and testing should be conducted or supervised by a qualified acoustician. Note that site testing is imperative since flanking sound is a continuing concern.

2.9 SPECIFIC ROOMS/AREAS

Privacy and quiet are desirable during the day as well as the night. Areas requiring privacy are bedrooms, study, and bathrooms. Unfortunately, typical residential dwelling design and construction has ignored this lack of acoustical privacy. This often occurs because builders use thin or hollow-core doors, leave open air spaces under doors, and provide lightweight stiff interior walls. As a very minimum, bedrooms, studies, and bathrooms should be designed to provide a 35 STC sound attenuation. This requires solid core doors with tight air seals, thick carpeting, and the elimination of airborne and structureborne flanking paths.

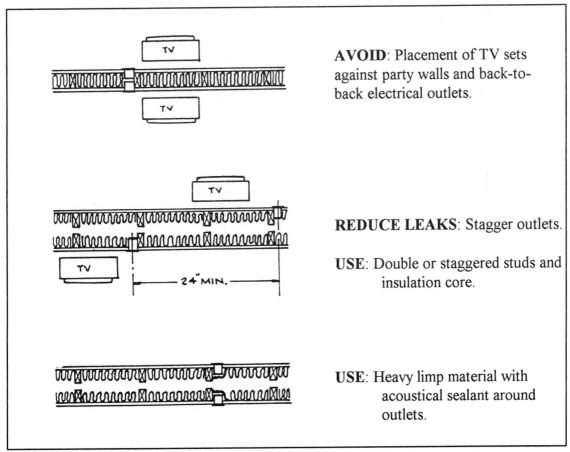

Figure 2-29 Party walls for TV rooms.

71

In rooms where there will be a TV, stereo or radio, especially when this room shares a common wall with another dwelling unit, special care should be taken to assure the common wall will adequately attenuate noise (Figure 2-29). Sound barriers should be at least 45 STC, even in the inner city where background noise levels may be high. In the suburbs or country, this wall may require STC 55 to properly isolate normal TV and stereo sounds. In these instances flanking paths become a paramount concern. Things like back-to-back electrical outlets must be eliminated and all wall penetrations carefully sealed with permanently elastic acoustical sealant.

Special listening rooms are a relatively new phenomenon in the home environment. New electronic equipment with high power capabilities has provided the user with sound levels far beyond normal human sound levels. Not only can this equipment produce loud sound, it can reproduce frequencies well above and below normal human response. The low-frequency capability, brought about with high-power amplifiers and phenomenal woofer speakers, is particularly troublesome. These units are capable of shaking the whole building down to 25 Hz.

Barrier designs and noise elements of building codes have not even considered the noise levels being generated by present day electronic equipment. Furthermore, the listeners seem not to be happy until the sound level activates their whole body.

Solutions for containing sound/noise[29] levels produced by modern electronic systems are expensive. Sound barriers may have to provide sound attenuation of 60 STC or higher. Barriers may require sound transmission loss values of 40 or more in the low-frequency range between 25 and 125 Hz. Achieving this degree of barrier performance may require measures far beyond the scope of this text. If a listening room is contemplated, locate the room well away from any room designated as a quiet space. Construct a "room within a room" where walls, floors, and ceilings are carefully isolated from the surrounding structure. Inside surfaces of the listening space should be rendered as sound absorptive as is feasible. Sound absorbing material 2 to 6 in. thick is recommended. Since modern equipment can faithfully reproduce any sound signal, including computerized reverberation, sound-reflecting surfaces should be nonexistent. Lastly, all potential flanking paths must be eliminated. It is recommended that a qualified acoustician with architectural design and engineering experience design this type of room. There are many nuances to be considered.

An obvious solution for the connoisseur of quality electronically reproduced sound is ear phones. Not only will the noise be eliminated for neighbors but others within the living unit will benefit. One word of caution. Those who use earphones tend to play them louder than realized. This is a well-established and documented way to ruin your hearing. Permanent hearing loss due to loud stereo earphones is at an alarming level in our society. As we move into the next generation of electronic equipment, we are now able to purchase and experience "virtual sound." Fortunately, at present, a "virtual reality" experience is conveyed via personal headsets. Given historical progression, we soon will be building rooms, like a Star Trek "holodeck," to experience this new technology using construction technology not yet perfected. Acoustical consultants will relish the challenge and opportunity.

---

[29] The term <u>sound/noise</u> is used here because the sounds produced by electronic equipment will be a beautiful sound to the listener but will likely be a terrible noise (unwanted sound) to neighbors.

2.10 AIRBORNE SOUND FLANKING PATHS

Airborne sound leaks, or flanking, is the most insidious problem in resolving sound transmission. Airborne sound is like water or air under pressure. It will find even the most minute hole or crack and seep through the intended sound barrier. Unless stopped, it will erupt into the adjacent space like a cork in a champagne bottle.

Even the smallest crack, one that can hardly be seen, will reduce a good sound barrier to a nearly worthless one. Furthermore, as sound barrier efficiency is improved, sound leaks become more prevalent and can cause greater reduction in barrier efficiency. For example, a standard wall of 2 x 4's and wallboard with a ⅛-in. crack along the floor may have STC 30 when properly sealed and STC 28 with the crack. A barrier with STC 55 fully sealed might drop to STC 35 with the same small crack.

Finding an airborne sound flanking path is more an art than a science. Flanking paths may be one or more of a multitude of paths, some obvious, others not easily identified (Figure 2-30).

Obvious flanking paths between a source and receiving room are:

- Open doors
- Open windows
- Air vents (supply and return) that serve both the source and receiving rooms

These types of flanking paths will reduce the performance of a demising partition to the point where it will be nearly worthless as sound barrier. Open windows and doors tend to also be reflectors, causing sound to bounce like a cue ball into the adjacent space. If the door or window is not in the position to reflect sounds, then sound attenuation of about 10 dB_A can be anticipated. Large air ducts made of sheet metal act like a speaking tube. They will even amplify sounds under some circumstances. A grill cover may cover the flanking path visually but generally will not attenuate airborne sound.

Detection of obvious flanking paths is a simple matter of observation. Listen to the noise and determine where the sound originates. In this case your own ears and perceptions are all that is necessary.

Design solutions generally involve the obvious - close the door or window. If this doesn't work, then a major redesign or acoustical solution must be found.

Less obvious flanking paths are:

- Joints between walls and floors, other walls or ceilings
- Back to back cabinets
- Pipe or other service line penetrations
- Unsealed panel joints
- Power and signal outlet boxes
- Joints between walls and window or door frames
- Poor seals on windows or doors (especially undercut doors)

A type of flanking path that is not readily apparent but nonetheless can substantially reduce the effectiveness of the barrier is a composite of several paths. Composite flanking paths are very difficult for an observer to identify, since the sound leak may be coming from several locations. An observer will hear the sound after it has been mixed by reflections in the receiving room. This mix of sounds make it difficult, if not impossible, to identify the precise location of the sound source. In fact, the sound may appear to be coming from a different location because of reflections etc.

Detection of the not-so-obvious airborne sound leak is where the art of noise control is fine-tuned. Since normal observations are not sufficient, other means must be employed to detect the sound leak. Electronic devices and sound test equipment can be used to identify and verify the fact that a sound leak exists, but none of the equipment is capable of locating the source of the leak or leaks. Detection becomes a matter of training. An acoustician with lots of field experience will likely commence a methodical search of the room.

Experience will tend to focus on areas of construction that are easily overlooked by the installing contractor. For example, drywall contractors typically will cut the gypsum wallboard short of joints, walls, and particularly at the floor line because they know that a cover molding will hide their loose-fitting joint. A common crack occurs along the juncture between the wall and floor. It is covered and out of sight when the base molding is installed. However, even a nice-appearing installation will have a large airborne sound path. Similarly, an outlet cover plate has holes for sound leaks. HVAC ducts are a favorite path since the most efficient distribution network will connect adjacent rooms on the same duct line. Verification of the leak may be a simple matter of close observation. With a radio tuned between the stations (static produces a broadband sound source) in one room, and listening carefully and intelligently in the adjacent room, sound leaks can be quite easily detected.

The solution for airborne sound leaks caused by cracks is to liberally install acoustical sealant at all building surface interfaces. This includes the joint between walls and floors, and all other dissimilar material joints, around penetrating pipes and outlet boxes (Figure 2-31).

Tough-to-Find Leaks. Considerable research has been done by several test laboratories and manufacturers of building products on sound leaks. In fact this research caused the introduction of materials called acoustical sealants. When the sound barrier is designed and tested in the laboratory to be 45 STC or higher, even minute sound leaks will cause a substantial degradation of the barrier performance. These leaks can be so small that even a trained ear cannot detect them. In order to verify that there may be sound leaks, the acoustician must first conduct a full scale sound transmission loss test in accordance with ASTM E336. This test is twofold. An initial test will be made to evaluate the overall sound isolation between the source and receiving room. Additional construction will isolate all potential flanking paths. If the two results differ, there is a good likelihood sound leaks exist. Tracing down the sound leaks requires persistence. For more detailed work, a stethoscope may be used. Caution must be exercised to avoid having localized effects, such as the "doubling effect" that occurs in a corner, lead to an incorrect assumption. For difficult circumstances, it may be beneficial to place a smoke bomb in the source room, pressurize the room slightly, and look for leaks.

Once the tough-to-find leaks are identified, the solution may be as simple as placing some acoustical sealant in the right spot or filling a nail hole with spackle compound. If the sound tests show that the wall is not performing as well as it did in the laboratory test, chances are that the construction details differ.[30]

---

[30] There are many reasons why a sound barrier will not perform as well in the field. A discussion of those experienced by the author and others is given in Chapter 5. A chart showing the effect of various size holes on the STC of a partition and research findings of typical field flanking, as evaluated in the laboratory is also given in Section 5.2.

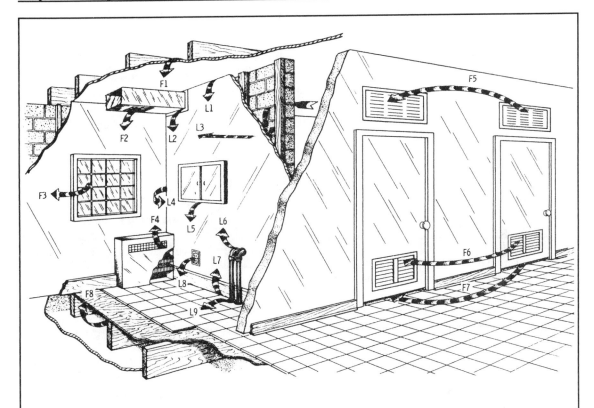

FLANKING NOISE PATHS

F1 OPEN PLENUMS OVER WALLS, FALSE CEILINGS
F2 UNBAFFLED DUCT RUNS
F3 OUTDOOR PATH, WINDOW TO WINDOW
F4 CONTINUOUS UNBAFFLED INDUCTOR UNITS
F5 HALL PATH, OPEN VENTS
F6 HALL PATH, LOUVERED DOORS
F7 HALL PATH, OPENINGS UNDER DOORS
F8 OPEN TROUGHS IN FLOOR-CEILING STRUCTURE

NOISE LEAKS

L1 POOR SEAL AT CEILING EDGES
L2 POOR SEAL AROUND DUCT PENETRATE
L3 POOR MORTAR JOINTS, POROUS MASONRY BLK
L4 POOR SEAL AT SIDEWALL, FILLER PANEL ETC.
L5 BACK TO BACK CABINETS, POOR WORKMANSHIP
L6 HOLES, GAPS AT WALL PENETRATIONS
L7 POOR SEAL AT FLOOR EDGES
L8 BACK TO BACK ELECTRICAL OUTLETS
L9 HOLES, GAPS AT FLOOR PENETRATIONS

OTHER POINTS TO CONSIDER, RE: LEAKS ARE (A) BATTEN STRIP A/O POST CONNECTIONS OF PREFABRICATED WALLS, (B) UNDER FLOOR PIPE OR SERVICE CHASES, (E) RECESSED, SPANNING LIGHT FIXTURES, (D) CEILING & FLOOR COVER PLATES OF MOVABLE WALLS, (E) UNSUPPORTED A/O UNBACKED WALL BOARD JOINTS (F) EDGES & BACKING OF BUILT-IN CABINETS & APPLIANCES, (G) PREFABRICATED, HOLLOW METAL, EXTERIOR CURTAIN WALLS.

Figure 2-30 Common airborne sound flanking paths in buildings.
(From U. S. Department of Commerce National Bureau of Standards, NBS Handbook 119, Brendt & Corlis, 1976.)

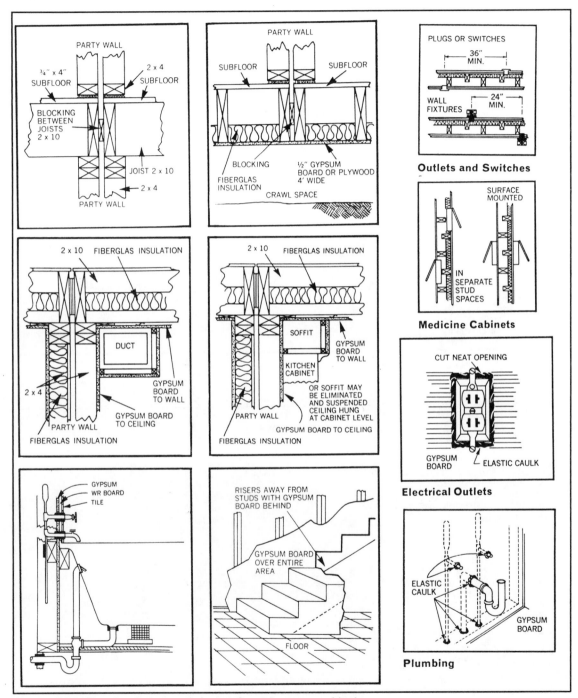

Figure 2-31 Sound isolation construction for airborne sound leaks.

76

2.11 SUMMARY

Following design guidelines for good sound control is imperative for occupant satisfaction. Implementing the elements shown in this chapter in the design stage is the most effective and economical solution to good noise control. It may increase the overall cost of the building by a very small percentage; the cost of a retrofit, on the other hand, can be devastating. Proper design corrections for noise control typically require major reconstruction with associated costs in time, labor, and project delay. In the worst scenario, potential buyers will disappear or buy elsewhere if the project gets bad press. An owner who cites poor noise control performance of the building prior to owner acceptance can cause financial and business ruin for the builder.

To avoid the potential for acoustical disaster, the owner and designer should insist on clear, concise specifications for the acoustical environment. Where noise is identified as a concern by the environmental impact analysis, detailed specifications and the requirement for field acoustical tests to demonstrate compliance are required.

This chapter discussed a wide range of design considerations. For additional information on how

Figure 2-32 Hand held sound level meter and frequency analyzer.

to estimate the acoustical properties of composite constructions, such as a wall with windows and doors, see Appendixes 2 and 3. Detailed sound ratings are given along with worksheets to estimate the expected results.

Chapter 3 RESIDENTIAL EQUIPMENT NOISE

3.1 INTRODUCTION

Control of residential equipment noise can be as easy as specifying a quiet appliance or as complex as controlling noise on the space shuttle. The term <u>quiet</u> has many meanings. Quiet to the residential occupant is relevant. With a high background sound level, the sound of an appliance may never be heard if it is masked by the ambient sound. Residents who live in center city with a constant din of noise will have a high ambient background. Those who live in the country well away from highways and airports who enjoy a very low background sound level may consider the same appliance to be highly intrusive. In both situations the first piece of information required to resolve an appliance noise problem is to know the sound level it produces. Unfortunately, this information is difficult to attain without measurement. Few manufacturers provide sound level data on their product. As a consequence, field measurements are usually required as the first step in the "quest for quiet."

In the late 1970s and early 1980s, sincere efforts by consumer advocates, manufacturers, and the United States federal government to bring forth acoustical labeling were caught up in a bureaucratic and political nightmare.[1] A few manufacturers, including members of the Noise Control Association (NCA), attempted to provide noise level and performance data. With no regulations or guidelines, there was little incentive to implement the program. As a result, we have no product labeling that provides the user with information on the noise level of a product, be it the acoustical benefits of a sound absorber, the sound attenuation characteristics of a noise barrier, or the sound level of a noise producer such as a dishwasher, lawn mower, furnace, or garbage disposal. While sound absorption and barrier performance may be found in most manufacturer's literature, the values are given in many different ways. Test data may be obtained by any of several different techniques and there is no dominant, or government-approved, standard for direct comparisons. When making a product selection based on noise levels, readers are cautioned to be sure they are making direct comparisons using data that is compatible.

The user and specifier of potentially noisy equipment are essentially at the mercy of the manufacturer. Traditionally, few manufacturers have placed quiet designs at, or near, the top of their list of priorities. Unless marketers insist that their equipment be quiet to be competitive, equipment designers opt to save a few dollars by limiting noise quieting designs. In many cases noise-quieting is a simple design oversight, or worse yet, based on misconception.[2] As a result, equipment made available to consumers is much noisier than need

---

[1] Consumer advocates convinced the Environmental Protection Agency (EPA) and Occupational Safety and Health Administration (OSHA) to prepare and implement noise-labeling legislation. Manufacturers fought the concept, along with other issues, on grounds that this type of legislation would pose an undue burden on them. After several losses in court, EPA and OSHA did not receive funding to police noise control legislation. This is the present status. The Noise Control Act of 1972 regulations are law but not policed. A few states such as California have noise control agencies with extensive regulations but they receive little funding.

[2] The classic misconception is the IBM typewriter. An IBM Selectric typewriter could be made very quiet for about $15 more than a standard machine. Asked why it does not make and market this quiet machine, the manufacturer's representative responded that the operator must hear the keys striking the paper. Those who use a computer keyboard have never missed the sound of keys striking the paper and find the low noise levels very pleasant. Chapter 5, covers such misconceptions in greater detail.

be. Only recently has this situation been changing. Consumer awareness of noise has been aroused by the automobile industry, which has latched on to lowering noise levels in vehicles and promoting quiet in its mass media advertizing.

Finding a solution for noisy equipment requires a basic understanding of noise control techniques. There are three basic elements to be considered in controlling noise:

- Treat the **source**
- Attenuate noise along the **path** from source to listener
- Control noise at the **receiver** (listener)

Called SPR, for source, path and receiver, this noise control process is prevalent in every noise control situation. While most of this book deals with how to control or attenuate noise along the **path**, it is the control measures at the **source** that provide the greatest decibel-per-dollar. In fact, the acquisition of quiet equipment is the paramount consideration in the economics of a noise control problem. Source control measures will yield but a pittance compared to path quieting measures.

It behooves the owner and specifier to "buy quiet." Many manufacturers of household equipment will provide a quieter version of their product. Chances are that the price tag will be only a few dollars more than their standard product. These few dollars may provide 10 to 30 dB$_A$ noise reduction. To effect this degree of sound isolation at the receiver or along the path may cost 10 to 100 times more. The decibel-per-dollar value of selecting quiet equipment should be obvious to all, yet is overlooked by most.

3.2 EQUIPMENT NOISE LEVELS (SOURCE)

Sound levels of typical residential appliances may range from a low of 25 dB$_A$ for a well-designed refrigerator to 75 dB$_A$ for a garbage disposal (Figure 3-1). Most of the typical residential appliances have a broad-spectrum frequency content. Any sound level that exceeds 60 dB$_A$ will tend to be disruptive since it interferes with normal speech. People must elevate their voices to be heard. For example, if occupants are listening to a TV or stereo, they will turn up the volume when the dishwasher is running. The extra sounds add up acoustically to a higher source sound level. In turn this level must be reduced in the adjacent living unit by sound barriers.

SOUND LEVELS, dB$_A$	
Refrigerator (very quiet)	25–30
Refrigerator (noisy)	30–50
Dishwasher (quiet)	30–50
Dishwasher (noisy)	50–70
Garbage disposal	55–75
Forced-air furnace (quiet)	25–40
Furnace (typical)	40–60
Central air conditioner	40–65
Room air conditioner	45–75
Washing machine	60–70
Clothes dryer	55–60
Television, stereo, radio	60–75
Exhaust fan	60–70
Typical speech level	60–70

Figure 3-1 Sound levels of common appliances

Quiet designs for residential appliances typically incorporate low revolutions per minute (rpm), well-balanced rotational equipment, resilient mountings, sound barriers, dampers, sound-absorbing material, and mufflers. When selecting an appliance for quiet operation without a noise label, or specified noise level, look for:

- **Low rpm**. Select equipment that has a low speed during normal operation. As the speed increases so does the noise level. Follow the rule: lower and slower. Since smaller motors will do the same amount of work if operated at a higher rpm, manufacturers of cheaper equipment tend to select high-rpm motors. These small, high-speed units are not only noisy, but they will wear out quicker. The added cost for higher-horsepower motors that allow the equipment to run at much lower speeds provides an excellent decibel-per-dollar ratio.

- **Well balanced** equipment. This machinery is nearly always quieter. Not only is the unit itself quieter, but the noise transmitted and amplified by the surrounding structure and closure panels will be reduced substantially. Starting with a well-balanced motor will reduce the need for expensive vibration mounts and extend the operational lifetime of the unit.

- **Resilient mountings.** These are imperative on nearly all motors, pumps, and dynamic devices. The purpose of a resilient mount is to isolate vibrations to the operational unit. When the vibrations are transmitted to structural elements and closure panels, the vibrations are amplified and cause noise. Mountings may be springs or resilient pads. Resilient mount designs take into account many issues including the frequency or rpm, mass of the vibrating equipment, operational temperatures etc.[3]

- **Sound barriers.** Barriers are used when the noise cannot be reduced by any or all of the foregoing measures. They may be sheet metal, lead, or composite materials of neoprene loaded with heavy material, or conventional impervious products such as gypsum wallboard.[4] As with any barrier material, the enclosure works best if totally airtight. If it is not feasible for it to be airtight, the enclosure may be equipped with air vents lined with a sound-absorbing material or a sound trap.

- **Sound-absorbing materials**. Such materials will effectively reduce high-frequency noise when placed near the sound source. In designing a barrier, it becomes efficient to add sound-absorbing material to the inside of the barrier, where sounds may be absorbed and quieted before passing through the barrier. Sound-absorbing material is not as effective as other quieting measures. Adding sound absorption rarely reduces noise levels by more than 5 to 7 dB. If not contained within an enclosure, the sound-absorbing material must be placed close to the sound source where it will receive direct sound. Typical sound-absorbing materials are fiberglass boards or blankets and open-

---

[3] For more information on resilient mounts, see Noise Control Manual by D. A. Harris, VNR, 1991.

[4] More information on sound barrier, absorber, and damping materials, mufflers, and associated design guidelines may be found in the Noise Control Manual by D. A. Harris, VNR, 1991. Acoustical data on barriers and absorbers is given in Appendix 3.

cell foams. More exotic materials are used in high-operating-temperature locations. [5]

- **Damping materials.** These are an effective way to reduce or eliminate resonating panels. Soft clay-like materials that adhere to a panel surface will cause damping and in most instances will eliminate the annoying buzz or ringing of a thin metallic membrane. Soft, resilient materials such as weather stripping or rubber washers will also provide damping. Damping materials are generally known as viscoelastic. Some are polymers formed in tape-like products that are designed to adhesively attach to the resonating panel. Locate these materials at the maximum point of vibration.[6]

- **Mufflers.** These devices are designed to quiet the sounds of engine noise where gases must escape. The most common mufflers are mounted on internal combustion engines and turbines. Mufflers also are used effectively on escaping high pressure air nozzles.[7]

Quieting at the source is by far the most effective way to reduce appliance noise levels. The watch word is Buy Quiet. When these techniques have been exhausted, then, and only then is it appropriate to investigate other measures outlined in this chapter.

3.3 HEATING, VENTILATION, AND AIR-CONDITIONING (HVAC) UNITS

3.3.1 Noise Control at the Source
Control of HVAC noise should be a simple matter of selecting quiet equipment. Many of the principles outlined in Section 3.2 above apply. Reduction of 10 to 30 dB_A can be accomplished in the design of the equipment with only minor increases in equipment cost. The rule of thumb in selecting or designing air handling systems is "lower and slower." In other words, lower air volumes per size of opening and slower fan speeds yield quieter equipment. The judicious placement of damping materials to reduce resonating panels and resilient or spring mounts is also vital to the achievement of low noise levels at the source.

It is difficult to acquire noise level data on residential HVAC equipment. However, most major manufacturers have conducted sound tests on their equipment. Some will publish the information in their technical literature. Others will make it available on request. Those who do not provide the information tend to have noisy equipment. The specifier should make an extra effort to acquire this data and use it to make equipment selections.[8]

Special attention should be given to the new "energy-efficient" gas furnaces. Their pulsating burner

---

[5] Same as note 4.

[6] Same as note 4.

[7] Same as note 4.

[8] If for no other reason, the data should be requested to impress on manufacturers that the information is desired by the consumer. Eventually, they will get the message and will provide the sound level data needed. For a more detailed discussion of HVAC noise see the American Society of Heating, Refrigeration and Air-Conditioning Engineers (ASHRAE) publications.

design produces a particularly annoying noise that approximates the sound of a small air-cooled internal combustion engine. Since these units require no chimney, their location should be selected to mitigate their propensity to make noise. For example, intake and outlet pipes should be located away from outside living areas. Major manufacturers of this type of equipment are aware of this problem and have designed noise suppression packages. Make sure this package is specified and properly installed. Essentially a muffler in design, these units should be installed on both the intake and exhaust lines. To avoid vibrations, these lines should be mounted with resilient devices in a space that has good sound barrier properties, and includes sound-absorbing materials.[9]

3.3.2 Airborne Sound Isolation in Wood Frame Construction

The most practical way to confine noise from existing HVAC systems is to enclose them within solid walls that have a high sound transmission class (STC) rating with the service entry to the outside. This allows for longer supply and return air duct runs which in turn make it easier to reduce the sound of noise down the ducts. Where the installation is in an interior location, the enclosure should be serviced by a heavy, solid-core door that is fully gasketed and sealed to be airtight. The fresh air supply for combustion and ventilation should be brought in through a port in the attic space or through a duct to the outside wall.[10] Inlet air duct size should be equal to the equipment manufacturer's recommendations. Flanking paths must be sealed and airtight.

Noise levels, measured as sound pressure level, in a typical mechanical room can vary from around 75 to 105 dB_A. There are many kinds of HVAC equipment, which makes for a wide variability in noise. Unless the sound levels are known to be very low, the mechanical room provided for the heating and cooling units should be located judiciously and be constructed of barriers that will provide a minimum of 55 STC. A buffer zone is sometimes necessary. Ducts and pipes that pierce these walls and doors must be acoustically lined.[11] Make certain that all sound leaks are properly caulked and sealed.

Machinery rooms should be treated with sound absorbing-materials on the interior wall surfaces to lower the noise level by preventing reverberant sound buildup. The interior surface treatments should be materials with a high noise reduction coefficient (NRC) such as low-density fiberglass board or open-cell foam.[12] In most cases a 2-in. thickness should suffice. Secure sound-absorbing material using manufacturer's recommendations. Finish coverings must be designed to allow sound to penetrate the foam or fiberglass material. Most acceptable materials are perforated or easily allow air to flow through the cover. Vinyl films are acceptable only if 1 mil or less thick. Fabricate enclosures of gypsum wallboard, plywood, lead, or other

---

[9] Additional information is given in a case study of a typical single family residence in Section 7.2.

[10] Caution; This outlet may be a noise source for neighbors or for those enjoying an outside porch or patio. Locate it judiciously and/or provide sound-attenuating measures.

[11] Acoustically lined ducts have a sound-absorbing material exposed to the interior of the duct. Sounds are absorbed as they travel down the duct. Fiberglass ducts are the most common. Where considerable sound attenuation is required, use special sound traps available from several manufacturers of noise control equipment. See Section 3.3.5 for more details. Duct attenuation data is given in Appendix 3.

[12] Open cell foam is an excellent sound absorber. However, most foam materials are combustible and should not be used where flammability is a concern.

sound barrier material and acoustically seal joints. Provide air vents with sound attenuators if necessary.

3.3.3 Structureborne Sound Isolation in Wood Frame Construction

When heating and cooling equipment is placed on wood frame construction, it **must be isolated** to minimize vibration transmission into the house structure. If the HVAC equipment is installed on a concrete base, as in a basement, crawl space, utility room, or attached garage, then vibration isolation is generally not of concern. [13]

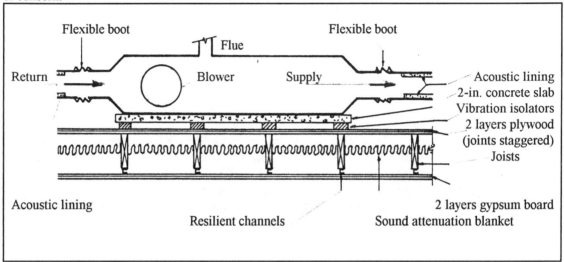

Figure 3-2 Attic installation of furnace/air conditioner.

Attic installations should be handled with extra care. The recommended location is generally determined by the location of the interior interconnecting hallway where the extra studs not only give the support required but form a more rigid supporting base. Because of the flexibility of ceiling joists, a more massive slab or inertia base may be required. In this instance, the inertia base may be composed of a 2-to4-in. thick poured or precast concrete pad over a plywood container that rests on resilient pads. The pads should be designed for this purpose. A typical pad consists of a 2-to-3-in. thickness of 5-lb-density fiberglass board or equivalent. Be sure the pads are large enough to support the total load of the inertia pad and the equipment it supports without compressing the material significantly. Compressed isolation pads are acoustically ineffective and may provide inadequate structural support. All HVAC lines, including piping, refrigerant lines, condensation lines, electrical connections, and duct systems, should be flexibly mounted to the unit to prevent transfer of vibrations to walls and ceilings. Construct a resilient ceiling of gypsum wallboard with a core of sound-absorbing material to improve the airborne and impact sound transmission loss (sound barrier)

---

[13] The greater mass of concrete provides an "inertia base." If the equipment is very large, however, even a concrete floor must be isolated. For most single-family HVAC units, a pad of concrete or lightweight concrete and gypsum concrete may be a sufficient inertia base to be acoustically effective.

characteristics (Figure 3-2).

3.3.4 Vibration Isolation in Commercial Construction

Commercial construction is typified by the use of steel or concrete structural elements and a poured concrete floor. While this type of construction may be sufficient as an inertia base for single-family-unit appliances, the equipment tends to be much larger. Care should be taken to install all HVAC equipment on vibration isolators such as steel springs, isolation pads or equivalent means to keep equipment vibrations from exciting the building structure. The design of the vibration system can become very complex and is best handled by an acoustical expert familiar with HVAC systems (Figure 3-3). Whole machinery rooms can be isolated or floated from the building structure, or special foundations can be constructed to isolate individual pieces of equipment.

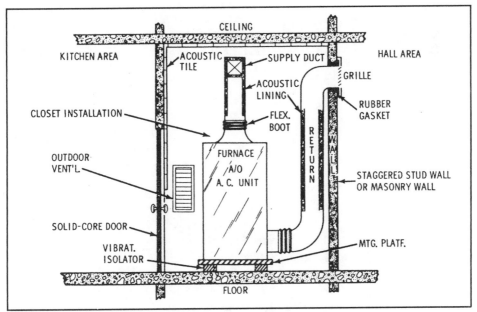

Figure 3-3 Furnace room design for commercial construction.

3.3.5 Locating HVAC Units

In general, HVAC units should be located as far as practical from the bedrooms. Unfortunately, the HVAC system designers and price-conscious builders tend to place the units in a closet or alcove near the center of the house with a single air return located near the center of the floor plan. All of these situations place the unit, acoustically, right in the center of the living space, thus causing undue annoyance (Figure 3-4).

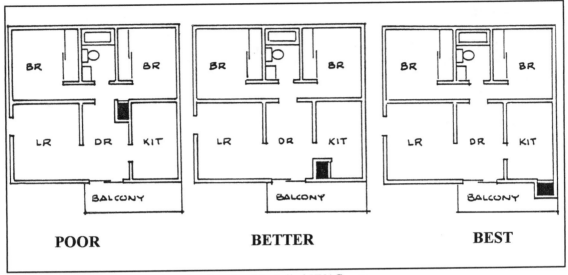

Figure 3-4 Selecting proper location for closet-installed HVAC units.

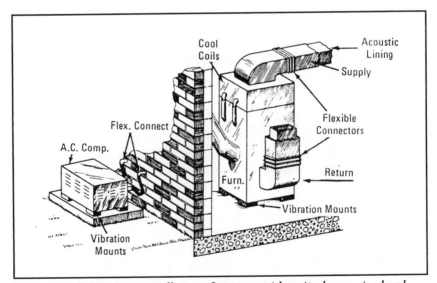

Figure 3-5 Split system installation. Locate outside unit where noise level will not be intrusive or enclose it.

Split systems, such as a heat pump where the heating and cooling elements are located over the furnace plenum and the water or air-cooled condensing or heat exchange unit is outside, require special design (Figure 3-5). Favorite spots for the outside units, which tend to be quite noisy, are the back patio or a narrow reverberant space next to a bedroom window and/or the neighbor's bedroom. Worse yet, these units are sometimes mounted on the roof where the noise can be broadcast to many neighbors. Heat pumps are particularly bothersome, since they operate both winter and summer. Their noise is less of a problem during winter, since most residences have closed windows and no one is outside. During the hot summer months when air conditioning cooling is needed, the outside unit will create a noise level that is totally unacceptable to the occupant enjoying a quiet relaxing nap on the back patio or a cool drink with friends in the evening air. The outside part of a split system unit should be carefully sited to avoid direct noise to a sensitive location. Typically this will require that the unit be relegated to a far corner of the lot and placed in an enclosure designed to attenuate the sound.[14]

A particularly disturbing practice is to place a condenser or air-cooling unit on the patio. Since an outdoor living area is intended as a place where you can experience the great outdoors and hear the birds, whispering pines, and other pleasant sounds, it is quite inappropriate to upset this setting with intrusive fan and compressor noise. Invariably this unit will be operating at full capacity during the same time frame that you would use the outdoor living space. Correcting the problem is difficult since the installer will insist that the patio is the most practical site for the unit. Preplanning with noise considerations in mind is a must. Once the unit is installed, there is little course of action except turning the unit off or constructing an expensive acoustical enclosure. The best preplan is to purchase a quiet operating unit. Unfortunately, HVAC manufacturers have taken an attitude of indifference to this vexing situation. Some will even be so bold as to insist that the unit provides "masking" for other unwanted sounds. Until manufacturers realize that quiet units will sell, there is little alternative left but to turn the unit off or build an expensive enclosure.

Locations of HVAC and associated mechanical equipment will vary depending on the type of construction. Examples of key considerations for commercial construction, single-family dwellings, and multifamily dwellings are:

- In **commercial construction** the mechanical equipment can usually be isolated from the occupants by locating equipment rooms in basement areas. Where basements are not practical, locate the equipment room away from the living units, especially the bedrooms, and buffer them from the rest of the building. This can be accomplished by locating corridors, stairwells, storage rooms, repair and maintenance shops, garages, and other noncritical areas adjacent to or as part of the mechanical room. When practical, a partial subbasement is warranted. In most cases, the increased cost of isolating the noisy equipment will offset the cost of acoustical treatment of ducts and party walls. Cooling tower noise control is discussed in a following paragraph on multifamily dwellings.

- In **single family dwellings** it pays to purchase better-quality and oversized equipment. The smallest size heating unit that will "just do the job" is typically installed as a result of economic pressure on the subcontractor. A higher-quality unit, at a slight increase in cost, can help a builder and owner obtain a quieter house package. Location of the base units should be in the basement and/or carefully

---

[14] An example is addressed in the worksheet for source quieting in Appendix 2. Also see case study in Chapter 7, Section 7.1.

isolated from the bedrooms.

Most types of furnaces are made in two sizes with the same Btu output. One is for heating only, and the other has provision for air conditioning. The one which includes air conditioning is somewhat larger and has a larger blower to accommodate the greater air volume required for cooling. Since larger fans run more slowly, there is a good possibility that it will be quieter.

An alternative is to use a furnace "the next size up." Again with a larger fan that runs slower, it will probably be quieter. A similar effect can be made by fully insulating the house, thereby allowing a smaller heating system. If an HVAC unit designed for a less-insulated-structure is used, it will run more efficiently, and hopefully, quieter.[15]

- **Multifamily dwellings** utilize HVAC equipment that ranges from the cheapest models on the market for individual units to large units serving the entire complex. As a result, noise quieting must be designed to fit the particular circumstances. If individual units are chosen, utilize the techniques outlined above for single-family residences. For larger units, serving many living units, the noise levels may over a wide range. Motors, and motor-driven compressors, blowers, and pumps all create flowing air and liquids which are all sources of noise and vibration. Start by determining the potential noise levels produced by each of the individual noise elements and/or the complete system. If the information is not readily available, and the project is of sufficient size, a performance specification that spells out the maximum sound level allowed will prompt manufacturers to test their equipment.[16]

Most responsible manufacturers of HVAC equipment have qualified personnel working in the areas of equipment noise control. Reliable sound level data can be obtained from the manufacturer for design purposes. However, it may take some extra effort by the manufacturer's representative to obtain noise level information. In the design stage it is wise to select equipment that will more than just do the job. This is particularly true with fans and pumps. Reducing fan speeds only slightly can produce noise level reductions far in excess of the lost electrical power utilization. In residential construction the greatest air-conditioning load is generally around 5:00 to 7:00 p.m. when exterior and interior conditions are at their extremes. Some systems barely get through this part of the day even with the maximum allowable temperature swing. Therefore, they are operating at full capacity late into the night when the background sound levels are greatly reduced. The result is a wide signal to noise ratio that causes undue annoyance. Designs that result in quicker recovery (e.g. an overdesigned unit) will be operating more quietly, at the precise time when occupants are most sensitive to noise.[17]

Cooling towers can be a major noise source in an air conditioning system. In order to operate efficiently, they are required to be out in the open, and the noise from the fan, water, motors, and driving mechanism can reach a high and disturbing level. The noisiest element is the fan. A cooling

---

[15] There is no substitute for specifying a quiet unit with a known sound level. These guidelines are only a rule of thumb.

[16] Noise criteria and related specifications are established by ASHRAE. Consult ASHRAE Handbook for guidelines and standards.

[17] Studies have confirmed the obvious, that noise intrusion will delay sleep.

tower design that will just do the job requires top fan speeds to keep the temperature of the water at a proper level. Slowing the fan to reduce noise will result in a less efficient refrigeration plant. For this reason, designs should have oversize fans that provide sufficient air movement at a slower speed. When slowing the fan is not possible, the cooling tower can be enclosed and noise-absorbing units (silencers) employed on both the inlet and outlet. An effective enclosure may be much more expensive than removing the high speed fans and replacing them with larger, slower units.[18]

3.3.5 Air Distribution System Design

Attention to air duct design is an important factor in noise control design. Metal ducts transmit noise much like speaking tubes used on a ship for communications. There are three areas of noise control in duct design namely: attenuation down the duct, through duct sidewalls, and at the outlet. Duct attenuation is designed to reduce the propensity for sound to travel down the duct, including cross talk between rooms with common ducts. Duct attenuation techniques are also used to reduce the noise of the fans or HVAC equipment reaching the occupied space. Duct outlets or grill covers may be a noise source, especially if high-speed air is forced through small openings. Ducts that penetrate sound barriers are also subject to noise traveling through the duct sidewall, traveling down the duct, and exiting through the duct sidewall to the other side of the sound barrier, thereby short circuiting an otherwise good construction. Sound will travel down a duct equally no matter which way the air is flowing. Therefore, it is vital to **treat both the delivery and return ducts equally**. Considerations relevant to each of these elements include:

- **Duct attenuation** is required when sound travels down the duct. The noise source may be sounds in another occupied space or the noise of the HVAC equipment. Called <u>cross talk</u> by some, the transmission of sounds from a source location to a receiving room via an air distribution duct is one of the most common forms of flanking.[19] Reducing this flanking path is vital to the success of the sound barrier. Sound attenuation down a duct can be accomplished by adhering to two design principles. The first is to separate the openings between the source and receiving spaces by as long and circuitous a route as practical. Each time sound turns a corner, it is reduced slightly. Likewise, a long route will tend to reduce the level of a sound source. If the duct is made of highly reflective material, like sheet metal, the amount of attenuation may be minimal. The second principle is to add a highly sound absorbent material to the **inside of the duct** so that sounds are attenuated much more rapidly. A highly effective sound-absorbing material for this application is fiberglass. Sound attenuations of 5 dB per foot length at certain frequencies are possible, depending on the duct size and absorptivity of the interior lining (Figure 3-6).

---

[18] Cooling tower noise situations are complex and are best handled by an acoustician.

[19] Flanking is discussed in Sections 2.10, 5.2 and 5.4.

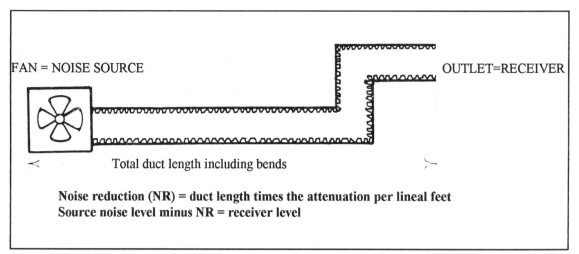

Figure 3-6 Sound attenuation/noise reduction in supply or return air ducts.

SHEET METAL DUCT – 12 by 24 in. (Duct liner as shown)	FREQUENCY (Hz)					
	125	250	500	1000	2000	4000
1-in. thick fiberglass, 1 lb density	0.5	0.5	1.5	2.8	4.0	2.7
1-in. thick fiberglass, 5 lb density	1.1	0.9	2.3	3.3	3.8	2.1
12-in. dia., 1¼-in. fiberglass flex duct	0.8	1.4	2.7	3.1	2.9	1.3

Figure 3-7 Typical noise reductions, in decibels per lineal foot for a 12 by 24-in. duct.

It is possible to calculate the amount of noise reduction (NR) for a duct with a given size and length. The formula, based on considerable research, requires knowledge of the sound-absorbing characteristics of the interior lining of the duct and the number and type of bends (Figure 3-7). Duct attenuation, sometimes called noise reduction in ducts is given in decibels. This value is a product of the duct length times the sound attenuation per lineal foot.[20] To determine the resultant sound level in a receiving room subtract the NR from the sound level in the source room. Note that duct attenuation is directly related to the duct configuration and size. Duct attenuation is usually

---

[20] Duct attenuation per lineal foot for various duct designs and sizes is given in Appendix 3.

given in terms of noise reduction per lineal feet for each size and type of duct.[21]

Installation of acoustically lined ducts in residences is highly desirable. Not only does the sound insulation provide excellent sound attenuation, it also is an excellent thermal insulation. Air ducts are very often routed through unheated spaces making it imperative they be insulated for thermal efficiency. Acoustically, the insulation provides an excellent decibel-per-dollar value. Caution: Be sure the acoustical insulation is on the **inside of the duct**. Thermal insulation placed on the outside of a sheet metal duct will have essentially no acoustical value.[22] Several manufacturers make and market prefabricated ducts made of fiberglass board for this application.

- **Sound transmission** through the duct sidewalls and into the occupied space is a secondary concern. Most ducts are poor sound barriers. Sound that is generated in one room can penetrate the duct sidewall, travel down a duct that penetrates a sound barrier, such as a party wall or floor/ceiling assembly, and exit through the duct wall into an adjacent receiver location. This is particularly likely in two-story multifamily dwellings. A typical practice is to place HVAC units in the basement and supply the second floor through duct systems that run through party walls and floor/ceiling systems. Some ducts may even be routed through interior walls of the lower living unit. The section of the barrier through which the ducts pass should be acoustically treated to assure the HVAC distribution is not a major flanking path. An example is HVAC ducts that are routed through a furred-down hallway ceiling. Reducing sound transmission through the duct wall can be handled in several ways:

1. Make the duct wall a better barrier by sealing any sound leaks in sheet metal joints, using heavy duty foil on the outside of fiberglass ducts, and, if necessary, encase the duct in a gypsum wallboard enclosure with sound absorbing-material placed between the duct and the wallboard. The degree of sound transmission loss will be governed by the overall attenuation required between the source and receiving rooms.
2. Finish the ceiling and walls in upper hall area with gypsum wallboard before installing the HVAC duct in a furred-down ceiling.
3. Install ducts on resilient mounts and be sure all penetrations through common sound barriers are made airtight with permanently elastic acoustical sealant.
4. Increase the sound attenuation down the inside of the duct by adding extra turns and sound-absorbing baffles inside the duct (as shown in Figure 3-8)
5. Make system layouts so the ducts do not have common outlets to adjacent occupancies, especially if ducts are routed through mechanical equipment or laundry rooms. This may require extra duct work but is well worth the expense.

---

[21] Duct attenuation data for different types and sizes of ducts is given in Appendix 3.

[22] HVAC contractors tend to avoid using fiberglass-lined ducts. Many were trained in the art of sheet-metal ducts and supported boycotts of fiberglass ducts when they appeared. They will argue that fiberglass ducts do not fit in the allocated space and are more expensive. In fact they are less expensive, and cheaper to install, and provide considerable fuel/thermal efficiency. Space is a concern and should be handled in the design stage. Retrofit installation of fiberglass duct in a project designed for sheet-metal ducts is difficult. See Chapter 7 for case study examples.

- **Grill noise** is caused primarily by turbulent air (e.g., high-velocity air moving through a restricted zone or being altered in direction). The result may be whistling, roaring, or vibrations. Whistling is a high-frequency sound caused by high-velocity air moving through a small constriction or orifice. Roaring is broad-band sound in the mid- and high-frequency zone caused by an air being compressed and accelerated while moving past a constriction. In some instances the grill may act like the vibrating reed of an instrument.

  Grill noise can be quieted in several ways. First, the grill should have more open area or be designed to change air flow direction by slow easy turns rather than abrupt, constricted paths. It is essentially a matter of choosing a grill designed for the proper air quantity. Another effective means is to reduce the air velocity coming out of the duct. Reduce the fan speed and/or increase the duct size to allow the air flow speed to be reduced without adversely affecting the air quantity. Use smooth gentle turns in the air ducts to reduce air turbulence. If the grill is vibrating like a reed, place some damping material or tape on the vibrating surface. The most effective and preferred method is to select the proper grill for the circumstances. Most major manufacturers have tested the acoustical performance of their grills and will provide recommendations based on air quantity and velocity.

- **Air return** systems must be handled in the same fashion as air supply. At the air speeds found in air distribution systems, sound doesn't care if it is traveling upstream or downstream in the air flow. Many tests using active systems have demonstrated this fact. Use sound absorbers, or sound traps if needed, inside the return air duct and isolate sidewalls. Grill noise is rarely a problem because HVAC engineers design air return flows to have less resistance and speed than the air supply.

Full-ducted air return systems are acoustically preferred over other methods. As discussed earlier, undercut doors, door air louvers, and common air returns for several rooms are potential significant sound flanking paths.[23] If a completely ducted air return is not used, it may be possible to utilize the floor of a linen closet or other area common to several rooms and/or the center hallway as an air return plenum. This plenum space should be treated like a large duct, with sound absorbing-material on all inside surfaces. Make a circuitous route for the air between inlets and add acoustical baffles to reduce cross talk sound paths. The furnace can also be raised above a sound absorbing box to reduce the spread of furnace noise. This system works well in slab construction (Figure 3-8).

---

[23] See Section 2.10 and Chapter 5 for more on flanking.

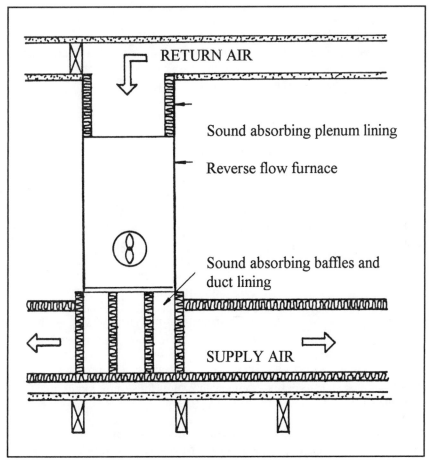

RETURN AIR

Sound absorbing plenum lining

Reverse flow furnace

Sound absorbing baffles and duct lining

SUPPLY AIR

Figure 3-8 Supply and return air duct quieting of furnace noise.

During construction, return air ducts can be installed between studs by simply lining the inside surfaces of the wallboard with fiberglass board. A similar version may be surface mounted (Figure 3-9). Surface-mounted ducts should be fabricated from gypsum wallboard to reduce sound transmission through the sidewall of the duct. Commercial sound traps are similar in design and will maximize sound attenuation down the duct over a short distance using sound baffles.[24]

---

[24] Air duct design may be made somewhat easier by using prefabricated fiberglass-lined ducts. There are two versions available from several manufacturers. One is a nominal 4 to 5-lb density 1-in. thick fiberglass board faced with a heavy metallic (usually aluminum) foil. These products are easily formed into square or rectangular ducts by a duct board fabrication machine that cuts a vee grove in the fiberglass board, leaving the foil intact. The board is then folded with the foil on the outside. Fiberglass lined ducts in 8-to 12-ft lengths are joined at butt ends with an overlapping design and foil-reinforced tape. A second design is a fully fabricated round and flexible duct with of a wire wound vinyl outer covering with light density fiberglass on the interior. Like a slinky toy in design, the duct is light, flexible and has good sound attenuation down the duct but poor sidewall transmission loss.

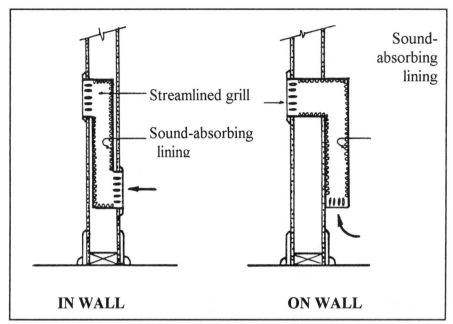

Sound-absorbing lining

Streamlined grill

Sound-absorbing lining

IN WALL

ON WALL

Figure 3-9 Design and installation of a quiet air return duct.

- **Makeup air system** designs that bring in outside air should be handled with caution. If neighborhood noise is of concern this system should be designed with the same parameters used for return or supply air.

- **Combustion air systems** are of specific concern in the same way as makeup air. If the intake air is for a pulse-type furnace, particular attention should be given to installing a muffler. You don't want to be bothered by this unit when you are on your patio or to disturb your neighbor who is sleeping with an open window next to the intake or outlet.[25]

- **Ventilating exhaust systems** for kitchens and bathrooms need acoustical treatment in the same fashion as HVAC ducts. The first rule of thumb is to always provide each occupancy with separate exhaust ducts. This will eliminate cross talk between living units and reduce the chance that a noisy exhaust fan becomes intrusive to the neighbor. It is particularly important to select quiet exhaust fans. These units are notoriously noisy. Most are made with very high speed motors and small fan blades. Larger and slower fans will be quieter. In kitchen units or where grease may accumulate, special designs for sound absorption may be required. Most sound-absorbing materials are good filters and a fire hazard if exposed to exhaust fumes.

---

[25] See Section 7.2 for a case study.

■ **Quieting sheet-metal systems** usually involves lining the **interior** of the duct with a sound absorbing material. Fiberglass duct board is the most typical material. Fiberglass boards for interior duct lining are available in board form having a density of 2 to 5 pounds per square foot and ⅜ to 1-in. thick. The amount of material required is determined by the amount of noise reduction needed between the source and receiver location (Figure 3-10).

Duct attenuation values are given in terms of decibels per lineal foot and will vary, depending on the size of the duct, and the thickness of the sound-absorber.[26] Note that the sound absorbing material must be placed on the inside of the duct to be effective in reducing sound transmission down the duct. A common mistake is to place the material on the outside. While it may provide thermal insulation on the outside, it is worthless acoustically. Also note that film facings designed to reduce fiber erosion will tend to reduce the sound absorption properties of the acoustical liner. If a facing is required, it should be film that is no thicker than 1 mil. Thicker films will tend to reduce the sound absorption characteristics at higher frequencies. Polyolefin (Mylar) is used because it has great strength in thin films. Be sure interior facings satisfy local fire code criteria.

A secondary noise complaint in sheet-metal ducts must be addressed. Sheet metal ducts will pop or squeak as the duct expands and contracts under different temperature conditions. An easy solution is to use fiberglass ducts.

DUCT ATTENUATION GUIDE

● **Provide 6-ft length of lined duct for each room.**

● **Place lined duct as close to grill as possible.**

● **Where more than one duct serves the same room, treat <u>all</u> ducts, including return ducts.**

● **Provide as many turns as feasible, especially where lining is used.**

● **Where space does not allow a 6-ft section of duct lining, provide extra baffling.**

Figure 3-10 Typical duct attenuation guidelines for a single-family residence.

---

[26] Duct attenuation values are given in Appendix 3 for a range of duct sizes and lining materials. Also see Figure 3-7.

REDUCING AIR FLOW NOISE

"STREAMLINE THE FLOW"

- **Low flow velocities**

- **Smooth boundary surfaces**

- **Simple layouts**

- **Long radius turns**

- **Flared sections**

- **Streamline transitions**

- **Remove obstacles**

Figure 3-11 Air flow noise considerations.

■ **Air flow noise** reduction can be a complex and confusing task. Since ducts are designed to deliver air as their first priority, any effort to treat them for noise control must be compatible with their air-handling capabilities. The key advice for effective air or liquid flow noise is "streamline the flow." Issues and design considerations for reducing air or liquid flow noise are common to all ducts whether they are in HVAC systems, vacuum cleaners, or plumbing systems (Figure 3-11).

1. Low flow velocities avoid turbulence, one of the main causes of noise. Flow velocities should be on the order of 8 to 10 feet/second (3 meters/second) in domestic forced-air HVAC systems.

2. Smooth boundary surfaces also avoid air turbulence. Provide duct interiors with smooth interior walls, edges, and joints.

3. A simple layout will likely be better than one with lots of turns, fittings, and connectors. While this may appear to conflict with duct attenuation design, note that this process is designed to reduce the source of the noise. Eliminating a source will always be better than trying to suppress one as a retrofit operation.

4. Using long-radius turns or changing the flow gradually and smoothly will reduce air turbulence. Turns having a curve radius equal to 5 times the pipe diameter or major cross-sectional dimension are suggested.

5. Flaring of intake and exhaust openings, particularly in a duct system, tends to reduce flow velocities, often with substantial reduction in noise output.

6. Streamline transition in flow path. Changes in flow path dimensions or cross-sectional areas should be made gradually and smoothly with tapered or flared transition sections to avoid turbulence. A rule of thumb is to keep the cross-sectional area of the flow path as large and uniform as possible throughout the system.

7. Remove unnecessary obstacles. The greater the number of obstacles in the flow path, the more

torturous, turbulent, and hence noisier the flow. All other required and functional devices in the path, such as structural supports, deflectors, and control dampers, should be made as small and as streamlined as possible to smooth out the flow patterns. Parts with perforations, slots, or other openings permit air leakage through or around the part, and may help eliminate pressure buildup on light-gauge parts that cause popping and cracking sounds.

WHERE TO USE SOUND ATTENUATING DUCTS

- ■ **Entire air supply system**

- ■ **Entire air return system**

- ■ **All interconnecting air ducts**

WHY TREAT ENTIRE DUCT SYSTEM? To reduce noise from:

- ● **Furnace and air conditioner – especially fans**

- ● **Cross talk between rooms with common air supply/returns**

- ● **Popping or cracking from expansion/contraction**

Figure 3-12 Where to use sound-attenuating ducts.

- ■ **Entire system control** is a very effective way to assure duct noise is reduced to a minimum. An easy and inexpensive technique for entire system control is to specify fiberglass ducts for all air distribution systems in the building. Equipment noise, cross talk, air flow noise, and popping or creaking noise will be eliminated or reduced to very acceptable levels. When coupled with energy-saving considerations, a fully insulated duct system provides excellent performance and economy[27] (Figure 3-12).

---

[27] Fiberglass ducts are the solution to quieting a single-family residence in a case study presented in Section 7.2. This solution is so effective and economical it is incomprehensible why all houses are not constructed with fiberglass ducts.

3.4 APPLIANCES

3.4.1 Buy Quiet
The key to having quiet appliances is to "buy quiet." Considering the decibel-per-dollar, acquisition of quiet equipment is heavily favored over retrofit noise suppression. Retrofit costs have been known to exceed the value of the appliance itself. By contrast, it may cost a manufacturer but a few dollars more to substantially reduce the noise level of a unit.

Most major appliances perform the job they were designed to do in a very satisfactory manner but are notoriously noisy. Making a quiet appliance is not difficult or expensive. It is simply a matter of marketing. Builders who can purchase in volume demand low prices. Low price generally means noise. To reduce costs and provide competitive prices, manufacturers ignore those elements that provide quiet. Removing quiet performance criteria from specifications, or ignoring them, allows manufacturers to utilize smaller, higher-pm motors and pay less attention to parts balancing, resilient mounts, damping, and sound absorption.

This self-serving attitude is enhanced by marketing managers who have determined that the sooner the appliance wears out, the more units they will sell. If the worn-out unit is not thrown out and replaced, they revel in the opportunity to participate in a large and lucrative aftermarket for parts and repairs. It is indeed unfortunate that well-designed, balanced, and durable equipment, that also yields quiet, has low sales volumes.[28]

Most of this situation is enhanced by a lack of knowledge. It is a well-established fact that consumers and users want quiet appliances. However, since the user rarely has specification control over builders' appliances, noise control attributes are lost. Furthermore, users have been led to believe that nothing can be done, or worse yet, that more noise means more powerful equipment. Quite the contrary, a noisy appliance usually means that it will break down or wear out quicker.

In the 1970s there was a concerted effort by the U. S. federal government to inform the consumer by requiring noise labels on all appliances. The reaction by manufacturers was as expected. They did not want regulation and promptly set about to kill the legislation. Consumers still have no way to determine the noise level of most appliances except to measure it in the field. Be assured that most manufacturers have tested the noise level of their equipment, and a few will provide the information eventually, but generally they make it difficult for the consumer to obtain. If you are in the position to control the specification, it is highly desirable to include noise level performance criteria.

3.4.2 Quieting Major Appliances
Major appliances that create the most noise in residential dwellings are the automatic dishwasher, garbage disposal, exhaust fans, washing machines, and vacuums.[29] Noise levels range from a low of 25 to

---

[28] Even the creative advertising, with the lonely repairman, has not increased the demand for quality and quiet appliances.

[29] HVAC and furnace noise is covered in Section 3.3.

30 dB$_A$ for a quiet refrigerator to 75 dB$_A$ for noisy dishwashers, disposals, and other household appliances[30] (Figure 3-1). Look for the following quiet attributes:

- **Dishwashers** are best quieted by purchasing quiet equipment. If you cannot obtain the noise level, look for:
  - Larger and slower motors
  - Oversized pumps
  - Well-balanced motors and pumps
  - Quiet-operating solenoids in water supply valves
  - Oversized water supply and waste lines
  - Quiet recirculating pump with large throat design
  - Resilient mounts for motors and pumps
  - Resilient pads on machine supports
  - Damped panels – foam or damping compound
  - Sound-absorbing material located near noise source
  - Flexible coupling on water and electrical lines.

- **Garbage disposal units** can be made quieter by the manufacturer. Purchase units with lower noise level. If noise level is not available, look for:
  - Larger and slower motors
  - Well-balanced motor
  - Oversized waste lines
  - Resilient mount for unit attached to sink
  - Damped panels– foam or damping compound
  - Sound-absorbing material located near noise source
  - Flexible coupling on water and electrical lines
  - Extra enclosure lined with sound-absorbing material

- **Exhaust fans** are notoriously and needlessly noisy. They are a classic case of neglect by the manufacturer and builder. Bathroom and kitchen exhaust fans are nearly always designed with high-speed motors and blade fans. For just a few dollars more, the fan can have a larger, slower motor and a quiet fan blade design. Squirrel cage fans are quieter than blade fans. Units with larger air volume ratings can be operated at lower blade speeds, which in turn means a quieter fan. It is much cheaper to remove the noisy fan and replace it with a quieter one than trying to quiet an existing unit. Install separate exhaust vents to avoid cross talk and fan noise intrusion between living units. If the unit squeals or squeaks, the fan blade may be touching the cowl or just needs oil on a bearing[31].

---

[30] The American Society for Testing and Materials (ASTM) Committee E33 has recently adopted test procedure E1574-95, <u>Standard Test Method for Measurement of Sound in Residential Spaces</u>. Designed specifically to measure sound levels of appliances in the field, test results are in dB$_A$.

[31] If the unit cannot be replaced, follow the recommendations of duct quieting in Section 3.3.

■ **Washers and dryers** are normally placed side by side or one over the other and are usually considered together. Washing machines are typically the noisier equipment. Quieting a dryer usually means the fan should operate at low rpm and air speeds. Vent with separate ducts to avoid noise in adjacent units. Quieting a washing machine is very similar to quieting a dishwasher. Again, buy quiet. If you cannot determine the sound level, look for the same attributes listed above for dishwashers. In addition, look for a tub that is mounted resiliently so it does not bang on mounts when in an unbalanced spin cycle.

■ **Vacuum cleaners** are a major source of broadband noise. Fortunately, their use is short and usually during waking hours. Sound levels near 80 dB$_A$ are commonplace.[32] Some concern has been expressed by manufacturers but few have made any attempt to quiet their units. About the only way to handle this situation is to install a central vacuum system and place the mechanical unit in a closet with good noise barrier characteristics and lined with sound-absorbing material. Outlet sound is a minor problem.

■ **Installation of appliances** requires special considerations, for example:

● All supply and drain lines should be flexible. Install air chambers in plumbing supply lines to reduce banging due to quick shutoff valves (Figure 3-13).

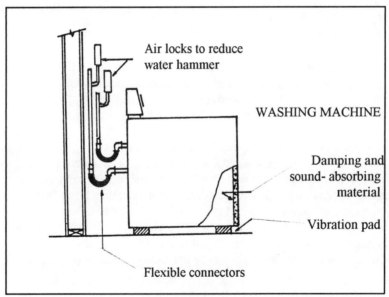

Figure 3-13 Quieting a washing machine.

---

[32] A vacuum cleaner has been found to be an excellent sound source for field sound measurements because of the loud broadband character it produces.

- In multifamily units, do not locate the dishwasher, disposal, kitchen sink, and exhaust fan parallel to the party wall. This arrangement will tend to resonate the cabinets. If the cabinets are affixed to a party wall, the potential exists for direct sound transmission of structureborne noise to the adjacent living unit. If possible, it is desirable to locate these units perpendicular to a party wall or on an outside wall.

- Be sure garbage disposals are mounted with flexible connections to the sink and drain line.

- Locate laundry equipment in a closet, garage, or basement. It is best to isolate them in their own space with tightly closed doors. For added quiet, install sound-absorbing material on the walls and ceiling. Concrete slabs are preferred over wood frame floors to improve vibration isolation.

- Do not interconnect exhaust vents or plumbing lines with adjacent living spaces. Be sure any pipes that penetrate party walls are sealed airtight and resiliently mounted.

3.4.3 Electrical

A poor electrical installation can raise havoc with good sound barrier construction by permitting sound to leak from area to area through holes cut for wiring or by "shorting" a construction by tie-through. This phase of construction demands careful job control for best results[33]. For example, avoid back-to-back outlet boxes on party walls and make all penetrations airtight.

In designing multifamily units, locate the television and stereo speakers in a predetermined spot that will not cause annoyance in other living units. For example, place cable outlets in desired locations but avoid placement on a party wall. If cabinets are part of the unit, place them on a wall that backs up to a kitchen, hall, or bathroom rather than an adjacent living unit or bedroom (See Figure 1-11).

Built-in intercom systems can be used to provide background sound levels and mask other intrusive noise. However, care should be taken in locating the speakers so the sound does not intrude on a neighbor. Installation, speaker layout, and wiring should be carefully analyzed to maintain the acoustical integrity of sound conditioned walls or floor/ceilings.

Most minor appliances are the choice of the tenant or occupant and will not be discussed here. If in doubt, determine the sound level of the unit before purchasing.

3.4.4 Room Air Conditioners

Room air conditioners, like many built-in appliances, are often supplied by builders. Noise levels may exceed 80 dB_A. Their noise can be intrusive to the occupant, neighbors and people outside. When multiple units open into a courtyard or confined space, the exterior noise level can be highly obtrusive. Motels and multifamily residences are particularly prone to noise from room air conditioners. Like many other motor-driven units, room air conditioners can be obtained in a wide range of cooling and air volume capacities with

---

[33] See discussion of flanking paths in Section 2.10.

an even wider range of operating noise levels. When not operating, these units are a major noise flanking path for an exterior wall that may otherwise be a good exterior noise barrier.

Major sources of noise in a room air conditioner are the fan, compressor, and motor. Fans are a major cause of high noise levels. Squirrel cage fans are quieter than the blade type. But because of cramped space and cost they are rarely used. Fans are generally designed to run at high speed, even when given several settings, and move air at high speeds. It is almost as if the manufacturer intended to provide cooling by high-speed air movement rather than cool air. In quality units, fans run at lower speeds, are well balanced, and are refined for smooth, vibration-free operation. Motors that hum may have poor field coils and may vibrate due to an out-of-balance condition. Compressors that are overtaxed and improperly mounted can vibrate the entire wall of a building, thereby amplifying the potential noise level.

Establishing a maximum noise level is obviously the most effective means to quiet room air conditioners. Room air conditioners are the most likely appliance to have a noise label. Multipurchase contracts should contain noise level performance criteria. Responsible manufacturers will provide the data. If no data is available, there is a good likelihood that the equipment is inordinately noisy. Acceptable noise levels should be limited to 45 or 50 dB_A. Units higher than 60 dB_A will interfere with speech. Because of the broadband spectrum of sound produced by these units, they can sometimes be used as sound masking to cover intrusive sound from an airport or nearby freeway.

Occupant room noise levels from room air conditioners can be reduced in several ways. Install the units on vibration pads and enclose them in a muffle box. The enclosure should be fabricated of a good sound barrier material and lined on the inside with sound-absorbing material. Place a sound-absorbing duct with several bends on all air intakes and outlets. For exterior quieting, construct a muffle box and/or muffler as described above.

If the room air conditioner is installed in a window or exterior wall that was designed to be a barrier to community noise sources, the units must have special attention paid to their sound transmission loss characteristics. This attribute has been addressed by many manufactures, and data on STL is available from responsible system offerors. This is a complex issue, and it is recommended that noise mitigation efforts be handled by a qualified acoustician.

3.4.5 Lighting Fixture Noise

Incandescent lights are rarely noisy. However, flourescent and high intensity discharge (HID) fixtures require a ballast that causes an annoying and noisy hum. Quiet ballasts can be obtained at increased cost but are well worth the price. A good light fixture manufacturer will isolate the ballast from the fixture with rubber grommets or similar vibration isolators. A fixture with an external ballast is a good choice. Noise levels are generally available from most responsible light fixture manufacturers. Choose levels below 35 dB_A if possible. Installation of recessed light fixtures notoriously cause sound flanking paths. Since most fixtures require air ventilation to remove excess heat, sound flanking paths result. Cover the back of the fixture with a box having good sound barrier performance and install ducts with sound-absorbing liners if the path becomes a noise problem.

3.5 PLUMBING NOISE

In normal residences, there are four, liquid piping systems that require attention for noise control: hot water, cold water, the drainage system including the vent lines, and piping for hydronic heating and cooling.

Piping noise problems are relatively simple to mitigate when one realizes that pipes themselves, being round, are a good acoustical design. Liquids generally flow smoothly, with little turbulence, down a pipe. The major effort in resolving pipe noise problems is to isolate the pipe from the building structure.

Pipes tend to act like a tuning fork with the structure as a resonance box. When one taps a tuning fork and holds it in a hand, it is almost impossible to hear it. Yet, when the base of the tuning fork is in contact with a wall, box or other resounding surface, it is very audible. Generally, pipes need to be in contact with a resounding surface to be noisy. Therefore, most pipe noise issues may be resolved with resilient mounts and connections (Figure 3-14) .

Noise can be created in piping by water rushing through the pipes. If the pipes are too small, a whistling sound is created. If there is air in the lines, a gurgling sound is created. And, when water rushes down a drain line, it will cause the pipes to vibrate. Pipes can expand or contract because of temperature change, particularly hot water and drain lines, causing creaking at the connections. Or a valve may whistle if opened only part way. If a pipe is in contact with, or connected to, a vibrating source, as with a garbage disposal, the vibrations of the unit will travel down the pipes and could be amplified by the structure a substantial distance away.

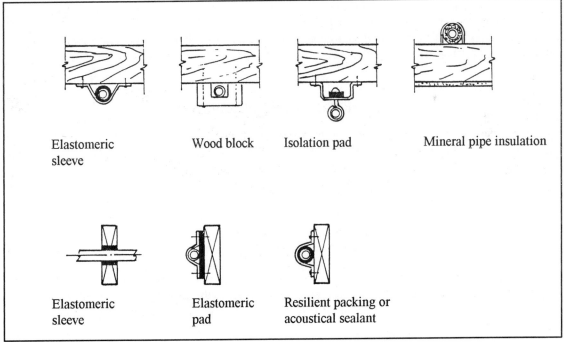

Elastomeric sleeve	Wood block	Isolation pad	Mineral pipe insulation
Elastomeric sleeve	Elastomeric pad	Resilient packing or acoustical sealant	

Figure 3-14 Resilient mounts for pipes.

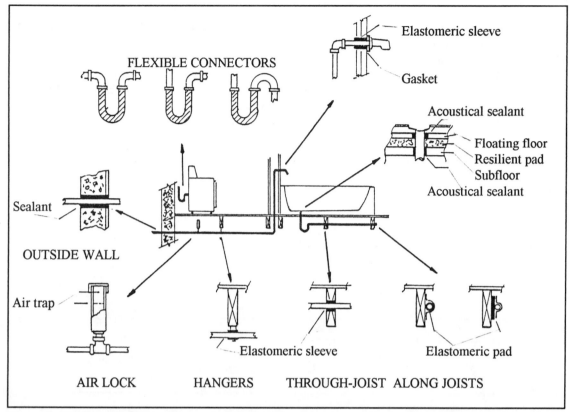

Figure 3-15 Pipe quieting techniques include flexible connectors, sealed penetrations, air locks, and resilient mounts.

To reduce whistling, reduce the water pressure or install a larger line or valve. To eliminate banging,or "water hammer," install an air lock or air pocket as a cushion for the rapid pressure change. Most other pipe noises can be resolved by utilizing resilient connectors and mounts (Figures 3-14 & 3-15).

All the above problems can be answered by proper design and installation techniques at minimal cost. If in doubt, lower the pressure, and use oversize pipes, resilient mounts, and resilient connectors.

See the checklist in Table 3-1 for plumbing noise problems and specific solutions.

TABLE 3-1 Plumbing Noise – Problems/Solutions

Problem	Solution
Noise flanking near pipes	Seal all holes where pipes pass through common walls with a permanently elastic acoustical sealant.
Pipes pop, snap or creak due to expansion/contraction from heating or cooling. (Note: 100° F differential can cause expansion of ⅜-in. in 10-ft pipe or 1¼-in. in 100 feet.)	Provide a swing arm at the end of a pipe run to permit movement. Design pipe supports to allow for movement without binding. Add collars of pipe insulation and resilient mounts.
Water gurgling from water closet or bathtub.	Drain lines may be rigidly attached to structural members. Encase pipe in pipe insulation, use resilient mounts and/or connectors.
Water noise at bottom of vertical stack.	Tilt stack slightly to adhere water to pipe when falling. Otherwise, install pipe insulation, and resilient mounts and isolate the pipe from the structure with a resilient pad of fiberglass or foam.
Water rushing through pipes, possibly causing whistling sounds.	Pipes and/or valves are undersize for the water pressure. Install oversize faucets and valves. Reduce water pressure. Install oversize pipes and encase in pipe insulation. Install resilient mounts.
Noise coming through medicine cabinets mounted back to back (Figure 3-16).	Locate cabinets in separate stud spaces. Better yet, use surface-mounted cabinets and seal all penetrations.
Water turned on and off quickly causes pipes to bang (water hammer) and vibrate. Problem is particularly acute with dishwashers and washers with electric shutoffs.	Provide air chambers at each outlet to provide a shock absorber of air to bring water to a slow halt without noise. If air locks already exist, recharge waterlogged air chambers by draining system at its lowest point with taps open. Close taps and refill system. Chambers with a diaphragm will eliminate waterlog.
Pipes carry sound from pumps or other vibrating equipment throughout the building. Problem may be associated with well pump, sump pump, heat exchanger, and hot water/steam lines.	Provide flexible connector in pipe connection to pump. Install pipe on flexible mounts. In some cases the pipe might need to be encased with pipe insulation having an outside noise barrier. Water pumps and special equipment may require an inertia base (Figure 3-17).

Toilet operation during flush and refill causes annoyance and embarrassment.	Reduce water flow rate with lower water pressure and/or oversize pipes and valves. Install a quiet action fill device. Mount all water lines on resilient attachments and install flexible connectors.
Neighbor can hear someone in bathtub or shower.	Back-to-back installation of bathtubs and showers requires a full wall behind both units. Builders tend to cut corners and install units prior to installation of wallboard. This causes a large noise flanking path. In multifamily units, utilize staggered wood stud separator walls with full wallboard faces and a core of sound insulation. Be sure these walls are fully constructed and caulked airtight prior to installation of tub and/or shower. Do not allow any pipes to penetrate the separating wall. Lines serving both units require a flexible coupling. Install all pipes on flexible mounts.
Garbage disposal vibrations are heard elsewhere in building.	Vibrations are being transmitted and amplified by structure and/or sink. Install flexible connector between pipe and disposal. Be sure disposal is attached to sink with resilient mount. Attach all pipes with resilient mounts.

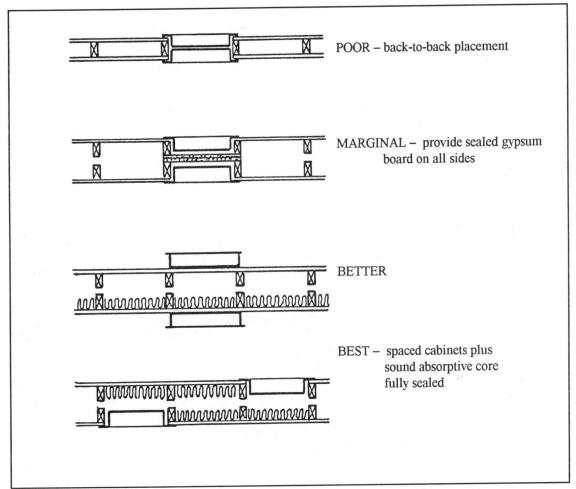

POOR – back-to-back placement

MARGINAL – provide sealed gypsum
board on all sides

BETTER

BEST – spaced cabinets plus
sound absorptive core
fully sealed

Figure 3-16   Reducing noise transmission through back-to-back bathroom cabinets.

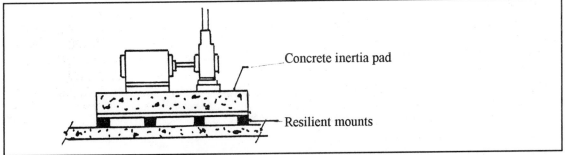

Concrete inertia pad

Resilient mounts

Figure 3-17   Inertia block/vibration isolation for special pumps and equipment.

3.6 SUMMARY: SOURCES, SOUND PATHS, AND REDUCTION METHODS

The installation of mechanical equipment creates two types of noise problems, namely;

- Mechanical vibration or impact-type sounds due to rotating parts

- Airborne sounds caused by pulsating equipment, fans, and blowers.

These noises are disturbing to occupants of a building, particularly if the equipment operates intermittently or is loud. Some heavy equipment may operate in sympathy with the building structure and create mechanical/structural vibrations that occupants actually feel as well as hear.

Solutions should be implemented in the following order[34]:

1. Locate the furnace room, air-conditioning equipment, cooling towers, fans, pumps, and like noise sources where the noise is least objectionable.

2. Buy quiet. Specify and purchase equipment that has been properly designed and creates the lowest noise level consistent with economy.

3. Isolate the resulting noise source by placing it in an acoustical enclosure composed of a good noise barrier that is lined with sound-absorbing material and providing resilient mounts and/or inertia bases.

4. Implement noise control measures along the path by eliminating flanking paths, building efficient noise barriers, and using sound-attenuating ducts, resilient mounts, and sound-absorbing wall/ceiling materials.

Almost any degree of quiet can be achieved for a price. In most cases quiet is merely a matter of good planning. A systems approach that involves the equipment supplier, mechanical engineer, HVAC designer, plumber, contractor, architect, and owner has been demonstrated to be very effective in selecting efficient practical solutions and assuring that the completed project provides the desired acoustical environment.

---

[34] Implement 3 and 4 only **after all** efforts to quiet the source have been diligently pursued. See Table 3-2 for additional details.

**TABLE 3-2 Check List for
Common Equipment Noise Paths**

Sound path	Noise reduction technique
Direct sound radiation from source to receiver. Sound reflections cause room echoes and poor speech intelligibility.	Add sound-absorbing material to the room, such as acoustical ceiling tile/panels or acoustical wall panels. Carpet and pad will reduce footfall and scraping sound. Select quiet equipment.
Noise radiating from ducts and plenums.	Locate ducts and terminals in noncritical areas, design ducts and fittings for low turbulence, and install sound-absorbing lined ducts or sound attenuators.
Noise coming from supply and return ducts.	Select quiet fans. Install sound-attenuating ducts.
Noise transmitted through equipment room walls and floor/ceilings.	Separate equipment room from critical areas. Install high STC walls and floors. Provide resilient mounts and isolators. Seal all cracks and penetrations with permanently elastic acoustical sealant.
Building structure transmitting sounds.	Balance equipment. Design equipment supports with isolation mounts, pads, inertia bases. Enclose in room with high STC walls and floor/ceiling. Seal all cracks, holes with acoustical sealant. Line interior surface of walls, ceilings, and floors with sound-absorbing material.
Transmission of noise along pipe and ducts.	Install flexible connectors and resilient mounts on all pipes and ducts.
Outside noise enters through air conditioner mounted in window.	Install window air conditioner that has high STC or sound attenuation rating. Install sound trap over inside and/or outside of unit. Locate window air conditioner away from noise source. Install high STC windows.
Outside noise coming through walls and/or roof/ceiling.	Look for and eliminate flanking paths. Eliminate all ducts and air exhausts or provide sound-absorbing ducts or attenuators.
Noise transmitted via connecting ducts.	Install sound absorbent lined ducts or duct attenuators.

Noise transmitted through room partitions or over and around them.	Look for sound flanking paths. Eliminate all ducts, pipes, and electrical connections that penetrate the wall. Provide sound attenuating ducts or sound traps for those air ducts that cannot be eliminated. Extend partitions to ceiling slab and seal tightly. Install high STC wall and ceiling systems.
Noisy Equipment	Buy quiet. Install mufflers. Enclose in acoustical envelope composed of a good barrier lined with sound-absorbing material and provide lined ducts/mufflers for air intake and exhaust. Use resilient mounts and inertia base.

Chapter 4 REGULATIONS, CODES, TEST METHODS

4.1 INTRODUCTION

Noise and noise control regulations appear in many different forms in our society. In the early days a city noise ordinance may have been as simple as stating that "No person, animal, or equipment shall produce noise that intrudes on a neighbor." Known by some as "barking dog ordinances" these early noise regulations had no criteria. They were purely subjective. Many attempts have been made by many different agencies and parts of society to regulate noise. As a result, our acoustical environment has become a part of our regulations and rules at all levels of government. It involves nearly all technical disciplines and is a concern of each individual.

Noise regulations may be found in many places. We have city, county, state, regional, and federal ordinances, regulations, and requirements. Noise control requirements show up in building codes at all levels of bureaucracy. These codes regulate construction practices for single-family homes, multifamily buildings, condominiums, apartments, high-rise residential buildings, hotels, motels, medical care centers, and all manner of residential buildings. Other codes regulate community noise, construction noise, environmental noise, transportation noise, and nearly any other type of noise imaginable. The most recent requirements are those that are generally implemented as part of an Environmental Impact Study (EIS). If done properly, every EIS will have a "Noise Element" section that addresses the present and predicted noise impact on the project along with "guidelines" and "requirements" for "noise mitigation." In effect the "Noise Element" of an EIS becomes a regulation and requirement.

"Noise control by means of laws and regulations involves balancing the rights and remedies of the individual against the rights and remedies of society."[1] Considering the breadth and scope of the noise control regulations in our society, it behooves the reader to keep this statement in mind when evaluating the merits of an individual code requirement, ordinance, criteria or regulation that governs our acoustical environment. The definition of noise may appear to vary, but in all cases must be reduced to the simple version, namely, **noise is unwanted sound.** Note that noise is always subjective. In classic terms, "One man's music is another man's noise." This simple premise must be retained by the reader in analyzing the myriad of noise regulations in our society. When criteria are set, they always are arbitrary. When performance requirements are implemented on a project, they are always selected to satisfy what the specifier perceives to be needed for the project. These criteria will change and should change depending on the circumstances and the overall acoustical environment.

This chapter is dedicated to providing a source for information about noise control codes, regulations, building codes, and test procedures. By no means is it complete. Furthermore, the laws and codes are continuously being updated.

As an individual, you are encouraged to apply the principles of noise control presented in the previous chapters and the information contained in the appendixes. These measures are the ones that will protect you from harmful and irritating noise. Yet, there are some noise sources over which you may have no direct

---

[1] From <u>Noise as a Health Hazard</u>, Proceedings of the American Speech and Hearing Association, paper titled "Control of Noise through Laws and Regulations", by James J. Kaufman, 1969.

control. What can be done in these cases?

You might pursue simple, informal remedies to reduce noise sources over which you have no control. For example: talk with your neighbors and ask them to quiet their dog or late-night parties, contact the construction workers down the street and request they stop pounding their pilings at night, call the peace officers in your town and request they enforce an existing law, call your council representative, public health officer, licensing bureau, planning department, or building inspector to determine if there are rules covering your situation.

If personal and local contacts do not provide a means to mitigate the intruding noise, then it behooves you to investigate regulations implemented by higher authorities, bureaus, associations, and the like. The following portions of this chapter are devoted to listing those agencies that have implemented some form of noise control documentation, code, criteria, or test procedures. Commentary about the agency will assist you in determining if the organization has measures that will impact your particular situation.

Builders, contractors, architects and property owners must be aware of all these regulations. It is very likely that you will be asked to comply. Anyone who has been served with a stop construction order will probably concur, that it is considerably less expensive to comply with the regulations during the planning stage than to implement retrofit noise control measures.

4.2 FEDERAL NOISE REGULATION AGENCIES

**OFFICE OF NOISE ABATEMENT
AND CONTROL (ONAC) – Responsibilities**

- **Regulate noise emissions**
- **Product labeling**
- **Low-noise-emission product purchase (by government)**
- **Coordinate government noise reduction activities**
- **Assist state and local noise control**
- **Noise education and research**

Figure 4-1 EPA/ONAC responsibilities/actions.

4.2.1 Environmental Protection Agency (EPA)

The EPA has legislation that covers many aspects of environmental noise. In 1972, the Ninety-Second Congress directed EPA to be the controlling agency for the Noise Control Act. The Act, approved by Congress, requires the U. S. Environmental Protection Agency to develop and publish information concerning noise levels that jeopardizes health or welfare. It required EPA to regulate noise emissions from new products used in interstate commerce, coordinate the noise abatement efforts of other agencies, and provide information to the public concerning the noise "Noise Control Act" of 1972.[2] The Act requires the U. S. Environmental Protection Agency to publish information concerning noise levels that jeopardize health or welfare. "It required EPA to regulate noise emissions from new products used in interstate commerce, coordinate the noise abatement efforts of other agencies, and provide information to the public concerning the noise emission of products."[3] In effect, EPA was provided overall authority to set standards defining permissible noise levels for products that have been identified as major noise sources, establish test procedures, require noise labeling, and assure compliance. EPA started by conducting an extensive noise study. After many hearings and years of public comment, EPA established the Office of Noise Abatement and Control (ONAC). ONAC subsequently implemented a broad scope of rules, criteria, regulations, and enforcement measures (Figure 4-1). Examples of the EPA requirements include acceptable noise levels for many different circumstances, noise labeling requirements, and processes for enforcing compliance. **These rules and measures are still in effect.**

Industry lawyers challenged many of the regulations in several significant court cases. As a result the White House Office of Management and Budget (OMB) ended funding for the EPA Office of Noise Abatement and Control in 1981. This effectively stopped EPA from enforcing its regulations. Funds already earmarked for ONAC were placed in a special account. Unfortunately, only a small portion of these funds earmarked for enforcement have been released. In fact, they probably do not exist, having been absorbed in general fund programs. As a result, EPA noise control regulations are unique in U. S. legislative history. "Of the 28 environmental and health an safety statutes passed between 1958 and 1980, the Noise Control Act of 1972 (NCA) stands alone in being stripped of budgetary support."[4] The law is in effect with no funding for enforcement.

To resolve the funding and enforcement dilemma, Congress adopted legislation that turned over much of the EPA program, in particular the enforcement, to the states. Several states, including California, set up their own version of the EPA along with similar regulations. Unfortunately, only a small portion of the funds were released and most states gave up organizing noise control agencies.

While not dead, the EPA noise control regulations were in limbo during the Reagan/Bush years. Recent efforts to rekindle the earlier efforts are being considered. The Clinton administration and Congress have succeeded in placing the issue on the agenda.[5] However, no one is predicting a favorable outcome as yet.

In the interim, several of the Noise Control Act elements have been taken over by other federal agencies.

---

[2] The Environmental Noise Control Act of 1972, Senate report no. 1160, House report no. 842, Ninety-Second Congress.

[3] "The Dormant Noise Control Act," Sidney A. Shapiro, Administrative Conference of the United States, 1991.

[4] Ibid.

[5] Ibid.

4.2.2 Department of Transportation (DOT), Federal Aviation Administration (FAA), Federal Highway Administration (FHWA), Federal Railroad Administration (FRA)

DOT assumed responsibility for all transportation noise from the EPA mandate. Aircraft noise control items are controlled by FAA, which has established criteria for aircraft flyover noise and established many programs, complete with funding, for airport noise mitigation. For example, after many years of public comment, FAA has established a set of engine noise limits for aircraft operating in the USA. Similar but less well-known regulations have been promulgated by FHWA for highway noise and FRA for railroad noise.

Federal Aviation Agency

Federal Aviation Regulations (FAR) Part 36 sets noise certification levels for all aircraft designed after 1970. The initial goal of FAR Part 36 was to reduce existing noise levels by 10 dB. Implementation of quieter airplanes is in stages. In 1989, the quieter stage III airplanes comprised nearly 40% of the domestic fleet (Air Transportation Association., 1991). By the year 2004, all of the noisier stage II aircraft must be phased out (Airport Noise and Capacity Act, 1990). This requirement should promote a quieter environment around airports, but the growth of air transportation and the pressing need for airport expansion threatens to offset the benefits of the quieter aircraft.

Nowadays, the problem of low-flying aircraft has added a new dimension to community annoyance, as the nation seeks to improve its "nap of the earth" warfare capabilities. In addition, the issue of aircraft operations over national parks, wilderness areas, and other areas previously unaffected by aircraft noise has claimed national attention over recent years.[6]

Airport Noise Compatibility Planning is contained in FAR Part 150. In addition to quieting the aircraft, FAA, in conjunction with local airport authorities, has implemented and funded extensive programs to mitigate noise intrusion into the local community. To aid the airport operator in attaining nose/land compatibility, the FAA promulgated Part 150, "Airport Noise Compatibility Planning" in 1981 and updated the document in 1988. Part 150 contains standards for airport operators who voluntarily submit noise exposure maps and airport noise compatibility planning programs to FAA. Included in the regulation is the establishment of a single system for determining the exposure of individuals to airport noise, and a single system for measuring airport (and background) noise. The regulation also prescribes a standardized airport noise compatibility planning program, which includes:

1. Development and submission of noise exposure maps and noise compatibility programs to the FAA by airport operators.
2. Standard noise methodologies and units for use in assessing airport noise.
3. Identification of land uses that are normally compatible (or incompatible) with various levels of airport noise.
4. Procedures and criteria for FAA evaluation, and approval or disapproval of noise compatibility programs by the FAA Administrator.

FAR part 150 contains a table entitled "Land Use Compatibility with Yearly Day-Night Average Sound Levels," identifying land uses that are "normally compatible" or "noncompatible" with various levels of noise exposure. The levels of noise exposure, in yearly day-night average sound level (L_{dn}) correspond to the contours developed for each airport. All land uses may be considered as normally compatible with noise

---

[6] " Noise and Effects," Dr. Alice H. Suter, Administrative Conference of the United States, 1991.

levels less than 65 L_{dn}.[7]

In some instances, the local airport authority, with mostly federal funds, has implemented programs to upgrade the exterior envelope sound attenuation properties of buildings within the 65 or 70 L_{dn} contour. Examples are Los Angeles (LAX), San Francisco (SFO), and Dayton, Ohio. A typical residential sound insulation program, such as that solicited by the noise abatement coordinator for the City of Dayton, included 156 residential homes located inside the 70 L_{dn} noise contour of aircraft noise. The project offeror was asked to perform tasks ranging from identification of those residences that need upgrading, preparing the plans and specifications, testing before and after upgrade, and managing the entire construction program (Figure 4-2).

The DOT/FAA programs are the most active part of the Noise Control Act. As is evident, FAA has taken a proactive role in quieting flyover noise. It's efforts appear commendable on the surface. However, it has created considerable controversy. The most notable constructive critic of FAA has been the National Organization to Insure a Sound Controlled Environment (NOISE).[8] Composed primarily of airport operators and community officials concerned about airport noise, the organization has been instrumental in guiding the FAA programs. Some members have been critical of FAA for bowing to pressure from the airlines and taking a cavalier attitude about noise implementation. For example, FAA has mandated that all airports comply with the FAA regulations on stage II aircraft and local noise regulations. In the case of Orange County Airport, FAA threatened to withdraw federal funds if Orange County continued to enforce noise limits on takeoff that were more stringent than the federal regulations. The same thing caused LAX airport officials to relent on their intent to require quieter aircraft than the FAA rules.

AIRPORT RESIDENTIAL SOUND INSULATION PROGRAM
(Example)

1. Prioritize areas within the 70 L_{dn} contour for sound insulation implementation.

2. Inventory units in the designated area and recommend candidate units for participation (i.e., conduct fact finding sound transmission loss tests on exterior envelope).

3. Develop and implement a public information program.

4. Specify criteria for evaluating applications (i.e., develop specific test procedures and establish noise limits above which require mitigation is required.)

5. Perform noise measurements and analysis.

6. Provide feasibility and cost/benefit analysis of alternative treatments and recommend strategy for implementation.

7. Plan and design the acoustical treatments.

8. Prepare detailed plans and work specifications to be included in bid documents for general contractors.

9. Provide acoustical insulation training to the selected general contractors and assure compliance with design criteria.

Figure 4-2 Work program for Dayton Airport noise mitigation.

---

[7] See Section 6.2 for more on L_{dn} contours and Section 7.8 for an example of a residential sound insulating program in Dayton.

[8] NOISE sponsors several symposiums per year where the positions of aircraft manufacturers and owners, airport managers, city officials, communities impacted by the noise, and noise-conscious citizens are aired. If airport noise is your prime concern, you will want to keep close tabs on the proceedings of NOISE. Staff offices are at 1225 19th St. NW, Washington, D.C., (202) 452-1487.

Federal Highway Administration

In 1982 the Federal Highway Administration established noise standards for traffic noise on federal highways. When these standards or noise abatement criteria (NAC), are approached or exceeded, noise impact occurs. The NAC for most sensitive receptors such as parks, residences, schools, churches, libraries, and hospitals is 67 dB$_A$ at the receiver location or the property boundary. Trucks are more stringently regulated than autos.

FHWA's implementation of the NAC prompted several states to adopt highway noise control programs that require highway noise barriers through sensitive noise corridors. California has the most active program. Similar programs are evident in Washington, D.C., Seattle, and Minneapolis. The usual process is initiated with an Environmental Impact Study. An EIS including a "Noise Element" has been required for all new highway construction in recent years. Those areas that required noise mitigation now have highway noise barriers constructed with federal funds.[9] In the late 1980s, Congress added an extra 5 cent per gallon gasoline tax to pay for the new highways with a significant portion of the funds allocated for noise control measures.[10] Extensive computer programs have been developed to predict the noise control performance of highway noise barriers. These programs have been used extensively by those designing highway noise barriers and researching truck noise. [11]

FHWA "has likewise de-emphasized enforcement of EPA's noise standards claiming high compliance rate and the burden of other inspection duties. The Government Accounting Office (GAO) reports, however, that older trucks may be making excessive amounts of noise because of inadequate maintenance."[12]

Federal Railroad Administration

The FRA is responsible for enforcing EPA's railroad noise standards. Like the FHWA, FRA imposes noise limits on sensitive neighbors. FRA also "has discontinued routine noise inspections because the rate of compliance has been extremely high, but the GAO found that high compliance rates may be explained, in part, by the FRA's practice of not citing any railroad that has made a good faith effort to correct the violation, even if the railroad is still in violation of the standard after the correction is made."[13]

There are reports that the railroads will try to comply if pressed. For example, a small town in Oregon succeeded in getting the railroad to curtail unnecessary whistle signals during the night. New rail lines are subject to the EIS process. All new light rail and mass transit construction is being very carefully studied for noise impact prior to community acceptance. This has caused contractors to implement stringent noise

---

[9] See Section 4.4 for more on an EIS "Noise Element."

[10] It is doubtful if very much, or if any, of this fund has been used for noise control.

[11] FHWA programs have spurred several truck manufacturers to research techniques for truck quieting. While this program limits outside truck noise, a similar program to quiet the interior or truck cab comes under the jurisdiction of the Occupational Safety and Health Administration (OSHA) and is discussed later.

[12] "The Dormant Noise Control Act," Dr. Alice H. Suter, Administrative Conference of the United States, 1991.

[13] Ibid.

mitigation efforts at the design, construction, and in-use phases. The San Francisco Bay Area Rapid Transit (BART) District was one of the first systems to implement quiet designs. The BART process has been emulated for several subsequent rail systems such as those in Washington, D.C. and Los Angeles.

Construction Impacts

In all the EPA-instigated noise control regulations, construction noise is a key issue. The noise of earth moving equipment, pile drivers, and increased truck traffic are all potentially intrusive to nearby noise-sensitive receivers. In those projects covered by an Environmental Impact Study, the "Noise Element" will address construction noise with specific noise limitations on the various equipment anticipated on the project. All too often this aspect of noise control is overlooked. A work stoppage due to noise can be an expensive result.[14]

4.2.3 Air Force Regulations

The United States Air Force has land use regulations that are similar to the FAA regulations. Thirteen Compatible Use Districts (CUDs) are used to classify noise zones from an L_{dn} of 65 to 70 (CUD-13) to 85 and above (CUD - 1). For example, it is recommended that no residential uses such as homes, multifamily dwellings, hotels, and mobile home parks should be located where the noise levels are expected to exceed L_{dn} 65. Some commercial an industrial uses are considered acceptable where noise levels do not exceed L_{dn} 75. However, in such instances a 25 to 30 dB_A noise level reduction should be incorporated into the design of noise-sensitive structures.[15]

4.2.4 Department of Labor – Occupational Safety and Health Administration (OSHA)

OSHA noise criteria are well known by those who must abide by noise control regulations in the workplace. Under the Occupational Safety and Health Act of 1970, OSHA enforces regulations designed to protect the hearing of workers. Authority is limited to the regulation of companies engaged in interstate commerce. Under this Act, the Department of Health, Education, and Welfare performs research to define occupational noise limits needed to safeguard the hearing of workers. The Department of Labor relies upon results of this research in setting noise standards. Present OSHA limits are a sliding scale depending on time of worker exposure. The maximum exposure for a worker during an 8-hour workday is 90 dB_A. Every worker exposed to 85 dB_A or above must have audiometric tests done yearly[16] (Figure 4-3). Compliance may be via engineering controls where the noise level is lowered by quieting the equipment, management controls where the worker is limited to the time of noise exposure or temporary measures. Temporary measures include ear protectors. However, they are acceptable only for 60 days or until engineering or management controls can be implemented.

OSHA criteria for noise control are widely accepted by the industry and compliance is fairly well

---

[14] Additional information about the EIS " Noise Element" is provided in Sections 6.3 and 7.9.

[15] An example of a" Noise Element" for an Air Force base that also serves as a commercial airport is given in Section 7.9.

[16] A more comprehensive review of OSHA criteria and the techniques utilized in the industry to comply is given in *Noise Control Manual*, D. A. Harris, VNR, 1991.

implemented. OSHA inspectors are a bit lax on those industries that have shown that the cost of compliance is not reasonable. A notable example is the airlines, whose baggage handlers and maintenance personnel are allowed to wear ear protectors on a permanent basis. Of interest to contractors and builders is the fact that OSHA regulations apply to all equipment operators and construction workers.

4.2.5 Department of Health, Education, and Welfare (HEW)

Under the authority of the Federal Coal Mine Health and Safety Act of 1969, the National Institute for Occupational Safety and Health is responsible for the development of noise standards to protect the hearing of miners. Enforcement of these standards is the responsibility of the Mining Enforcement and Safety Administration, Department of the Interior. Regulations are similar to those from OSHA.

NOISE EXPOSURE LEVELS ALLOWED	
Noise level, dB_A	Allowable daily exposure
85	8 hours
Audiometric testing of worker required	
90	8 hours
92	6 hours
95	4 hours
97	3 hours
100	2 hours
105	1 hour
110	30 minutes
120	1 minute or less

Figure 4-3 OSHA Requirements.

4.2.6 Department of Housing and Urban Development (HUD), Federal Housing Administration (FHA), Veterans Administration (VA)

The Department of Housing and Urban Development has developed broad criteria for the sound insulation characteristics of many types of buildings as well as noise considerations for the site. Concerned about the economic value and marketability of residential programs, the Federal Housing Administration , an agency within HUD, has minimum property standards for housing that will be guaranteed by an FHA loan. Likewise the Veterans Administration has minimum property standards for residences guaranteed by a VA loan. For those who qualify, guaranteed low interest loans are made available by both VA and FHA provided the construction satisfies Noise Assessment Guidelines.[17] Current policy is defined in regulation 24 CFR (Code of Federal Regulations) Part 51-B, "Noise Abatement and Control,"[18] and covers site acceptability standards.

FHA and VA have extensive noise control elements that must be implemented by the builder in order to be acceptable for a low-interest loan. Included are criteria for sound insulation characteristics of walls, floor/ceilings, and exterior envelope in multifamily dwellings of all types, row houses, nursing homes, and single family residences. These criteria must be met in order to qualify for HUD/FHA and VA mortgage insurance. Excerpts from the FHA Minimum Property Standards that apply to acoustical control are given in Table 4-1.

---

[17] *Noise Assessment Guidelines*, HUD-PDR-735, Office of Policy Development and Research, U.S. Department of Housing and Urban Development, June 1983.

[18] *Noise Abatement and Control*, 24 – Code of Federal Regulations (CFR), Part 51, Subpart B, *Federal Register*, Vol. 44, No. 235, July 1979.

TABLE 4-1 FHA Minimum Property Standards For Acoustical Control

Reference	Description
401.1	<u>General:</u> Living units shall be designed to provide an acoustically controlled environment in relation to exterior noise and noise from adjacent living units and public spaces. If a site is unacceptable according to the Interim Standards of HUD Circular 1390.2, the design of new buildings for that site must consider exterior noise and provide for the appropriate sound attenuation.
404.2	Sound Transmission Limitations
404.2.1	Mechanical equipment shall be located and installed to minimize transmission of objectionable sound.
404.2.2	Sound transmission class (STC) shall be determined in accordance with ASTM E90 and ASTM E413.
404.2.3	Impact insulation class (IIC) shall be determined in accordance with ASTM E492.
404.2.4	Living units shall be provided with acoustic separation according to Table 10 of *A Guide to Airborne, Impact, and Structure Borne Noise Control in Multifamily Dwellings.*[19]
515.2.3	<u>Sound Levels</u> of kitchen exhaust and range hood fans shall not exceed 8 sones. Bathroom fans shall not exceed 6.5 sones.

FHA TABLE 10-1 KEY CRITERIA OF AIRBORNE AND IMPACT SOUND INSULATION BETWEEN DWELLING UNITS

	GRADE I	GRADE II	GRADE III*
Wall Partitions	STC ≤ 55	STC ≤ 52	STC ≤ 48
Floor/Ceiling Assemblies	STC ≤ 55	STC ≤ 52	STC ≤ 48
	IIC ≤ 55	IIC ≤ 52	IIC ≤ 48

---

[19] *A Guide to Airborne, Impact, and Structure Borne Noise Control in Multifamily Dwellings* was prepared for FHA by Brendt, Winzer, and Burroughs of the National Bureau of Standards (now National Institute of Standards and Technology) and published by the U. S. Department of Housing and Urban Development, September 1967.

(Notes to FHA table 10-1)

GRADE I is applicable primarily in suburban and peripheral suburban residential areas, which may be considered as the "quiet" locations and lower, as measured using the "A" weighting network of a sound level meter which meets the current standards. The recommended permissible interior noise environment is characterized by noise criteria of NC 20-25 (*Noise Reduction*, L. L. Beranek, McGraw-Hill, 1960). In addition, the insulation criteria of this grade are applicable in certain special cases such as dwelling units above the eighth floor in high-rise buildings and the better class or "luxury" buildings, regardless of location.

GRADE II is the most important category and is applicable primarily in residential urban and suburban areas considered to have the "average" noise environment. The nighttime exterior noise levels might be about 40-45 dB(A); and the permissible interior noise environment should not exceed NC25-30 characteristics.

GRADE III criteria should be considered as minimal recommendations and are applicable in some urban areas which generally are considered as "noisy" locations. The nighttime exterior noise levels might be about 55 dB(A) or higher. It is recommended that the interior noise environment should not exceed the NC-35 characteristic.

STC = Sound transmission class

IIC = Impact isolation class

FHA TABLE 10-2   CRITERIA FOR AIRBORNE SOUND INSULATION OF WALL PARTITIONS BETWEEN DWELLING UNITS

Partition Function Between Dwellings			Sound Transmission Class		
Apartment A		Apartment B	Grade I*	Grade II*	Grade III*
Bedroom	to	Bedroom	55	52	48
Living room	to	Bedroom	57	54	50
Kitchen	to	Bedroom	58	55	52
Bathroom	to	Bedroom	59	56	52
Corridor	to	Bedroom	55	52	48
Living room	to	Living room	55	52	48
Kitchen	to	Living room	55	52	48
Bathroom	to	Living room	57	54	50
Corridor	to	Living room	55	52	48
Kitchen	to	Kitchen	52	50	46
Bathroom	to	Kitchen	55	52	48
Corridor	to	Kitchen	55	52	48
Bathroom	to	Bathroom	52	50	46
Corridor	to	Bathroom	50	48	46

* See definitions for FHA Table 10-1.

FHA TABLE 10-3   CRITERIA FOR AIRBORNE & IMPACT SOUND INSULATION OF
FLOOR/CEILING ASSEMBLIES BETWEEN DWELLING UNITS

Barrier Function Between Dwellings			Grade I*		Grade II		Grade III*	
Apartment A		Apartment B	STC	IIC	STC	IIC	STC	IIC
Bedroom	above	Bedroom	55	55	52	52	48	48
Living room	above	Bedroom	57	60	54	57	50	53
Kitchen	above	Bedroom	58	65	55	62	52	58
Bathroom	above	Bedroom	60	65	56	62	52	58
Corridor	above	Bedroom	55	65	52	62	48	58
Bedroom	above	Living room	57	55	54	52	50	48
Living room	above	Living room	55	55	52	52	48	48
Kitchen	above	Living room	55	60	52	57	48	53
Family room	above	Living room	58	62	54	60	52	56
Corridor	above	Living room	55	60	52	57	48	53
Bedroom	above	Kitchen	58	52	55	50	52	46
Living room	above	Kitchen	55	55	52	52	48	48
Kitchen	above	Kitchen	52	55	50	52	46	48
Bathroom	above	Kitchen	55	55	52	52	48	48
Family room	above	Kitchen	55	60	52	58	48	54
Corridor	above	Kitchen	50	55	48	52	46	48
Bedroom	above	Family room	60	50	56	48	52	46
Living room	above	Family room	58	52	54	50	52	48
Kitchen	above	Family room	55	55	52	52	48	50
Bathroom	above	Bathroom	52	52	50	50	48	48
Corridor	above	Corridor	50	50	48	48	46	46
Mechanical/ Equipment room	above	sensitive area	65		62		58	

* See definitions for FHA Table 10-1.

SELECTED NOTES TO FHA TABLE10-3

The sound insulation between living units and other spaces within the building requires special considerations. Placement of living units vertically or horizontally adjacent to mechanical equipment rooms should be avoided whenever possible. If such cases arise, the following is applicable. Generally the recommended airborne sound insulation criteria between mechanical equipment rooms and sensitive areas in dwellings are STC ≥ 65, 62 and 58 for grades I, II, and III respectively. Mechanical equipment rooms include furnace-boiler rooms, elevator shafts, trash chutes, cooling towers, garages, and the like. Sensitive areas include bedrooms and living rooms. Similarly, the recommended criteria between mechanical equipment rooms and less sensitive areas in dwellings are STC ≥ 60, 58 and for grades I, II, and III, respectively, where less sensitive areas include kitchens and family or recreation rooms. Double-wall construction is usually necessary to achieve adequate acoustical privacy. (Note: This was true in the 1970s. Later developments have produced sound barriers that provide the degree of isolation with resilient furring channels, damped wallboard, and other constructions.) Where living units are above noisy areas, the airborne sound insulation is important and impact insulation becomes a moot point as long as structureborne vibration is minimal. However, where mechanical equipment rooms are above living areas, the airborne sound insulation must be maintained, but in addition the impact insulation becomes extremely important and elaborate steps must be taken to assure freedom from intruding vibrations and impact noise. It is not advisable to ascribe impact insulation criteria values to this case, but rather, such structures should be designed to assure quiet living spaces.

Placement of dwelling units vertically or horizontally adjacent to business areas such as restaurants, bars, and community laundries should be avoided whenever possible. If such situations arise, the recommended airborne sound insulation criteria between business areas and sensitive living areas are STC ≥ 60, 58, and 56 for grades I, II, and III respectively. If the living areas are situated above business areas, impact insulation criteria of IIC ≥ 60, 58, and 56 should be adequate; however, if the relative locations are reversed, i.e., business areas above living areas, the impact insulation criteria values should be increased at least by 5 points.

If noise levels in mechanical equipment rooms or business areas exceed 100 dB, as measured on the linear or C scale of a standard sound level meter, the airborne insulation criteria given above must be raised 5 points.

FHA TABLE 10-4 SUGGESTED CRITERIA FOR AIRBORNE SOUND INSULATION WITHIN A DWELLING UNIT.

Partition function between rooms			Sound Transmission Class		
			GRADE I*	GRADE II*	GRADE III*
Bedroom	to	Bedroom	48	44	40
Living room	to	Bedroom	50	46	42
Bathroom	to	Bedroom	52	48	45
Kitchen	to	Bedroom	52	48	45

* See definitions in notes to FHA Table 10-1.

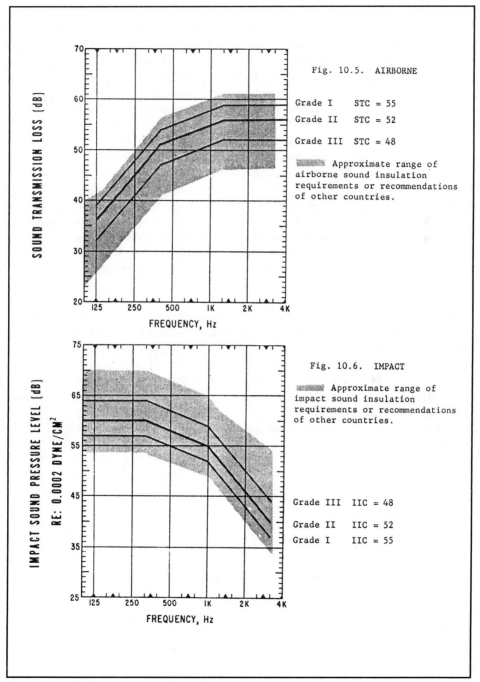

Figure 4-4 FHA recommended sound insulation criteria.

123

Table 4-1 and Figure 4-4 represent recommendations by FHA in 1967. These criteria are still valid for today's environment. For upscale living units, Class III or better ratings are recommended. They are also representative of the criteria used by VA to review their projects for financing. These criteria are recommended for any type of dwellings. Note that FHA Table 10-4 is also applicable to single family residences.

FHA criteria are based on single-number ratings developed by ASTM Committee E33 Environmental Acoustics. STC and IIC evolved from considerable research and have been adopted by the industry as the preferred rating systems. Graphical presentation (Figure 4-4) of individual frequency values required by the rating system for Classes I, II and III are similar to rating systems in Europe.[20]

Satisfaction of the IIC and STC ratings in the design may be achieved by reviewing sound test data for a wide range of systems of construction. Many typical systems are given in Appendix 3. Other systems and their ratings are available from building materials manufacturers. Members of the National Council of Acoustical Consultants (NCAC)[21] can identify and select systems that will meet the criteria, establish that they are appropriate for the type of construction anticipated, and within the project budget.

FHA and VA guaranteed financing was utilized by a major portion of the construction industry in the 60s and 70s. While losing ground in the 80s this form of financing appears to be increasing in the 90s since it allows low down payments. Construction that meets VA and FHA minimum property standards is easier to market because of the type of financing, and the acoustical controls required. It also enhances the resale value and desirability of the property. Coupled with community noise controls and property noise assessments, property that satisfies all the criteria for the acoustical environment will maintain greater value throughout its useful life.

It is important to note that all the FHA criteria are based on laboratory test ratings in accordance with ASTM E90 for airborne sound and E492 for impact sound.[22] It is expected that field conditions such as flanking paths will not jeopardize the achievement of these criteria using standard field test methods. If the ratings achieved under actual field conditions are 5 points lower than the laboratory values there is good reason to believe that flanking paths exist or construction of the system of construction differs from that in the laboratory description. [23]

---

[20] ASTM E33 procedures are outlined in Section 4.6. STC and IIC ratings for typical systems are given in Appendix 3.

[21] The author is a member of NCAC. For more information call Dave Harris at (360) 437-0814.

[22] E90 and E492 procedures are found in ASTM Volume 4.06, Environmental Acoustics. Descriptions of these and other ASTM acoustical standards are given in Section 4.2.

[23] Flanking paths are described in Chapter 2. Field versus laboratory results are discussed in Section 5.5.

4.3 STATE AND LOCAL REGULATIONS

4.3.1 Introduction

For almost a century and a half, the control of noise by regulation within the United States has been a matter of concern for city governments. It is only within the last decade that some states have seriously entered the field of noise regulation. The most notable is the state of California. By 1980, over 400 municipalities, representing a combined population in excess of 60 million people, or 30% of the population, enacted noise regulations. It is estimated the coverage now includes over 75% of the population.

Most of the expansion of state and local acoustical regulations have been prompted by local concern for noise as an environmental issue, considerable press about EPA, OSHA, FAA and other federal noise programs and the adoption of local building codes. Environmental Impact Statements and the federal programs discussed earlier have raised the level of awareness for noise mitigation on local projects. Considerable technical information and guidelines made it possible for local authorities to implement noise control regulations for their community. In addition, three model building codes have made it easy for local authorities, who admit to having little expertise, to adopt up-to-date and technically responsible building requirements similar to those adopted by HUD/FHA and VA minimum property standards.

Local and state noise control laws are usually classed as general nuisance laws which include a prohibition of "unnecessary," "excessive," or "unreasonable" noise. these criteria are often subjective and difficult to enforce. When vigorously challenged, this type of noise control law tends to be struck down by the court because it is too broadly worded for objective enforcement. What is "unnecessary" or "unreasonable" for one person may be "acceptable" to another.

Noise control laws which are based on performance standards appear to be far superior to nuisance laws if they are properly written and correctly applied. Performance standards typically specify the maximum allowable noise level at a given point and this requires monitoring equipment and trained personnel for enforcement.

Who has control of the noise ordinances in your area? A good question, but there is no simple answer. It varies widely. In the cases of a large municipality, it may be the city. However, because of local politics noise control may be governed by the county or state. And, to make matters complicated, noise control may be adopted by one or all of the authorities with considerable overlap and/or inconsistency between them. The agency within the public authority having noise control responsibility presents an even greater maze. It may be the police, zoning board, building official, or, if you are lucky an "Office of Noise Abatement." The root of a noise control ordinance is elusive, and tracing it can be a frustrating experience. Even finding the ordinances is a bureaucratic challenge. Interpreting the varied language requires both an acoustician and lawyer.

The balance of this section is devoted to providing a few examples of local and state ordinances that cover noise. They are provided primarily to guide you in your search for who has responsibility for your particular project.

4.3.2 Model Building Codes

Model building codes available for use by local municipalities are:

- Uniform Building Code (UBC)
  Sponsor: International Congress of Building Officials, Inc. (ICBO), Whittier, California.

- Standard Building Code (SBC)
  Sponsor: Southern Building Code Congress International, Inc., Birmingham, Alabama

- BOCA National Building Code (BOCA)
  Sponsor: Building Officials and Code Admin. Int., Inc., Country Club Hills, Illinois

- Life Safety Code (NFPA 101)
  National Fire Protection Association, Quincy, Mass

These building codes are formatted for adoption in whole or in part by the state, county, city, township, or village as part of its ordinance. Each code contains a significant section on acoustics and noise control. However, each governing agency has adopted portions applicable to its circumstances and supplemented it with considerable local flair.

For example, California and Washington have written their own noise criteria into law. The Revised Code of Washington, Chapter 173-60, Washington Administrative Code, establishes maximum permissible noise levels for various environments, or class of use, called environmental designation for noise abatement (EDNA). Class A covers a residential, school, or hospital area; Class B applies where interference with speech is of concern, such as an office complex or park; and Class C covers industrial land use. Enforcement of the Code is assigned to the State of Washington Department of Ecology by Title 70 Revised Code of Washington Chapter 70.107. Each day of noncompliance is considered a separate violation with the penalty not to exceed $100. Fees are paid to the State General Fund. Implementation varies depending on local assignment of duties but generally is the responsibility of the local building inspector.

Almost every structure and building system must meet the provisions of a building code. A building code or standard, when adopted through legislative action, mandates minimum requirements for the design and construction of buildings. Adoption of codes tends to follow a geographic pattern; the Uniform Building Code is favored in the west; the Standard Building Code in the south; and the BOCA National Building Code in the east and midwest (See Figure 1-25). Each code covers similar aspects of the building such as fire safety and acoustics. However, because of divergent writing groups and their experiences, each code is different. Most are revised or supplemented to satisfy local needs.

Codes apply primarily to new construction but can also apply to the remodeling, alteration, or repair of an existing building, depending on the value of the alteration in relation to the building value and the time frame of alteration. Changing the occupancy or use of a building will generally require the building to satisfy the current building code.

Codes establish only minimum requirements for acoustical, safety, and other environmental issues. It is often necessary for a building owner, or the planning and design team, to go beyond the code guidelines to meet the particular needs of the building or users.

The UBC is typical and requires all residential units with separating assemblies, such as party walls and floor/ceiling systems, to have a minimum sound rating. Typical ratings are STC 50 for airborne sound transmission and IIC 50 for impact sound isolation. The section referencing sound transmission control in UBC is as follows:

Uniform Building Code - Appendix to Chapter 35
SOUND TRANSMISSION CONTROL

Sound Transmission Control

Sec. 3501. (a) General. In Group R Occupancies (Residential of all types), wall and floor-ceiling assemblies separating dwelling units or guest rooms from each other and from public space such as interior corridors and service areas shall provide airborne sound insulation for walls, and both airborne and impact sound insulation for floor-ceiling assemblies.

(b) **Airborne Sound Insulation**. All such separating walls and floor-ceiling assemblies shall provide an airborne sound insulation equal to that required to meet a Sound Transmission Class (STC) of 50 (45 if field tested) as defined in U. B. C. Standard No. 35-1.

Penetrations or openings in construction assemblies for piping, electrical devices, recessed cabinets, bathtubs, soffits, or heating, ventilating or exhaust dusts shall be sealed, lined, insulated or otherwise treated to maintain the required ratings.

Entrance doors from interior corridors together with their perimeter seals shall have a laboratory-tested Sound Transmission Class (STC) rating of not less than 26 and such perimeter seals shall be maintained in good operating condition.

(c) **Impact Sound Insulation**. All separating floor-ceiling assemblies between separate units or guest rooms shall provide impact sound insulation equal to that required to meet an Impact Insulation Class (IIC) of 50 (45 if field tested) as defined in U. B. C. Standard No. 35-2. Floor coverings may be included in the assembly to obtain the required ratings and must be retained as a permanent part of the assembly and may be replaced only by other floor covering that provides the same sound insulation required above.

(d) **Tested Assemblies**. Field or laboratory tested wall or floor-ceiling designs having an STC or IIC of 50 or more as determined by U. B. C. Standard No. 35-1, 35-2, or 35-3 (ASTM E90/413 for airborne sound, E492/989 for impact sounds, E336 or E597 for field measurements) may be used without additional field testing when, in the opinion of the building official, the tested design has not been compromised by flanking paths. Tests may be required by the building official when evidence of compromised separations is noted.

(e) **Field Testing and Certification**. Field testing, when required, shall be done under the supervision of a professional acoustician who shall be experienced in the field of acoustical testing and engineering and who shall forward certified test results to the building official that minimum sound insulation requirements stated above have been met.

(f) **Airborne Sound Insulation Field Tests.** When required, airborne sound insulation shall be determined according to the applicable Field Airborne Sound Transmission Loss Test procedures of U. B. C. Standard No. 35-3 (ASTM E336 or 597/413). All sound transmitted from the source room to the receiving room shall be considered to be transmitted through the test partition.

(g) **Impact Sound Insulation Field Test**. When required, impact sound insulation shall be determined in accordance with U. B. C. Standard No. 35-2 (ASTM E492/989)

Sound Transmission Control Systems

Sec. 3502. Generic systems as listed in the Fire Resistance Design Manual, October 1984, Eleventh Edition, as published by the Gypsum Association may be accepted where a laboratory test indicates that the requirements of Section 3501 are met by the system. (Otherwise, a specific sound test report from an approved acoustical testing laboratory or field testing agency is required.) *****

To determine what building code applies in your area, consult your local building code official. If the local authority has adopted a particular code in its entirety, you may contact the sponsoring organization for a copy of the code. Most will also provide some technical assistance on interpreting the code for your application. However, they may charge a fee for the service. Unfortunately, local building officials are not well trained in acoustics and acoustical requirements or procedures. There have been many cases of misleading or missing information. Since noncompliance may cause litigation and/or expensive retrofit it is best to consult a member of the National Council of Acoustical Consultants who is familiar with the building code covering your location.

4.3.3 City and Local Building Codes and Noise Ordinances

Most major municipalities have adopted one of the model building codes addressed in Section 4.3.2. A few have expanded and/or upgraded the requirements for STC and IIC ratings of occupancy separators. To determine what is applicable for your building, contact the building inspection official responsible for your location.

Local noise ordinances vary widely. They may range from the simple nuisance-type ordinance where the "barking dog" must be quieted during certain hours or they may be so sophisticated that they include "day/night" limitations for different land use, vehicle noise limits, and a whole range of items too numerous to cover in this text. In some cases, a local ordinance will go so far as to limit the sound level of your car radio.[24] An example of a fairly common nuisance ordnance was adopted by the City of Carson, Calif.. Excerpts of the "Prohibited Conduct; Offenses" section of Article IV – Public Peace that relate to noise follows:

CITY OF CARSON NOISE ORDINANCE (excerpts)

4101. <u>Unnecessary Noises.</u> No person shall make, or cause or suffer, or permit to be made upon any premises owned, occupied or controlled by him, any unnecessary noises or sounds which are physically annoying to persons of ordinary sensitiveness or which are so harsh or so prolonged or unnatural or unusual in their use, time, or place as to occasion unnecessary discomfort to any persons within the neighborhood from which said noises emanate or so as they interfere with the peace and comfort of the resident or their guests, or the operators or customers in places of business in the vicinity, or which may detrimentally or adversely affect such residences or places of business. The following acts, among others, are declared to be prima facie evidence of violations of this section:

(a) <u>Unnecessary Noises.</u> The unnecessary making of, or knowingly and unnecessarily permitting to be made, any loud, boisterous or unusual noise, disturbance or commotion in any hotel, motel apartment house, court, rooming house, auto court, trailer camp, dwelling, place of business or other structure, or upon any public street, park or other place or building, except the ordinary and usual sounds, noises or commotion incident to the operation with the usual and normal standard of practice applicable thereto and in a manner which will not disturb the peace and comfort of adjacent

---

[24] A Redondo Beach, California ordinance requires any noise emanating from a vehicle shall not exceed 80 dB_A at a specified distance (nominally 20 ft). The purpose is to reduce the noise of vehicles, mostly trucks and vans with a "boom box." Police will arrest violators and confiscate the vehicle. Since this level is easily exceeded by most automobile radios being played at a level to be heard above the breeze, in an open vehicle the ordinance is very stringent. It has been upheld in court.

residences or which will not detrimentally affect the operators or customers of adjacent places of business.

(b) <u>Radios, Phonographs, Etc.</u> The using, operating or permitting to be played, used or operated between the hours of 11:00 p.m. and 7:00 a.m. of any radio, musical instrument, phonograph, television set, or instrument or device similar to those heretofore specifically mentioned for the production or reproduction of sound in volume sufficiently loud as to disturb the peace, quiet or repose of persons of ordinary and normal sensitiveness who are in the immediate vicinity of such machine or device.

(c) <u>Band or Orchestral Rehearsals.</u> The conducting of or carrying on of band or orchestral concerts or rehearsals or practice between the hours of 11:00 p.m. and 7:00 a.m. in any area, building or structure wherein the adjoining occupied building used for human habitation is nearer than 200 feet.

(d) <u>Loudspeakers and Amplifiers for Advertising.</u> The playing or operating, or permitting the playing or operating of any musical instrument, radio, television set or phonograph, or the operation and use of any loudspeaker or amplifying device on any public street, or in any public place, or outside of any doorway of any building, facing upon a business street in the City of Carson for the purpose of advertising or in conjunction with any commercial venture, without a permit being granted therefore by the City Administrator upon application filed with him and hearing held, after at least 48 hours notice to applicant of such hearing, unless notice is waived; provided that if said application is denied, applicant may within 15 days of the date of said order of denial, appeal from said decision of the City Administrator to the City Council, which body shall hold a hearing thereon upon the notice to applicant in the same manner as the hearing before said City Administrator.

(e) <u>Engines, Motors and Mechanical Devices Near Residential District.</u> The operation between the hours of 11:00 p.m. and 7:00 a.m. of any electric motor or engine or the use of operation of any automobile, motorcycle, machine or mechanical device is enclosed within a sound insulated structure so as to prevent noise and sound from being plainly audible at a distance of 50 feet from such structure, or within 10 feet of any residence.

(f) <u>Noises by Animals.</u> It shall be unlawful for any person having charge, care, custody, or control of any animal to permit such animal to emit any excessive noise which is disturbing or offensive. The City shall enforce this ordinance as follows: (deleted)

(g) <u>Amplifying devices for Hawkers and Peddlers.</u> The operation of any vehicle or instrumentality equipped with, or to which is attached and used any loudspeaker or sound amplifying device, or the operation of any device or instrumentality, either mobile or stationary, through which device the spoken word, or other sounds are produced or reproduced in such increased volume as to be clearly audible to a person of normal hearing under normal and ordinary conditions, for a distance of more than 200 feet, from the source of such sound, or between the hours of 9:00 p.m. and 8:00 a. m. without first obtaining a permit for such operation from the City Administrator, upon application filed with him, and hearing held, after at least 48 hours notice to the applicant of such hearing, unless notice is waived, and provided that if said application is denied, applicant may, within 15 days from the date of said order of denial, appeal from said decision of the City Administrator to the City Council, which will hold a hearing thereon upon notice to applicant in the same manner as the hearing before said City Administrator.

(h) <u>Motor Vehicles.</u> The racing of the engine of any motor vehicle or needless bringing to a sudden start or stop any motor vehicle.

(i) <u>Pile Drivers, Hammers, Etc.</u> the operation between the hours of 6:00 p.m. and 7:00 a.m. of any pile driver, steam shovel, pneumatic hammer, derrick, hoist or other appliance, the use of which is attended by loud or unusual noise.

(j) <u>Construction or Repair of Buildings</u>. The erection (including excavation), demolition, alteration, construction or repair of any building other than between the hours of 7:00 a.m. and 6:00 p.m. on weekdays, except in case of urgent necessity in the interest of public health and safety, and then only with a permit from the City Engineer, which permit may be granted for a period not to exceed three days while the emergency condition continues, and which permit may be renewed for periods of three days or less while the emergency continues; if the City Engineer should determine that the public health, safety, comfort and convenience will not be impaired by the erection, demolition, alteration or repair of any building or the excavation of streets, highways, or other sites within the hours of 6:00 p.m. and 7:00 a.m. or any part thereof, or on Sunday, and that substantial loss or inconvenience would result from any party in interest denied permission to do so, he may grant permission for such work, on any day, or at such times within such hours as he shall fix in accordance with such determination.

(k) <u>Yelling, Shouting, Etc.</u> Yelling, shouting, hooting, whistling or singing on the public streets or in a public place between the hours of 11:00 p.m. and 7:00 a.m. or at any time or place so as to annoy or disturb the quiet, comfort or repose of persons in any office, or in any dwelling, hotel or other type of residence, or of any persons in the vicinity.

****

Note that the entire foregoing document is subjective and contains no performance criteria. Enforcement is handled by the police. Construction requirements for noise control are covered in the Uniform Building Code.

By way of comparison, the City of Los Angeles adopted an ordinance that is a combination of nuisance and performance. All construction requirements for noise control are contained in the Los Angeles Building Code, which is based on UBC but has many exceptions and additions. The noise ordinance is contained within the Los Angeles Municipal code. Both the noise ordinance and the building code, including noise control provisions, are enforced by the Los Angeles Department of Building and Safety. A separate office called the Office of Noise Abatement is staffed with experienced personnel that coordinate noise issues with the police, building officials and planning department. The Los Angeles City Noise Ordinance 156,363, dated and approved February 9, 1982, is reprinted in part as follows:

CITY OF LOS ANGELES (NOISE) ORDINANCE 156,363

Section 111.01. DEFINITIONS

Unless the context otherwise clearly indicates, the words and phrases used in this chapter are defined as follows:

(a) "Ambient Noise" is the composite of noise from all sources near and far in a given environment, exclusive of occasional and transient intrusion noise sources and of the particular noise source or sources to be measured. Ambient noise shall be averaged over a period of at least 15 minutes at a location and time of day comparable to that during which the measurement is taken of the particular noise source being measured.

(b) "Commercial Purpose" is the use, operation, or maintenance of any sound amplifying equipment for the purpose of advertising any business, goods, or services, or for the purpose of attracting the attention

130

of the public to, advertising for, or soliciting patronage or customers to or for any performance, show, entertainment exhibition, or event, or for the purpose of demonstrating such sound equipment.

(c) "Decibel" (dB) is a unit of level which denotes the ratio between two (2) quantities which are proportional to power; the number of decibels corresponding to the ratio of two (2) amounts of power is ten (10) times the logarithm to the base (10) of this ratio.

(d) "Emergency Work" is work made necessary to restore property to a safe condition following a public calamity or work required to protect persons or property from an imminent exposure to danger, or work by private or public utilities when restoring utility service.

(e) "Impulsive Sound" is sound of short duration, usually less than one second, with an abrupt onset and rapid decay. By way of example, "impulsive sound" shall include, but shall not be limited to, explosions, musical bass drum beats, or the discharge of firearms.

(f) "Motor Vehicle" includes, but shall not be limited to, automobiles, trucks, motorcycles, minibikes and go-carts.

(g) "Noncommercial Purpose" is the use, operation, or maintenance of any sound equipment for other than a "commercial purpose". "Noncommercial purpose" shall mean and include, but shall not be limited to, philanthropic, political, patriotic, and charitable purposes.

(h) "Octave band Noise Analyzer" is an instrument for measurement of sound levels in octave frequency bands which satisfies the pertinent requirements for Class II octave band analyzers of the American National Standard Specifications for Octave, Half-Octave, and Third-Octave Band filters, S1.11-1966 or the most recent revision thereof.

(i) "Person" is a person, firm, association, co-partnership, joint venture, corporation or any entity, private or public in nature.

(j) "Sound Amplifying equipment" is any machine or device for the amplification of the human voice, music or any other sound, but shall not include:

1. Automobile radios, stereo players or television receivers when used and heard only by the occupants of the vehicle in which the same is installed.

2. Radio, stereo players, phonographs or television receivers used in any house or apartment within any residential zone or within 500 feet thereof.

3. Warning devices on emergency vehicles.

4. Horns or other warning devices authorized by law on any vehicle when used for traffic purposes.

(k) "Sound Level" (Noise level) in decibels (dB) is the sound measured with the "A" weighting and slow responses by a sound level meter; except for impulsive or rapidly varying sounds, the fast response shall be used.

(l) "Sound Level Meter" is an instrument including a microphone, an amplifier, an output meter, and "A" frequency weighting network for the measurement of sound levels which satisfies the pertinent requirements for Type S2A meters in American Standard Specifications for sound level meters in S1.4-1971 or the most recent revision thereof.

(m) "Sound Truck" is any motor vehicle, or any other vehicle regardless of motive power, whether in motion or stationary which carries, is equipped with, or which has mounted thereon, or attached thereto, any sound amplifying equipment.

(n) Supplementary Definitions of Technical Terms. Definitions of technical terms not defined herein shall be obtained from American Standard Acoustical Terminology S1-1-1971 or the most recent revision thereof.

SECTION 111.02. SOUND LEVEL MEASUREMENT PROCEDURE AND CRITERIA.

(a) Any sound level measurement made pursuant to the provisions of this chapter shall be measured with a sound level meter using the "A" weighting and response as indicated in Section 111.01(k) of this article. Except when impractical, the microphone shall be located four to five feet above the ground and ten feet or more from the nearest reflective surface. However, in those cases where another elevation is deemed appropriate, the latter shall be utilized.

Interior sound level measurements shall be made at a point at least four feet from the wall, ceiling, or floor nearest the noise source.

Calibration of the sound level meter, utilizing an acoustic calibrator, shall be performed immediately prior to recording any sound level data.

The ambient noise level and the level of a particular noise being measured shall be the numerical average of noise measurements taken at a given location during a given time period.

Section 3. Subsection (b) of Section 111.02 of the Los Angeles Municipal Code is hereby amended to read:

(b) Where the sound alleged to be offending is of a type or character set forth below, the following values shall be added to the sound level measurement of the offending noise:

1. Except for noise emanating from any electrical transformer or gas metering and pressure control equipment existing and installed prior to the effective date of the ordinance enacting this chapter, any steady tone with audible fundamental frequency or overtones have 200 Hz. +5.

2. Repeated impulsive noise. +5

3. Discontinuous noise lasting less than 5 hours. +5

4. Noise occurring five minutes or less in any period of 60 consecutive minutes between the hours of 7:00 a. m. and 10: p.m. of any day. -10

Section 4. Section 111.02 of the L. A. Municipal Code is hereby amended by adding Subsection (d) thereto, said subsection to read:

(d) For those cases where a sound level measurement has been made pursuant to the provisions of this chapter and two or more provisions of this chapter apply, the provision establishing the lower or lowest noise level, respectively, shall be used.

Section 5. Section 111.03. MINIMUM AMBIENT NOISE LEVEL

Where the ambient noise level is less than the presumed ambient noise level designated in this section, the presumed ambient noise level in this section shall be deemed to be the minimum ambient noise level for the proposes of this chapter.

TABLE II - SOUND LEVEL "A" DECIBELS

In this chart, daytime levels are to be used from 7:00 a.m. to 10:00 p.m. and nighttime levels from 10:00 p.m. to 7:00 a.m.

	PRESUMED AMBIENT NOISE LEVELS [dB(A)]	
ZONE (L. A. City land use)	DAY	NIGHT
A1, A2, RA, RE, RS, RD, RW1, RW2, R1, R2, R3, R4, and R5	50	40
P, PB, CR, C1, C1.5, C2, C4, C5 and CM	60	55
M1, MR1, and MR2	70	70

At the boundary line between two zones, the presumed ambient noise level of the quieter zone shall be used.

Section 6. Section 111.04 of the Los Angeles Municipal Code is hereby amended to read:

SECTION 1104. VIOLATIONS; ADDITIONAL REMEDIES. INJUNCTIONS.

As an additional remedy, the operation or maintenance of any device, instrument, vehicle, or machinery in violation of any provision of this chapter, which operation or maintenance causes discomfort or annoyance to reasonable persons or which endangers the comfort, repose, health, or peace of residents in the area, shall be deemed and is declared to be, a public nuisance and may be subject to abatement summarily by a restraining order or injunction issued by a court competent jurisdiction.

Section 7. Section 111.05 of the Los Angeles Municipal Code is hereby added and shall read:

SECTION 111.05, ENFORCEMENT, CITATIONS.

(a) The Department of Building and Safety shall have the power and duty to enforce the following noise control provisions of this Code: Section 12.14A-6(h), Section 12.19A-4(b)(1), Section 112.01(c), Section 112.02, Section 112.04, Section 112.05, Section 112.06, Section 114.01(c), and Section 114.02(a)3. However, the sound measurements described in Section 112.01(c), Section 114.01(c) and Section 114.02(a)3 need only be performed by the Department of Building and Safety in conjunction with a request from the Police Department or the City Attorney to provide them with information to help them in the enforcement of Section 112.01(a), Section 114.01(a) and Section 114.02(a)1 and 2.

(b) The Police Department shall have the power and duty to enforce the following noise control provisions of this Code: Section 41.32, Section 41.40, Section 41.42,, Section 41.44, Section 41.57, Section 63.51(m), Section 112.01(a) and (b), Section 112.03, Section 113.01, Section 114.01(a) and (b), Section 114.02(a)1 and (a)2, Section 115.02, and Section 116.01.

(c) Any Building Mechanical Inspector assigned to noise enforcement inspection shall have the power, authority and immunity of a public officer and employee, as set forth in the Penal Code of the State of California, Section 836.5, to make arrests without a warrant whenever such employee has reasonable cause to believe that the person to be arrested has committed a misdemeanor in his presence which is a violation of any provision set forth in Section 111.05(a) of this chapter. The provision of said Penal Code section regarding issuance of a written promise to appear shall be applicable to arrests authorized herein.

Section 8. Section 112.01 of the Los Angeles Municipal Code is hereby amended to read:

SECTION 112.01. RADIOS, TELEVISION SETS, AND SIMILAR DEVICES.

(a) It shall be unlawful for any person within any zone of the City to use or operate any radio, musical instrument, phonograph, television receiver, or other machine or device for the producing, reproducing or amplification of the human voice, music, or any other sound, in such a manner, as to disturb the peace, quiet, and comfort of neighboring occupants or any reasonable person residing or working in the area.

(b) Any noise level caused by such use or operation which is audible to the human ear at a distance in excess of 150 feet from the property line of the noise source, within any residential zone of the City or within 500 feet thereof, shall be a violation of the provisions of this section.

(c) Any noise level caused by such use or operation which exceeds the ambient noise level on the premises of any other occupied property, or if a condominium, apartment house, duplex, or attached business, within any adjoining unit, by more than five (5) decibels shall be a violation of the provisions of this section.

Section 9. Section 112.02 of the Los Angeles Municipal Code is hereby amended to read:

SECTION 112.02. AIR CONDITIONING, REFRIGERATION, HEATING, PUMPING, FILTERING EQUIPMENT.

(a) It shall be unlawful for any person, within any zone of the city to operate any air conditioning, refrigeration or heating equipment for any residence or other structure or to operate any pumping, filtering or heating equipment for any pool or reservoir in such manner as to create any noise which would cause the noise level on the premises of any other occupied property or if a condominium, apartment house, duplex, or attached business, within any adjoining unit, to exceed the ambient noise level by more than five (5) decibels.

(b) This section shall not be applicable to emergency work, as defined in Section 111.01(c) of this chapter, or to periodic maintenance or testing of such equipment reasonably necessary to maintain such equipment in good working order.

Section 10. Section 112.04 of the Los Angeles Municipal Code is hereby amended to read:

SECTION 112.04. OTHER MACHINERY, EQUIPMENT & DEVICES

Except as to the equipment and operations specifically mentioned and regulated elsewhere in this chapter, and except as to aircraft, tow tractors, aircraft auxiliary power units, trains and motor vehicles in their respective operations governed by state or federal regulations, no person shall operate or cause to be operated any machinery, equipment or other mechanical or electrical device in such manner as to create any noise which would cause the noise level on the premises of any other occupied property, or if a condominium, apartment house, duplex, or attached business, within any adjoining unit, to exceed the ambient noise level by more than five (5) decibels.

This section shall not be applicable to emergency work, as defined in Section 111.01(c) of this chapter, or to periodic maintenance or testing of such equipment reasonably necessary to maintain such equipment in good working order.

SECTION 112.05. MAXIMUM NOISE LEVEL OF POWER EQUIPMENT OR POWERED HAND TOOLS.

(a) No person shall operate or cause to be operated any power equipment or powered hand tool that produces a maximum noise level exceeding the following noise limits at a distance of 50 feet therefrom.

1. Construction and industrial machinery including crawler tractors, dozers, rotary drills and augers, loaders, power shovels, cranes, derricks, motor graders, paving machines, off-highway trucks, ditchers, trenchers, compactors, scrappers, wagons, pavement breakers, compressors and pneumatic powered equipment. 75 dB(A).

2. Agricultural tractors and equipment. 75 dB(A).

3. Powered equipment of 29 HP or less intended for infrequent use in residential area, including chain saws, log chippers, and power hand tools. 75 dB(A).

4. Powered equipment intended for repetitive use in resident areas, including lawn mowers, gardener's backpacks, small lawn and garden tools and riding tractors. 65 dB(A).

Except that the noise limits for particular equipment listed above shall be deemed to be superseded and replaced by noise limits for such equipment from and after their establishment by final regulation adopted by the Federal Environmental Protection Agency and published in the Federal Register.

Said noise limitations shall not apply where compliance therewith is technically infeasible. The burden of proving that compliance is technically infeasible shall be upon the person or persons charged with a violation of this section. Technical infeasibility shall mean that said noise limitations cannot be complied with despite the use of mufflers, shields, sound barriers and/or any other noise reduction device or techniques during the operation of the equipment.

Section 12. Section 112.06 of the Los Angeles Municipal Code is hereby added and shall read:

SECTION 112.06. PLACES OF PUBLIC ENTERTAINMENT

It shall be unlawful for any person to operate, play or to permit the operation or playing of any radio, television receiver, phonograph, musical instrument, sound amplifying equipment, or similar device which produces, reproduces, or amplifies sound in any place of public entertainment at a sound level greater than 95dB(A) at any point that is normally occupied by a customer, unless a conspicuous and legal sign is located outside such place near each public entrance stating: "WARNING: SOUND LEVELS WITHIN MAY CAUSE HEARING IMPAIRMENT".

Section 13. Section 114.01 of the Los Angeles Municipal Code is hereby amended to read:

SECTION 114.01. VEHICLE REPAIRS.

It shall be unlawful for any person, within any residential property located within any residential zone of the City or within 500 feet thereof, to repair, rebuild, reconstruct or dismantle any motor vehicle between the hours of 8:00 p.m. of one day and 8:00 a.m. of the next day in such manner:

(a) That a reasonable person residing in the area is caused discomfort or annoyance;

(b) That such activity is audible to the human ear at a distance in excess of 150 feet from the property line of the noise source;

(c) As to create any noise which would cause the noise level on the premises of any occupied residential property, or if a condominium, apartment house or duplex, within any adjoining unit, to exceed the ambient noise level by more than five (5) decibels.

Section 14. Section 114.02 of the Los Angeles Municipal Code is hereby amended to read:

SECTION 114.02. MOTOR DRIVEN VEHICLES.

(a) It shall be unlawful for any person to unreasonably operate any motor driven vehicle upon any property within the City or unreasonably accelerate the engine of any vehicle, or unreasonably sound, blow or operate the horn or other warning device of such vehicle in such manner;

1. As to disturb the peace, quiet and comfort of any neighborhood or of any reasonable person residing in such area;

2. That such activity is audible to the human ear at a distance in excess of 150 feet from the property line of the noise source;

3. As to create any noise which would cause the noise level on the premises of any occupied residential property, or if a condominium, apartment house or duplex, within any adjoining unit, to exceed the ambient noise level by more than five (5) decibels.

(b) This section shall not be applicable to any vehicle which operated upon any public highway, street or right-of-way or to the operation of any off-highway vehicle to the extent it is regulated in the Vehicle Code.

Section 15. Section (c) of Section 80.31 of the Los Angeles Municipal Code is hereby repealed.

Section 16. Paragraph (h) of Subdivision 6 of Subsection (a) Section 12.14 of the Los Angeles Municipal Code is hereby amended to read:

(h) Any automobile laundry or wash rack, in which power driven or steam cleaning machinery is used, shall be so sound-proofed, the entire development shall be so arranged, and the operations shall be so conducted that the noise emanating therefrom, as measured from any point on adjacent property, shall be no more audible than the noise emanating from the ordinary street traffic and from other commercial or industrial use measured at the same point on said adjacent property; provided, however, that in no event shall it be necessary to reduce the noise from such laundry or wash rack to below the level provided in Section 111.03 of this Code.

The comparison between the noise emanating from the automobile laundry or wash rack and from the street and commercial or industrial uses shall be made in the manner set forth in Section 111.02(a).

Every wash rack shall be so constructed or arranged so that entrances, exits, and openings therein shall not face any residential property within 100 feet thereof.

Section 17. Subparagraph (1) of Paragraph (b) of Subdivision 4, Subsection A of Section 12.19 of the Los Angeles Municipal Code is hereby amended to read:

(1) No crushing, smashing, baling or reduction of metal is conducted on the premises unless such is conducted without producing substantial amounts of dust and is so conducted that the noise emanating therefrom, as measured from any point on adjacent property shall be no more audible than the noise emanating from ordinary street traffic and from other commercial or industrial use, measured at the same point on said adjacent property; provided however, that such noise shall be permitted in the event it does not exceed the levels provided in Section 111.03 of this Code as measured from any point on adjacent property in an "A", "R", "C", "P", or "M" zone.

Section 19. The City Clerk shall certify to the passage of this ordinance and cause the same to be published in some daily newspapers, printed and published in the City of Los Angeles.

I hereby certify that the foregoing ordinance was passed by Council of the City of Los Angeles at its meeting of February 2, 1982.

REX E. LAYTON, City Clerk, by Edward W. Ashdown, Deputy.

Approved February 9, 1982.   JOEL WACHS, Acting Mayor

File No 77-5319   (DJG17216) Feb. 26
****

Comparison of Noise Ordinances

City and local municipalities noise ordinances vary widely. A survey of noise ordinances for a selected list of cities in southern California and Portland, Oregon, was conducted in 1991 by the author and is summarized in the table 4-6. Excerpts from the actual ordinance for the Cities of Carson and Los Angeles precede the noise ordinance summary.

The summary identified the following information given in the noise ordinance for comparison purposes. (<u>Note</u>: These are also the headings for the summary chart in Table 4-6.)

Descriptor -	This is the way the noise is to be measured. dB(A) means that the sound level is measured on the A scale of a sound level meter meeting International Standards Organization (ISO) criteria. Octave limits means that the ordinance also requires the sound to be measured in octave bands. Nuisance only indicates that the noise is subjective only and judged by someone listening to the noise. L(eq) means that the sound level loudness is measured by a technique known as "equivalent loudness" as measured over a specified period of time.
Residence -	The values are the maximum loudness levels allowed in a residential zone. The numbers indicated in this column represent a loudness measurement during the daytime and during nighttime. These are labeled "day/night" levels. For example, 55/45 means that a level of 55 dB(A) is the maximum allowed during the day between 8 a.m and 8 p.m. and 45 dB(A) is the maximum allowed between 8 p.m. and 8 a.m.
Commercial -	Maximum day/night noise levels in commercial zones such as retail stores, etc.
Industrial -	Maximum day/night noise levels in industrial zones such as factories, etc.
Heavy Industry -	Maximum day/night noise levels in a heavy industry zone such as a gas or steel plant, etc.
Construction -	Maximum day/night noise levels allowed on any construction site.

136

Vehicles - Most autos, trucks, trains and aircraft are governed by federal or state criteria. Cities usually enforce the State or Federal law that is applicable. However, some will enact their own legislation and supersede the higher authority.

Interior - The numbers given are the maximum sound level allowed <u>inside the building</u>, usually a residence, due to noise coming from the community. This means that to comply, someone must either reduce the noise in the community (i.e.; the source) or construct the exterior envelope of the building to have sufficient sound attenuation that it will reduce the sound level to or below the values listed. Usually, it is the party responsible for the noise source that must pay for the exterior envelope upgrade. In the case of airport noise, it is the airport authority that will pay for the upgrade of doors, windows, etc. to meet the requirements. This criterion is very interesting since it addresses the listener's environment within a building. However, the outside sound level may be allowed to run rampant.

Administration - Those responsible for enforcing the noise ordinance could be any of a number of different job titles and bureaucratic agencies.

Fines - The fine for noncompliance of $500 per day for each day of excessive noise can get monumental for continuing noise. Litigation and court judgments have upheld this type of regulation. While actual fines were not available, typically they are established by the nature and severity of the offence. In the case of noise from an auto in Hawthorne, California there is automatic seizure of the vehicle. In other instances, the penalty may be prison. By contrast some fines may be nothing more than a warning.

Adjustments are made by many communities (shown with * in Table 4-6) for noise levels of a short duration. For example, a noise that lasts more than 30 minutes may be rated at the same level required for long term exposure. If the sound is less than 15 minutes, the allowable limit may be increased by 5 dB(A). For sounds less than 1 minute in length the allowable limit may be increased by 20 dB(A). In essence these rules are relaxing the noise limits for short-term noise. Most ordinances with this type of variance allow the short noise to occur only once in any 24-hour period and may limit the time of day.

Another form of adjustment (shown with ** in Table 4-6) will adjust the allowable sound limit depending on the length of noise duration. For example, if the noise is steady the limit is reduced by 5 dB(A). If the noise is repetitive the limit is decreased by 5 dB(A). For noise that is "not continuous" over a 5-hour period the level is increased by 5 dB(A). If the noise lasts less than 90 minutes the limit is increased by 10 dB(A). For noise that is less than 30 minutes in duration, the limit is increased by 15 dB(A). If the noise occurs on Sunday in the AM the limit may be reduced by 5 dB(A). These variations become very complex. To be sure of what the actual noise limits are, one must determine the exceptions to the base criteria.

After reading a typical document it will become evident that the exceptions become the rule and are buried within other bureaucratic jargon. It becomes obvious that great care must be taken when reading the ordinances and that there are no rules of thumb to describe noise ordinance requirements. This unfortunate situation has led to unnecessary litigation over noise ordinances.

TABLE 4-6 NOISE ORDINANCE SUMMARY (NC = no criteria, NA = not available)

Municipality	Descriptor	Residence	Commercial	Industrial	Heavy Ind.	Construction	Vehicles	Interior	Administration	Fine
Rolling Hills Est.	dB(A)	44/45	65/55	55/45	7455	75/45	state	45/40	city mgr./police	$500
Los Angeles	dB(A)	50/40	50/55	65/65	70/70	70/70	comprehensive	NC	Buildg. & Safety	NA
Orange County	dB(A)	55/50	55/50	55/50	55/50	55/50	state/fed.	55/45	Sheriff/health officer	NA
El Segundo	dB(A)	--/45*	50/55*	65/70*	65/70*	65/none**	state/fed.	45 (cum./hr.)	Noise Officer	NA
Lawndale	none	NC	NC	NC	NC	none @ nite	state	NC	Planning Commission	NA
Manhattan Beach	dB(A)	45/50	50/55	60/55	60/65	none @ nite & weekends	state/FAA	NC	Dir. Community Dev./police	NA
Inglewood	dB(A)	45/55*	65/65*	75/75*	75/75*	not allowed nites & weekends	Aircraft - 90 eng.-50@ nite	NC	police/council	NA
Portland, OR	dB(A)	55/50	60/55	60/55	65/60	85 @ 50'	state/fed	NC	Noise officer	$500
Huntington Beach	dB(A)	55/50*	55/55*	60/60*	70/70*	70/70*	state/fed.	55/45	Health Officer/Noise Variance Board	NA
Torrance	dB(A)	50/40**	55/50**	60/55**	70/65**	70/65**	state/fed.	NC	Code/Police	NA
Rancho Cucamonga	dB(A)	60/55	60/55	60/55	60/55	60/55	state/fed.	45/50	Code Enforcement	NA
San Bernadino	Nuisance only	NC	NC	NC	NC	NC	state/fed.	NC	Code Enforcement	NA
Fontana	Nuisance only	NC	NC	NC	NC	NC	state/fed.	NC	Code Officer	NA
Chula Vista	L(eq)	55/45 res. 50/50 apt.	65/60	70/70	80/80	80/80	state/fed.	55/45*	Building Director	NA
Santa Barbara	dB(A)	NC	NC	NC	NC	5 over ambient	state/fed.	Sonic Boom Ordinance	Buildg. & Zoning	NA

* Adjustment for short duration noise: 30 min = 0, 15 min. = +5 dB, 5 min. = +15 dB, 1 min. = +20 dB ** Adjustment for steady, reptitive & not continuous noise + time of day

4.4 ENVIRONMENTAL IMPACT STUDIES/STATEMENTS (EIS)

Local projects of considerable magnitude will likely require an Environmental Impact Study. Published reports of an EIS are known as Environmental Impact Reports (EIRs) and generally provide considerable analysis of the impact the proposed project will have on environmental issues. A properly completed EIS or EIR will include a "Noise Element" and is required for all new or major revisions of federal highway or airport projects. States also have enacted legislation requiring an analysis of environmental issues on everything from state highway construction to community development. Likewise, an EIS may be required by the city, county, or any other municipality.

A federally mandated EIS usually stems from regulations enacted by Congress and is implemented by the Environmental Protection Agency (EPA). Examples of projects requiring a federal EIS are all new federally funded highways, airports, and power plants; military bases; federal buildings, and most federally funded construction programs.[25]

State and local EIS format and scope should be similar to the federal counterpart. However, local interest and funding are distinct limiters. Since an EIS is prompted by political action and implemented by a political body, the "Noise Element" may be comprehensive or nonexistent. Recommendations for noise mitigation may be well funded or left to the budget axe. An EIS should include specifics on how mitigation efforts will be paid for because an EIS without an economic plan becomes a political football with many strong feelings and no action potential. Examples of projects requiring a local EIS are a new community, hotel, resort, etc.[26]

The "Noise Element" of an EIS is usually written to assess the impact of the proposed project on the surrounding environment, be it a nature preserve or a residential community. Most "Noise Element" sections address the impact on a nearby residential community such as that of an airport expansion on a nearby town. Issues will include an assessment of the present community noise environment and a prediction of the resultant noise environment. Mitigation recommendations are made to assure compliance with all federal, state, and local noise ordinances as well as general well-being. Noise levels and their impact on sensitive receivers during construction are a major part of the overall noise environment, therefore noise limits are placed on construction equipment such as pile drivers, graders, and trucks. If the project will create increased traffic volume, limits may also be placed on general traffic noise.

When an EIS has been approved for a project, the recommendations become regulations. For example, if the EIS recommends that construction equipment not exceed a specified noise level, then that becomes a local and project-oriented regulation. Likewise, the EIS may recommend that a multiunit residential building have party walls with STC 60 and an exterior envelope with STC 50. These recommendations become requirements that must be satisfied in the completed project. Not meeting the requirements is a building code violation, a breach of promise, or noise ordinance violation and could become very costly.

---

[25] The "Noise Element" of an EIS for runways and auxiliary facilities at an Air Force base and commercial airport is presented in Section 7.9.

[26] A noise study for a new residential development in Palmdale, Calif., is presented in Section 7.6. The developer was required to submit a "Noise Assessment and Noise Control Recommendation" study and agree to comply with the noise mitigation recommendations in order to obtain a permit to proceed with development from the city Planning Department.

4.5 NOISE CONTROL LAWS — SUMMARY

Noise control laws that are based on performance standards appear to be far superior to nuisance laws if properly written and correctly applied. Performance standards typically specify the maximum allowable noise level at a given point. This requires monitoring equipment and trained personnel for enforcement.

A promising trend has been the adoption of noise control laws with performance standards and the EIS process including a "Noise Element." The advantages of such legislation and regulatory agency include:

1. An Office of Noise Abatement can act as a focal point for activities in noise control regulation, enforcement, and public education.
2. A staff of specifically trained personel can to respond to complaints and monitor noise levels.
3. An Office of Noise Abatement is authorized to use initiative in seeking out the worst offenders.
4. And, has legal authority to advise zoning commissions on the noise impact of different zoning plans.

The main disadvantage is the cost of such a program. Although this cost is not high relative to other kinds of pollution controls, it represents yet another demand on city, county, and state budgets.

What corrective action can be taken to provide a better acoustical environment? For openers, if you concur with the need for better noise control legislation implementation, you are encouraged to let local, state, and federal elected officials know about the need. It is indeed unfortunate that EPA has had its hands tied in the matter of noise control.

To be effective in solving community-wide noise problems, average citizens must make themselves heard. For example, in Chicago an organization called Citizens Against Noise persuaded the City Council to pass an effective antinoise ordinance in 1971. In New York City, Robert Baron and Citizens for a Quiet City, Inc., succeeded in establishing a Bureau of Noise Abatement.

Obviously, it is necessary to familiarize yourself with the various regulations (federal, state and local) which affect your noise environment. This familiarization need not be a close and detailed reading of all the applicable laws. At the local level, a telephone call to the police department, the building official, and the city or county council is a starting point for determining what laws exist, if any, relating to noise control and their applicability to your particular case. The matter of noise in a community could also be brought up at a local civic association meeting.

At the state level, you may contact your elected officials to determine what they are doing about an effective noise control law in the state. The governor's office should also be contacted. Again, it is a matter of being heard. If enough organizations and private citizens ask probing questions about noise pollution, elected officials will take notice and hopefully corrective action will be forthcoming.

It is at the federal level where your voice can be the most effective. The means and laws already exist. EPA regulations regarding noise control are stringent, comprehensive, and complete. All that is missing is funding for implementation and policing. With an administration that is environmentally sensitive it should not take too much of a ground swell of citizen concern to get the EPA noise control regulations rolling again. Your voice, individually and collectively, can be heard at the federal level through many channels of communication and especially via messages to your congressional representatives and senators and the Executive Office. An effective corrective action program must involve powerful political groups in the community. Since both government and industry ultimately respond to public demands, the only real solution to the overall noise problem is a rising public awareness of the dangers of noise and a persistent demand for effective noise control.

4.6 TEST METHODS AND STANDARDS

The American Society for Testing and Materials (ASTM) annually publishes a library of standards (68 volumes) covering a broad range of standards including:

- Test methods
- Specifications for materials
- Standard practices
- Guides for proper usage
- Classifications and single number ratings

Formed in 1898 as a scientific and technical organization for "the development of standards on characteristics and performance of materials, products, systems services and the promotion of related knowledge," ASTM currently has over 35,000 members and has published over 9000 separate standards. Acoustical standards are prepared by a consensus writing group called ASTM Technical Committee E33 on Environmental Acoustics and published in <u>ASTM Annual Book of Standards,</u> Volume 04.06. Members of the writing groups are all volunteers. Membership is balanced between those representing producers, users, and general interest. The ASTM standards are recognized at all levels of the United States government, states and municipalities, by the construction industry,and many foreign entities including Canada and Mexico. The ASTM Environmental Acoustics Standards provide the mainstay for standards in acoustics within America.

Many trade groups also publish their own set of voluntary standards. A few of the more notable organizations are the American National Standards Institute (ANSI), American Society of Heating, Refrigeration and Air-Conditioning Engineers (ASHRAE), Acoustical Society of America (ASA) and the American Institute of Architects (AIA). The following organizations publish ACOUSTICAL consensus standards.

<u>PREFIX</u>	<u>ORGANIZATION</u>/address
AMA	Acoustical Materials Association (defunct); Contact: Ceiling & Interior Contractors Association, 104 Wilmot Rd., Deerfield, IL 60015
ANSI	American National Standards Institute, 1430 Broadway, New York, NY 10018
ARI	Air Conditioning and Refrigeration Institute, 1815 N. Ft. Myer Dr., Arlington, VA 22209
ASA	Acoustical Society of America (Standards Secretariat/Journal - ASA & ISO) 335 East 45th St., New York, NY 10017
ASHRAE	American Society of Heating, Refrigerating and Air-Conditioning Engineers, 1791 Tullie Circle, N.E., Atlanta, GA 30329

ASTM   American Society for Testing and Materials (Book of Standards Volume 04.06, <u>Environmental Acoustics</u>), 100 Barr Harbor Dr., West Conshohocken, PA 19428-2959, (610) 832-9500

IEEE   Institute of Electrical and Electronic Engineers, 345 East 47th St., New York, NY 10017

ISA   Instrument Society of America, P.O. Box 12277, 67 Alexander Dr., Research Triangle Park, NC 27709

ISO/IEC   International Organization for Standardization, International Electrotechnical Commission, 1 Rue de Varembe, Case Postale 56, CH-1211 Geneva 20, Switzerland

NCA   Noise Control Association, 680 Rainier Lane, Pt. Ludlow, WA 98365, (360) 437-0814

NEMA   National Electrical Manufacturers Association, 2101 L Street N.W., Washington, DC 20037

SAE   Society of Automotive Engineers, 400 Commonwealth Drive, Warrendale, PA 15096

STANDARDS AND PROCEDURES (Related to Acoustics)

The following sections contain information and, where available, a brief synopsis of the standards relevant to residential architectural acoustics. Please refere to the standard for additional details. The descriptions for ASTM standards were excerpted from ASTM E1433, <u>Standard Guide for Selection of Standards on Environmental Acoustics</u> issued by ASTM Technical Committee E-33 on Environmental Acoustics. Other descriptions were prepared by D. A. Harris. For additional information or clarification, ASTM E-33 Subcommittee jurisdiction is as follows:

E33.01 – Sound Absorption
E33.02 – Open Plan Spaces
E33.03 – Sound Transmission
E33.04 – Application
E33.05 – Research
E33.06 – International Standards
E33.07 – Definitions and Editorial
E33.08 – Mechanical and Electrical System Noise
E33.09 – Community Noise

(NOTE: This listing is for information purposes only. Please refer to the individual standard for proper use, limits, and related concerns.)

4.6.1 Sound Absorption Standards

ASTM C384 – Test Method for Impedance and Absorption of Acoustical Materials by the Impedance Tube Method

Use: Intended primarily as a research screening tool, useful for manufacturers and/or researchers in evaluating the absorption of materials. It is also valuable for evaluating small units, such as anechoic wedges. It can be used to rank order the absorption and impedance characteristics of materials.
Result: Normal Incidence Sound Absorption Coefficients, Normal Specific Impedance Ratios
Discussion: A sound wave traveling down a tube is reflected back by the test specimen, producing a standing wave that can be explored with a probe microphone. The normal absorption coefficient is determined from the standing wave ratio. In addition, an impedance ratio at any one frequency can be determined using the position of the standing wave with reference to the face of the specimen (see also Test Method E1050). Values do not necessarily correlate with those of C423.

ASTM C423 – Test Method for Sound Absorption and Sound Absorption Coefficients by the Reverberation Room Method

Use: Primary method for evaluating sound absorption capabilities of building materials and systems. One can use the sound absorption coefficients and volume of a room, or sabins per unit, to determine how much material is needed to limit room reverberation and/or reduce noise to a desired level.
Result: Sound Absorption Coefficients, Noise Reduction Coefficient (NRC), Absorption figures in Sabins, Sabins/Unit
Discussion: Random noise is turned on long enough for the sound pressure in a reverberant room to reach a steady state. When the signal is turned off, the sound pressure level decreases. The rate of decrease (decay) in a specified frequency band is measured. The absorption of the room and its contents is calculated both before and after placing the specimen in the room. The increase in absorption due to the specimen, divided by the area of the specimen is the absorption coefficient. Noise Reduction Coefficient is the average of the four absorption coefficients of the third-octave bands centered on 250, 500, 1000, and 2000 Hz, rounded to the nearest 0.05. NRC is a single number rating and is convenient for ranking building materials and systems. However, in some critical applications, study of all available frequency data is advised to determine suitability.

ASTM C522 – Test Method for Airflow Resistance of Acoustical Material

Use: Indicates sound-absorbing properties in some materials where airflow resistance is related to sound absorption.
Result: Airflow resistance (R), Specific Airflow resistance (r), Airflow resistivity (r_0)
Discussion: The specific airflow resistance of an acoustical material is one of the properties that determine its sound-absorptive and sound-transmitting properties. The specific air flow resistance is given by the formula $R = P/U$, where P = air pressure difference across the specimen, U = volume velocity of airflow through it. The specific airflow resistance measured by this method may differ from the specific resistance measured by the impedance tube method in Test Method C384. *Caution: Materials exist that*

do not allow any airflow yet exhibit excellent sound absorption.

ASTM E795 – Practices for Mounting Test Specimens During Sound Absorption Tests

Use: Reference to specific mounting methods helps laboratory operators simulate expected field applications. It also helps specifiers by allowing comparison of materials tested in similar mountings.
Result: A letter designation describing the method of mounting a C423 test specimen.
Discussion: These practices cover test specimen mountings to be used during tests performed in accordance with Test Method C423. Sound absorption of a material covering a flat surface depends not only on the physical properties of the material, but also on the way in which the material is mounted over the surface. The mountings specified in these practices are intended to simulate, in the laboratory, conditions that exist in normal use.

ASTM E1042 – Classification for Acoustically Absorptive Materials Applied by Trowel or Spray

Use: This standard helps specifiers select materials by classifying certain characteristics.
Result: Classification of materials
Discussion: Acoustically absorptive materials are used for the control of reverberation and echoes in rooms. This standard provides a classification method for such materials applied directly to surfaces by trowel or by spray. Classification is made according to type of material: acoustical absorption determined by C423, flame spread determined by E84, and dust propensity determined by E859.

ASTM E1050 – Test Method for Impedance and Absorption of Acoustical Materials Using a Tube, Two Microphones, and a Digital Frequency Analysis System

Use: Intended primarily as a research tool. This is an alternative to C384 using digital instruments.
Result: Normal Incidence Sound Absorption Coefficients, Normal Specific Acoustic Impedance Ratios
Discussion: A broadband noise is produced on one end of a tube, the other end of which contains a test specimen. The plane wave produced is detected by two microphones located at different positions along the tube. A digital frequency analyzer measures output from the two microphones. Results match C384.

4.6.2 Sound Transmission Standards

ASTM E90 – Test Method for Laboratory Measurement of Airborne-Sound Transmission Loss of Building Partitions

Use: Primary method for evaluating transmission loss of materials and systems used in building construction, such as interior partitions, doors, windows, and floor/ceiling assemblies.
Result: Transmission Loss (TL) and Sound Transmission Class (STC)
Discussion: A test specimen is installed in an opening between two adjacent reverberation rooms, care being taken that the only significant sound path between rooms is by way of the specimen. An approximately diffuse field is produced in one room, and the resulting space-time average sound pressure

levels in the two rooms are determined at a number of one-third-octave band frequencies. In addition, the sound absorption in the receiving room is determined. The sound transmission loss is calculated from a basic relationship involving difference between the sound levels, the receiving room absorption, and the test specimen size. The TL data are used in E413 to determine Sound Transmission Class (STC).

ASTM E336 – Test Method for Measurement of Airborne Sound Insulation in Buildings

Use: Primary method for evaluating on-site noise reduction between two rooms or sound barrier performance of interior partitions. Can be used for acceptance of recent construction or improvement of existing buildings. It is not recommended to use test data in one facility to predict results in another.
Result: Field Transmission Loss (FTL), Noise Reduction (NR), Normalized Noise Reduction (NNR)
Discussion: The noise reduction between two rooms is obtained by taking the difference between the average sound pressure levels in each room at specified frequencies in one-third- octave bands when one room contains a noise source. The noise reduction may be normalized to a reference reverberation time of 0.5 seconds. When the rooms' size and absorption requirements are satisfied so that the sound fields are sufficiently diffuse and when flanking is not significant, the field transmission loss may be reported. Results will usually be lower than in E90 laboratory tests for the same specimen. Note that this test requires minimum room characteristics to be valid (see also Method E597). The data are used in E413 to Determine Noise Isolation Class (NIC), Normalized Noise Isolation Class (NNIC), or Field Sound Transmission Class (FSTC).

ASTM E413 – Classification for Rating Sound Insulation

Use: Permits specifiers to rank the transmission loss or noise reduction performance of similar materials or systems, using data from one of several test methods.
Result: Sound Transmission Class (STC), Field Sound Transmission Class (FSTC), Noise Isolation Class (NIC), Normalized Noise Isolation Class (NNIC)
Discussion: To determine the Sound Transmission Class (STC) of a test specimen, its transmission loss (as determined in accordance with Method E90), Field Transmission Loss (E336), Noise Reduction (E336 or E596), or Normalized Noise Reduction (E336) in a series of 16 test bands, are compared with those of a reference contour. When certain conditions are met, the class is found. It is recommended that the test data be presented in a graph together with the corresponding class contour. The single number rating is convenient for ranking building materials and systems. However, it is appropriate only for commonly found indoor sounds similar to speech. For critical applications, study of all available frequency data is advised to determine suitability.[27]

ASTM E597 – Practice for Determining a Single-Number Rating of Airborne Sound Isolation for Use in Multi-unit Building Specifications

Use: Determines the degree of acoustical isolation between and within dwelling units in apartment

---

[27] Additional discussion of STC and how it is determined may be found in Chapters 2 and 5. STC results are given in Appendix 3.

buildings, hotels, etc.

Result: Sound Level Difference (D), Normalized Sound Level Difference (D_n), Average Sound Level (L)
Discussion: The sound level difference between two rooms is measured by establishing a sound field with specified spectrum in a source room, of sufficient level that the corresponding sound in a receiving room predominates over the sound from all other sources. With the sound source in operation, the space-time average A-weighted sound level, L, in each of the two rooms, is measured. The difference between the levels is the sound level difference, D, for that room pair. (This value may vary from the Noise Isolation Class (NIC); see E336.) The sound level difference is a property of the two rooms and their contents, not of the dividing partition alone. Results will be lower than those found in E90, since flanking paths and imperfections are not eliminated from the test.

ASTM E756 – Test Method for Measuring Vibration-Damping Properties of Materials

Use: This method determines the vibration-damping properties of materials.
Results: Young's Modulus (E), Loss Factor (LF), Shear Modulus (G)
Discussion: This method is accurate over a frequency range of 50 to 5000 Hz and over the useful temperature range of the material being tested. It is useful in testing materials that have application in structural vibration, building acoustics, and the control of audible noise. Such materials include metals, enamels, ceramics, rubbers, plastics, reinforced epoxy matrices, and woods that can be formed to the test specimen configurations.

ASTM E966 – Guide for Field Measurement of Airborne Sound Insulation of Building Facades and Facade Elements

Use: Field test guide for measuring noise isolation of exterior walls and facade components.
Result: Outdoor-Indoor Transmission Loss (OITL), Outdoor-Indoor Level Reduction (OILR)
Discussion: Loudspeaker or traffic sound sources may be used. The outdoor sound field may be inferred from pre-calibration, or measured on site near the facade or at the facade surface. A fixed sound source is located at a specific angle, while traffic may move along a straight line in front of the facade. Indoors, a space average is taken in the room adjacent to the test facade. The difference between the two sound levels is OILR. (For uncontrolled sound sources and traffic, the outdoor and indoor sound levels are measured simultaneously.) To obtain OITL, OILR is normalized for room absorption, and flanking transmission paths must be blocked. If flanking transmission is present or unknown, the measurement is labeled the "apparent OITL" and represents the lower limit of noise isolation performance. Because of angle of incidence and flanking effects, results may not agree with those obtained with other test methods, such as E90 or E336.

ASTM E1332 – Classification for Determination of Outdoor-Indoor Transmission Class

Use: Provides a single-number rating to be used to compare building facade designs, including walls, doors, windows, and combinations thereof. The rating can be used by specifiers to rank-order building materials.
Result: Outdoor-Indoor Transmission Class (OITC)

Discussion: Using transmission loss data in the range of 80 to 4000 Hz, as measured in accordance with Method E90 or Guide E966, the OITC is calculated by applying A-weighting criteria to the reference source sound spectrum or source room sound levels, and subtracting the transmission loss. The resulting data are used in a provided formula to yield OITC. A sample manual worksheet and a computer program in the BASIC language are provided to help in applying the classification.

ASTM E1408 – Test Method for Laboratory Measurement of Sound Transmission Loss of Doors and Door Systems

Use: Procedure for installing doors and seals in E90.

Result: Establishes requirements for installation of operable door systems.

Discussion: This procedure is a supplement to E90 that extends the procedures for measuring sound transmission loss of doors. The method is used for laboratory measurement of the sound transmission loss of fully operable doors equipped with a particular combination of hardware and seals, the sound transmission loss of a laboratory sealed door panel, and the force or torque required to operate the door system. A nonmandatory test for assessing individual door components is given in the appendix.

ASTM E1414 – Test Method for Airborne Sound attenuation Between Rooms Sharing a Common Ceiling Plenum

Use: Measure the sound attenuation provided by a suspended ceiling in the presence of a continuous plenum space.

Result: The result of running this test is to obtain data to determine the CAC (Ceiling Attenuation Class) by using Classifiction E413 (previous results were STC and CSTC)

Discussion: This specification replaces AMA I-II. This test method utilizes a laboratory space so arranged that it simulates a pair of horizontally adjacent small offices or rooms separated by a partition and sharing a common plenum space. The only significant sound transmission path is by way of the ceiling and the plenum space. This procedure is one of two methods to evaluate the acoustical performance of ceiling systems in an open/closed plan design, the other being Test Method E1111.

4.6.3 Impact Noise on Floors

ASTM E492 – Test Method for Laboratory Measurement of Impact Sound Transmission Through Floor-Ceiling Assemblies Using the Tapping Machine

Use: This is the primary laboratory method for evaluating floor/ceiling assemblies as barriers to structure-borne rather than airborne noise. The standard tapping machine does not duplicate human footfall noise.

Result: Normalized Impact Sound Pressure Levels (L_n)

Discussion: A standard tapping machine is placed in operation on a test-floor specimen that forms a horizontal separation between two rooms, one directly above the other. The transmitted impact sound characterized by the spectrum of the space-time-average one-third- octave band sound pressure levels produced by the tapping machine is measured in the receiving room below. Since the spectrum depends on the absorption of the receiving room, the sound pressure levels are normalized to a reference absorption. Resulting data are used in E989 to determine Impact Isolation Class (ICC).

ASTM E989 – Classification for Determination of Impact Insulation Class (IIC)

Use: Provides single-number rating of the barrier capabilities of floor-ceiling assemblies against structure-borne noise.
Result: Impact Insulation Class (IIC), Field Impact Insulation Class (FLIC)
Discussion: The one-third octave laboratory impact noise data obtained in Method E492 or field data obtained in Method E1007, are compared with those of a reference contour. When certain conditions are met, the Class is found. It is recommended that the test data be presented in a graph together with the corresponding Class Contour. A single number rating is convenient for ranking building materials and systems. However, for critical applications, study of all available frequency data is advised to determine suitability.

ASTM E1007 – Test Method for Field Measurement of Tapping Machine Impact Sound Transmission Through Floor-Ceiling Assemblies and Associated Support Structures

Use: Measures transmission of impact sound generated by a standard tapping machine through floor/ceiling assemblies and associated supporting structures in field situations. Can be an acceptance or improvement tool for specifiers.
Result: Normalized Impact Sound Pressure Levels (L_n)
Discussion: Measurements may be conducted on all types of floor/ceiling assemblies, including those with floating-floor or suspended ceiling elements, or both, and assemblies surfaced with any type of floor surfaces or coverings. This field method does not distinguish between sound transmitted through the entire building and that transmitted solely through the floor/ceiling assemblies. The standard tapping machine does not duplicate human footfall noise. Because room sizes and shapes can vary widely, it is preferable to confine the use of test results to the comparison of closely similar floors and supporting structures. Resulting data are used in E989 to determine Impact Isolation Class (IIC).

AMA I-I – Impact Sound Transmission Test by the Footfall Method

Use: This procedure was developed specifically to evaluate actual footfall sounds being transmitted to the receiver room below.
Result: Footstep sounds transmitted to rooms located directly below are masked by an acoustical environment of not greater than $NC_{40} = 40$ or 35.
Discussion: Either a male or female walker may be utilized with size, weight, and footwear carefully identified. It is applicable to both the laboratory and field. AMA no longer exists, however, copies of the procedure may be obtained from Geiger and Hamme Laboratories, Ann Arbor, Mich., or the Ceiling and Insulating Systems Contractors Association (CISCA). This procedure evaluates the full spectrum of frequencies prevalent when a person walks. It does not correlate with the tapping machine tests leaving one to speculate which is appropriate. AMA I-I was designed to measure actual field conditions.

4.6.4 Application Standards

ASTM C367 – Test Methods for Strength Properties of Prefabricated Architectural Acoustical Tile or Lay-in Ceiling Panels

Use: This standard, when used in conjunction with tests of acoustic performance, helps specifiers select materials with the best combination of acoustic and strength properties for an intended application.
Result: Hardness, Friability, Sag, Transverse Strength
Discussion: Materials used for absorbing sound often have a porous, low-density structure and may be relatively fragile. These test methods cover procedures for evaluating those physical properties related to strength. The methods are useful to develop, manufacture, & select acoustical tile or lay-in panels.

ASTM C635 – Specification for Manufacture, Performance, and Testing of Metal Suspension Systems for Acoustical Tile and Lay-in Panel Ceilings

Use: This specification allows specifiers to evaluate and compare the physical characteristics of metal suspension systems.
Purpose: To aid in selecting metal suspension materials and systems.
Discussion: This specification sets forth suspension member tolerances, load tests, and finish tests, to guide manufacturers and specifiers on acceptable products, and to give users and designers comparative test data to choose appropriate products.

ASTM C636 – Practice for Installation of Metal Ceiling Suspension Systems for Acoustical Tile and Lay-in Panels

Use: This practice is intended to be referenced by architects, designers, and/or owners of buildings, or all of these.
Purpose: Identify significant installation requirements.
Discussion: This practice presents guidelines to designers and installers of acoustical ceilings and to other trades if their work interferes with ceiling components. Practices concerning hangers, carrying channels, main runners, cross runners, spline, assembly devices, and ceiling fixtures are discussed. Where seismic restraint is required, E580 should also be consulted, along with industry recommendations and code requirements.

ASTM E497 – Practice for Installing Sound-Isolating Lightweight Partitions

Use: Architects, designers, builders, and owners utilize this practice to assure fixed partition systems are free of major noise flanking paths and unnecessary leaks.
Purpose: To aid in design and specification.
Discussion: This practice details precautions that should be taken during the installation of gypsum board partitions to maximize their sound-insulating effectiveness. Potential problems with flanking sound transmission and sound leaks are discussed, and methods to avoid these are offered. A number of figures and drawings are included to illustrate the potential errors and to provide suggested precautions.

ASTM E557 – Practice for Architectural Application and Installation of Operable Partitions

Use: Architects, designers, builders, and owners utilize this practice to assure operable partition systems are free of major noise flanking paths and unnecessary leaks.
Purpose: To aid in design and specification
Discussion: This practice details precautions that must be taken before and during the installation of an operable partition to ensure that the maximum attainable sound insulation is achieved between the two spaces separated by the partition. Specific paragraphs refer to potential sound leakage through the partition joints, the seals, the ceiling and plenum, an HVAC system, and through hollow floors. Other paragraphs deal with deflection of the partition and the potential of focusing sound with curved surfaces.

ASTM E580 – Practice for Application of Ceiling Suspension System for Acoustical Tile and Lay-in Panels in Areas Requiring Seismic Restraint

Use: This practice is an extension of Practice C636 and is intended to be referenced by architects, designers, and/or owners of buildings. It is critical in areas affected by earthquakes or tremors. Refer to local codes.
Purpose: To aid in design and specification
Discussion: This practice presents guidelines for designers and installers to provide additional restraint required in areas deemed by local authorities to be subject to major seismic disturbance. Acceptable suspension system components, additional attachment points, and support elements for seismic restraint are described. Sketches show additional hanger wire locations and attachment. Specification C635 and Practice C636 cover suspension systems and their application, without regard to seismic restraint needs. Building codes and manufacturers recommendations remain applicable and should be followed when this practice is specified. They, plus building codes and manufacturers recommendations remain applicable and should be followed when this practice is specified.

ASTM E1264 – Classification for Acoustical Ceiling Products

Use: This classification is intended to serve a similar purpose to Federal Specification SS-S-118B. Fire endurance and physical properties are not covered.
Result: Classification by acoustical, light reflectance, and surface burning characteristics.
Discussion: This classification covers ceiling products that provide acoustical performance and interior finish to buildings. It serves to classify and aid in the selection of acoustical ceiling products. Products are categorized by type, pattern, light reflectance, acoustical properties, and surface burning characteristics.

ASTM E1433 – Guide for Selection of Standards on Environmental Acoustics

Use: Listing of all ASTM Standards for Acoustics.
Purpose: Aid in selecting standard
Discussion: Standard is organized like this listing.

4.6.5 Mechanical and Electrical System Noise Standards

ASTM E477 – Test Method of Measuring Acoustical and Airflow Performance of Duct Liner Materials and Prefabricated Silencers

Use: This method applies to heating and air-conditioning ducts in buildings with low pressure and air speed.

Result: Insertion loss (IL), Airflow-generated Sound Power Levels, pressure drop.

Discussion: The sound pressure level in a reverberation room is measured while sound is entering the room through a length of straight, empty duct and again, after a section of the empty duct has been replaced with the test specimen. The insertion loss is the difference between the two sound pressure levels. Airflow-generated noise is measured while air is passing through the system with the specimen installed. Pressure drop performance is obtained by measuring the static pressure at designated locations upstream and downstream of the test specimen at various air flow settings.

ASTM E1124 – Test Method for Field Measurement of Sound Power Level by the Two-Surface Method

Use: Provides an estimate of the normal sound power level of a specimen operating in situ.

Result: Sound power level (L_w)

Discussion: The average one-third or full octave band sound pressure levels are measured over two different surfaces which surround the specimen. These surfaces should be selected to consist of rectangular, cylindrical, and/or hemispherical surfaces so that the areas may be easily calculated. From the difference between the two average sound pressure levels and from the areas of the surfaces the sound power level may be calculated. The calculation accounts for both the effect of the reverberant field and the noise of other sources.

ASTM E1265 – Test Method for Measuring Insertion Loss of Pneumatic Exhaust Silencers

Use: This standard permits specifiers to evaluate and compare the performance of pneumatic exhaust silencers.

Result: Flow Ratio, Average Insertion Loss

Discussion: This method covers the laboratory measurement of both the acoustical and mechanical performance of pneumatic exhaust silencers designed for quieting compressed gas exhausts from orifices up to 3/4-in. NPT. The method is not applicable for exhausts performing useful work, such as part conveying, ejection, or cleaning. The method evaluates acoustical performance using A-weighted sound level measurements.

ASTM 1574 – Test Method for Measurement of Sound in Residential Spaces

Use: This test method produces measured sound data that may be compared with applicable criteria for the noise in residential spaces from built-in utilities and major appliances.

Result: Octave band sound pressure levels for continuous noise and A weighted, fast response, levels for transient noise.

Discussion: The location of the highest A-weighted sound level is identified for each continuous sound source of interest. The octave band sound pressure levels are then measured for each source. The highest A-weighted, fast response sound level is measured in the center of each space (that is, room) for each transient sound source.

ANSI S5.1 – Standard Test Code for the Measurement of Sound from Pneumatic Equipment (1971)

ASA 3-1975 – Test-Site Measurement of Noise Emitted by Engine Powered Equipment

ASHRAE 36-72 – Methods of Testing for Sound Rating Heating, Refrigerating, and Air-Conditioning Equipment (supersedes ASHRAE 36-72, 36A-63 and 36B-63)

IEEE 85 – Test Procedure for Airborne Sound Measurements on Rotating Electric Machinery (1973)

SAE J672a – Exterior Loudness Evaluation of Heavy Trucks and Buses (1970)

SAE J952b – Sound Levels for Engine Powered Equipment (1969)

SAE J986a – Sound Level for Passenger Cars and Light Trucks (1973) (ANSI S6.3)

SAE J88a – Exterior Sound Level Measurement Procedure for Power Mobile Construction Machinery (1973)

AMCA 300 – 67 - Test Code for Sound Rating

AHAM RAC-2SR – Room Air Conditioner Sound Rating (1971)

NEMA MG1-12.49 – Motors and Generators – Methods of Measuring Machine Noise (1972)

NEMA TR1-1972 – Transformers, Regulators and Reactors – Section 9-04, Audible Sound Level Tests.

4.6.6 Community Noise Standards

ASTM E1014 – Guide for Measurement of Outdoor A-Weighted Sound Levels

Use: Results from this guide may appropriately be used in conjunction with ordinances or land-use restrictions of noise by communities.

Result: A-weighted sound levels (dB_A)

Discussion: This guide covers measurement of A-weighted sound levels outdoors at specified locations or along particular site boundaries, using a general purpose sound-level meter. Three distinct types of measurement surveys are described: around a site boundary; at a specified location; and at a specified distance from a source (to find the maximum sound level). Since outdoor sound levels usually vary with time over a wide range, the data obtained using this guide may be presented in the form of a histogram of sound levels. The data obtained using this guide enables calculations of average or statistical sound levels for comparison with appropriate criteria.

ASTM E1503 – Test Method for Conducting Outdoor sound Measurements Using a Digital Statistical Analysis System

Use: This test method covers the measurement of outdoor sound levels at specific locations using a digital statistical analyzer and a formal measurement plan.
Result: L_{eq} (equivalent sound level) obtaned by integrating A-weighted sound level measured over a specific period of time, or in the case of un-weighted sound pressure and fractional octave bands, equivalent sound pressure level. Also, percentile exceedance level – a result of statistical analysis of data in a measurement set.
Discussion: This test method deals with methods and techniques that are well defined and that are understood by a trained acoustical professional. This test method has been prepared to provide a standard methodology that, when followed, will produce results that are consistent with requirements of government and industry, and which can be validated using information gathered and documented in the course of the measurement program.

4.6.7 Definitions of Environmental Acoustic Terms

ASTM C634 – Terminology Relating to Environmental Acoustics

Use: Many other standards rely on these definitions. The user may need these definitions to understand a specific standard.
Result: Understanding of definitions of acoustic terms
Discussion: Definitions of terms used in environmental acoustics standards, including, in those entries with physical properties, the symbol, dimensions, and units.[28]

4.6.8 Sound Power and Level Measurement

ASTM Proposed Method – Standard Method for Measurement of Sound in Residential Spaces

Use: Primary method for measuring sound in enclosed residential spaces produced by built-in-utilities and major appliances such as plumbing, heating, ventilation, air-conditioning systems, refrigerators, and dish washers. The measured values may then be used to assess compliance, design, or habitation suitability.

---

[28] Many of the terms are contained in Appendix 1, "Glossary of Acoustical Terms."

Result: Decibels on A scale of sound level meter (dB$_A$) and Decibels (dB) at octave or fractional band widths.

Discussion: This new procedure is in the final stages of approval. It is intended to fill an important gap in present measurement procedures and provide the contractor, homeowner and acoustician with a simple straightforward technique for measuring sounds in a residence under actual field conditions. This procedure should replace most other standards listed in this section.

ANSI S1.13 – Standard Methods for the Measurement of Sound Pressure Levels (1976)

ISO 3740 – Determination of Sound Power Levels of Noise Sources – Guidelines for Use of Basic Standards and Preparation of Noise Test Codes (1978)

ISO 3741 – Determination of Sound Power Levels of Noise Sources – Precision methods for broadband sound sources operating in reverberation rooms (1975)

ISO 3742 – Determination of Sound Power Levels of Noise Sources – Precision Methods for Discrete-Frequency and Narrow-Band Sound Sources Operating in Reverberation Rooms (1975)

ISO 3744 – Determination of Sound Power Levels of Noise Sources – Engineering methods for free-field conditions over a reflecting plane

ISO 3745 – Determination of Sound Power Levels of Noise Sources – Precision methods for sources operating in anechoic rooms

ISO 3746 – Determination of Sound Power Levels of Noise Sources – Survey method

ISO 3747 – Determination of Sound Power Levels of Noise Sources – Methods using a reference sound source

ANSI S3.17 – Method for Rating the Sound Power Spectra of Small Stationary Noise Sources

ANSI S1.21 – Standard Methods for the Determination of Sound Power Levels of Small Sources in Reverberation Rooms (1976)

ANSI S3.4 – Standard Procedure for the Computation of Loudness of Noise (1972)

ANSI S3.5 – Methods for the Calculation of the Articulation Index (AI) (1976). This single-number rating system provides a rating scale for the clarity or understandability of speech. The rating scale ranges from an AI of 0.0 -1.0 and is similar to a scale of percent speech intelligibility. An AI of 0.2 indicates a person may hear only 20% of the words spoken. An AI of 0.9 indicates one can hear 90% of the words.

4.6.9 Acoustical Measurement/Meter Standards

IEC (ISO) 179 – Precision Sound Level Meters (1973)

IEC (ISO) 179A – Additional Characteristics for the Measurement of Impulsive Sounds (1973)

IEC (ISO) 225 – Octave, Half-Octave and Third-Octave Band Filters Intended for the Analysis of Sounds and Vibrations (1966)

IEC (ISO) 327 – Precision Method for the Pressure Calibration of One-Inch Standard Condenser Microphones by the Reciprocity Technique (1971)

IEC (ISO) 402 – Simplified Methods for Pressure Calibration of One Inch Condenser Microphones by the Reciprocity Technique (1972)

ANSI S1.2 – Method for the Physical Measurement of Sound (1976)

ANSI S1.4 – Specification for Sound Level Meters (1971)

ANSI S1.6 – Preferred Frequencies and Band Numbers for Acoustical Measurements (1976)

ANSI S1.8 – Preferred Reference Quantities for Acoustical Levels (1974)

ANSI S1.10 – Method for the Calibration of Microphones (1976)

ANSI S1.11 – Specification for Octave, Half-Octave, and Third-Octave Band Filter Sets (1976)

ANSI S1.12 – Specifications for Laboratory Standard Microphones (1972)

4.6.10 SUMMARY OF TEST STANDARDS

The foregoing is a listing of the acoustical standards published by industry organizations and consensus standards writing groups that relate to residential and architectural acoustics. For other areas of interest, such as industrial noise control see the *Noise Control Manual* or *Planning and Designing the Office Environment*, by D. A. Harris, VNR, 1992. This list contains those standards known to the author and are well known in the United States of America. However, there are likely many others that are applicable. Revisions to the existing standards are ongoing and new standards are in the process of being adopted. Check with the latest edition of ASTM and other organizations for current listings.

4.7 ACOUSTICAL CONSULTANTS AND FIELD TESTING

In the interest of greater acoustical efficiency, the design, testing, and inspection of the acoustical environment should be done under the supervision of an acoustical expert. An acoustical consultant is trained and experienced in the preparation of project acoustical guidelines, specifications, testing and project follow-up for compliance. For example the acoustical consultant can efficiently and effectively provide:

- The "Noise Element" of an Environmental Impact Study (EIS).
- A clear definition of user needs and the acoustical environment appropriate for the project. User needs should be established during the initial design phase.
- Site analysis of the acoustical environment and community noise control recommendations.
- Recommendations in the design phase for building placement, orientation and layouts that will enhance the acoustical solution. These recommendations will likely save large amounts of design and construction time, and costly noise control mitigation efforts; avoid costly errors, omissions, and re-designs; and assure compliance with the acoustical criteria of building codes and ordinances.
- Acoustical detail designs and specifications for all building systems.
- Sound testing of new designs/systems both in the laboratory and the field using industry accepted procedures and standards.
- Compliance testing at the site and site review for good acoustical workmanship.
- Follow-up noise control analysis and recommendations.

To select an Acoustical Consultant for your project contact: National Council of Acoustical Consultants (NCAC), 66 Morris Ave., Suite 1B, Springfield, New Jersey 07081-1409, phone (201) 564-5859, for a list of members in your area. You may also contact Noise Control Association (NCA), 680 Rainier Lane, Port Ludlow, WA 98365, phone and fax (360) 437-0814. The Noise Control Association is a "not for profit association of Noise Control Products, Materials and Systems Manufacturers." Established in 1962 it states its mission as follows:

- Establish and encourage adherence to the highest standards of professional ethics and business practices.
- Inform the public of the existence of acoustical consultants and the services which they provide.
- Provide members with a forum for discussion and exchange of information on matters of common interest.
- Cooperate with representatives of other organizations on matters of mutual interest and concern.
- Preserve and protect the public welfare by encouraging accurate and proven representations concerning acoustical products, materials and services.
- Participate in the development of performance and measurement standards, and regulations.
- Encourage and promote continuing growth and education in the profession.

While there are organizations dedicated to scientific and engineering aspects of the field of acoustics, only the National Council of Acoustical Consultants is dedicated to management and related concerns of professional acoustical consulting firms and to safeguarding the interests of the clients and public which they serve. Members of the National Council of Acoustical Consultants adhere to a Canon of Ethics, a copy of which is available free of charge by contacting NCAC Headquarters.

Chapter 5   3 M's of ACOUSTICS:
MYTHS, MISCONCEPTIONS, AND MYSTERIES

5.1 INTRODUCTION

The premise of this chapter is not to be instructive. In fact it is the fun part of this book. However, the following stories, anecdotes, and opinions will provide a better understanding of principles advanced in other chapters by discussing them from a different angle.

In the past 35 years, the author has encountered many situations where well-meaning persons have conceived new ideas to solve nagging acoustical problems. Efforts have ranged from pure brilliance to just plain stupidity. Along the way some ideas have found practical use, others haunt those still searching for "a better way." Still others have spawned products and systems that fit in the category of "super duper sound sucker." The line between brilliance and stupidity is sometimes so fine that even the experts are fooled. Therefore, it is imperative that the reader of this chapter analyze the thoughts expressed herein very carefully.

The evidence presented is sometimes based on solid physical evidence that has been thoroughly researched. When this is the case, references are identified and it will be made clear if the conclusions have a reasonable and/or solid technological basis. In others, the conclusions are pure gut feelings based on circumstantial evidence. In still others, they are dreams of what might be or of research projects that deserve attention. The reader may also find a few whimsical stories. In all cases the material is pure opinion.

Please do your own research to arrive at facts. Do not jeopardize your project by utilizing quotes from this chapter alone. In legal jargon, you have just read appropriate boiler plate to absolve the author of any responsibility for the information herein. On the other hand, the material has been chosen with the hope it may spark some new and creative thought. If some new product idea is germinated from this chapter, it will have been a grand success. And, if someone were to ask, the answer is "Yes," the author would enjoy having additional communication on any of the subjects.

My first inclination was to organize this chapter into a section for each M. To do so would categorize the subject and stifle creativity. Next, I tried to put them in order of importance, but who is to say what is important to the reader. Consequently, there is no particular sequence except that the most widely misunderstood issues, causing the greatest decibel-per-dollar loss, are first. Last are the "ideas of the future." In between – who knows?

MYTH TO REALITY
yields
optimum dB per $ and ideas of the future

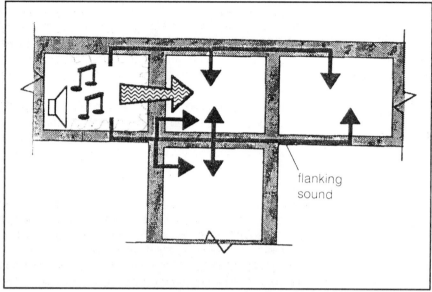

Figure 5-1 Use sound-absorbing materials in source room, sound barrier between source and receiving room, and eliminate flanking.

5.2 ABSORPTION, TRANSMISSION LOSS, AND FLANKING

Probably the most widely misunderstood issue in architectural acoustics is when to use sound absorbing and sound barrier materials and to find the flanking path. The natural inclination is to find a material that somehow has magical powers to correct <u>any</u> acoustical problem. In short, there is, and probably always will be, a market for a "super duper sound sucker." Since a practical version has not yet appeared, one must understand when to apply absorption principals, when to build a sound barrier, and when to use a combination. Flanking is then the nemesis that tends to render good acoustical designs worthless (Figure 5-1).

A simple rule of thumb is **absorption works best when placed in the source location**. That sounds simple enough, but it is amazing how many times expensive, highly efficient sound absorbers are placed in a receiving zone, over your own head, when it is the noise next door that is the problem. Efforts to quiet one's own space, the receiver, by adding absorption materials may actually increase intrusiveness of the noise from next door. By lowering the existing background sound in the vicinity of the receiver, the signal to noise ratio is adversely affected. The result is usually worse than before treatment.[1]

The first law of noise quieting is "Quiet the source." Adding sound absorption material to the source room is one means to quiet the source. Sound absorption will also reduce the echoes in the room, making it

---

[1] Signal to noise ratio is covered in Chapter 6 and addressed in the discussion of STC in Appendix 3.

more pleasant to converse. [2]

MATERIAL	NRC	STC
Fiberglass 3-lb, 1-in. board	0.75	5
above with foil back	0.65	10
3½-in. fiberglass insulation	0.95	3
above with FRK face to sound	0.80	3
½-in. Gypsum board	0.05	26
¼-in. Glass	0.05	26
½-in. Plywood	0.10	21
FRK = fiberglass reinforced kraft.		

Figure 5-2 NRC and STC of key building materials.

Materials that are **good sound absorbers are notoriously bad sound barriers.** Fiberglass, for example, is probably the most efficient sound absorber commercially available. Ratings of noise reduction coefficient (NRC) greater than 1 are easily attained with relatively inexpensive products that also provide thermal insulation. However, used only as a barrier, fiberglass insulation has poor sound transmission loss (TL) capability. Sound barrier ratings for fiberglass board or batts rarely exceeds a sound transmission class (STC) of 10. Adding facings with a good sound barrier attribute to fiberglass may improve barrier capability, but may reduce sound absorption or NRC values (Figure 5-2). Adding fiberglass to a partition cavity may improve the STC 8 to 10 points.

Facings on a sound absorber must be acoustically transparent or they will block the sound from entering the absorption material. A facing with good barrier characteristics placed over the sound-absorbing material has the same low NRC as the facing material. For example, placing a good absorber such as 1-in. fiberglass board on a wall and then covering it with gypsum wallboard will yield an NRC of 0.05 instead of 0.65 for the fiberglass. If the facing is to be transparent to sound it must have an open area of over 25%. Air flow through the face must not be impeded. In fact, a good way to determine if the facing is open enough is to blow through it. If you feel your breath is not impeded, then sound will likely flow through the facing. About the only exception to this situation is a thin film. A film that is less than 1-mil thick will be acoustically transparent to all normal frequency sounds and yet may not let air pass.

Good sound absorbers are used primarily to reduce reverberation time (i.e., echoes) within the room that contains the sound source. While sound absorption will reduce the sound level in the source room, the practical limit is approximately 10 dB$_A$. By reducing reverberation time, speech intelligibility is increasesed. A reverberation time of approximately 2 seconds is ideal for speech clarity. The amount and type of materials needed are explained in Chapter 4. However, if you like the effect of music by composers like Bach or Handel, played in an old stone cathedral, a long reverberation time is needed. In this instance, all interior surfaces should be good sound reflectors to achieve the reverberant effect. Conversely, electronic equipment that can generate reverberant sound should be used in a room that is very absorbent.[3] For example, loud rock music should be presented in a highly absorbent room. Rock music is generally played with electronically enhanced instruments and the hall will adversely affect the players' options to create the sounds desired.

---

[2] More on room acoustics is given later in the chapter. Also see Chapter 2.

[3] For the affecionado of cathedral music, an electronically reproduced reverberation may seem to be similar to the real thing. However, there are nuances to the sounds in an authentic stone cathedral setting that cannot be reproduced artificially. It is also interesting to note that Bach and the others of this era created their music for performance in stone cathedrals. They counted on the long reverberation time to enhance and mix the sounds. Performances of these classics in a room with short reverberation time will sound weak and without body or character. This may be the reason why many do not like the classics. They hear them differently than the composer intended solely because of the acoustics of the hall.

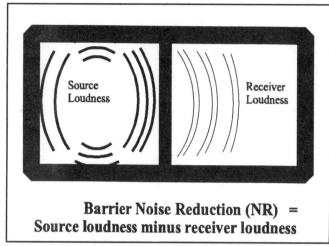

**Barrier Noise Reduction (NR) =
Source loudness minus receiver loudness**

Figure 5-3 Noise reduction of barrier equals loudness in the source room minus loudness in the receiving room.

Sound barriers should be installed between the source and receiver. Barriers may be walls and floor/ceiling assemblies, including all the penetrations such as windows and doors. A good barrier will have a high sound transmission loss, good sound attenuation, and a high sound transmission class. The first hang-up many have with sound transmission loss is the apparent double-negative connotation. As the barrier becomes a better barrier, the STC will increase. Sound transmission loss could also be called noise reduction. In some cases, such as a general field test, the resultant values are called just that: noise reduction (NR). TL, STC, and NR are all basically similar in that the value is the difference between a sound level in a source room minus the sound level in the receiving room after it passes through the barrier (Figure 5-3).

Sound barriers are never better than their weakest link. Barriers with good noise reduction, sound transmission loss, or sound attenuation, will contain the sound of a source, and block it from reaching the receiver location. A small crack, or flanking path can be a disaster. It is convenient to think of sound like air in a balloon; anywhere air can go, so goes sound. Paraphrasing an audiovisual presentation by the Bureau of Standards, a flanking path is a chink in shining armor. Stated another way, a flanking path is the weak link in the chain connecting the source with the receiver. Liberal use of caulking compound provides dramatic decibel-per-dollar results. Also, check out all the other potential flanking paths such as windows, doors, heating, ventilation, and air-conditioning (HVAC) ducts; back-to-back outlets; pipe penetrations; back-to-back bathtubs, showers, and cabinets, and all other possibilities listed in Chapter 2.

Only when all the flanking paths are identified and eliminated should one begin to upgrade the barrier characteristics of the wall or floor/ceiling. To do so requires knowledge of the basic principals of barrier design, namely, increase mass, decouple the surfaces to create a limp mass, and fill the decoupled void with sound-absorbing material. Why does sound absorbing-material work in this situation? Because it absorbs sounds within the cavity (i.e., it is a minisource control) and acts as a damper to reduce vibrations of the barrier surface materials.

In short, be sure you analyze your acoustical problem to determine whether you need a sound absorber to reduce echoes and noise levels in the source room or a good barrier to keep the noise away from the receiver.

5.3 SOURCE QUIETING

Source, path, and receiver (SPR) analysis of any noise problem can be done easily by anyone with just a bit of common sense. **Treatment of the source will always provide the most dramatic results at the**

least cost, unless of course you prefer ear plugs or are deaf.

SOURCE ⇒ PATH ⇒ RECEIVER

<u>Reduce source first</u>, receiver second, and path last.

It is amazing how the solution to a noise problem can be reduced to a simple statement: "Shut that damn thing off, or at least turn it down." Communication between the individual controlling a source and the receiver is the key to success. It may be that the person creating the noise is not aware of how intrusive it is. Most of us will be glad to cooperate if we only know. Don't spend thousands of dollars building a high TL barrier when a simple request to the source controller is possible.

The same goes for noisy equipment; ask for quiet equipment– " buy quiet." Manufacturers of appliances have made hundreds of surveys that indicate customers want low-noise items. Researchers and noise control engineers know how to make the equipment quieter. Yet, few bother to spend the effort to engineer for quiet. The cost to make quiet equipment may be nearly insignificant in many cases, but until the consumer demand is there, quiet equipment will be a long time coming.

Some simple features to look for in assessing whether equipment is as quiet as practical include:

- Well balanced parts. They are not only quiet but last longer.
- Proper lubrication also promotes quiet and longevity.
- "Slower is lower." Equipment that runs slower than a counterpart will be quieter. This is especially true for fans and motors.
- Streamlined air and liquid flows. Less turbulence means quiet.
- Vibration mounts. Isolate vibrations before they spread via the structure.
- Equipment enclosures that are tightly sealed and lined with sound-absorbing material.

Many manufacturers now provide sound level labels on their products or can provide acoustical performance values on request. Ask for the rating and keep asking. Maybe the manufacturer will eventually "see the light."

5.4 IMPORTANCE OF CAULKING

When building sound barriers such as walls or floor/ceilings with high sound transmission loss characteristics, not enough can be said about the importance of sealing sound leaks or flanking paths (Figure 5-4).

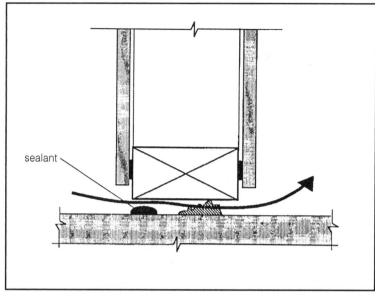

Figure 5-4 Caulking seals cracks at juncture of wall and floor caused by floor surface deviations or debris.

Caulking is every acoustician's best friend. The molding that conceals a crack at floor, ceiling, or corner gives one a false sense that the leak is gone simply because it is out of sight. **One must continuously think of the barrier as a container.** If necessary, think of a source room filled with water. Water leaks represent sound leaks. A neat way to find leaks is to fill the source room with smoke under pressure. Like the plumber who uses soap to find gas leaks by watching bubbles form, you can find sound leaks by watching smoke seeping through the cracks.

Finding sound leaks can be a difficult and frustrating experience. In some difficult cases, smoke may be the only way to find a sound leak. For a more practical initial search for leaks, turn on a radio in the source room and just listen in the receiving room. Try to identify where the noise is emanating. Tune the radio between stations. The static will provide a uniform broadband sound, sometimes called white noise since it has equal intensity at all frequencies. Listen with an inexpensive stethoscope. This device will aid in identifying sound leaks. Merely scan the wall and listen for major increases in levels. Do not place the stethoscope directly on the wall, keep it an inch away if you are listening for airborne sound leaks. If you are listening for structureborne noise, direct contact is necessary. Corners will always produce sounds slightly higher because of reflection doubling, so don't confuse this with a sound leak. Since most leaks occur in a corner, scan the corner in a uniform fashion, listening for sound variations.

Caulking must become a ritual when the sound barrier is designed for an STC 40 and higher. An excellent example is a wood-stud wall with resilient furring channels (Figure 5-5). Most manufacturers of resilient furring channels recommend placing a filler strip of ½-in. wallboard at the base for a base nailer. Sealant is vital to assure this vulnerable area does not leak. The permanently elastic sealant complements

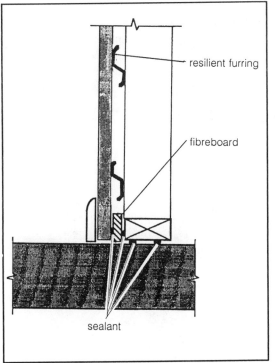

Figure 5-5 Caulking a wall with resilient furring channels and an STC > 40.

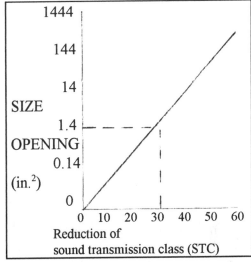

Figure 5-6 Cracks and holes create sound leaks that dramatically reduce the STC. e.g., 50 STC barrier with a crack ⅛-in.wide by 9.3 ft long (area 1.4-in.²) will be reduced to 20 STC.

the resilient facing and assures achievement of a high STC.[4] The size of sound leaks may be very small yet cause a considerable problem. For example a ⅛-in. wide crack that is 10 ft long, typical of a narrow crack at the juncture of two pieces of wallboard, is equivalent to a hole of 15-in². A hole of this size will reduce an STC 60 wall to STC 25 (Figure 5-6). In the case of an electrical outlet, the box will have leaks along all the edges and at the plugs. A total open area of 1-in.² is possible. This opening would reduce a 60 STC wall to nearly 30. It becomes obvious that the small cracks and holes add up quickly and can totally negate the sound transmission loss characteristics of high-performance sound barriers.

Eliminating sound leaks is the responsibility of the person constructing the acoustical envelope: the walls, floors, ceiling, windows, doors, and any other part that surrounds a noise source or the listener. While the acoustician can specify and test the flanking paths and the architect or contractor can specify and design for no flanking paths, it is ultimately the person doing the construction who will provide quiet. **It is imperative that the carpenter, drywall installer, and finish carpenter understand the principles of noise control.**

---

[4] Several manufacturers of gypsum wallboard have conducted extensive research into the proper use of caulking and sealants on wall systems. One notable study was conducted by U. S. Gypsum Co., the manufacturer of Sheetrock and acoustical sealants, demonstrated that poor sealant techniques could reduce the STC by 10. USG recommends four beads of caulking be placed at the base of the wall – two under the plate and one each under each piece of wallboard after the wallboard is in place and prior to installing the base.

Examples of how important it is for the person on the job to understand noise control principles abound in the construction industry. Probably the most flagrant noise flanking paths are caused by the drywall installer. Drywall's predecessor, lath and plaster, did not generate many noise complaints since the plaster was installed in a liquid or mud state. A good plasterer learned to press the gooey material into all cracks and crevices. This was done to assure that the plaster bonded to the lath. Excess plaster was forced into all the holes and crevices so that when it dried there was a strong mechanical key between the lath and plaster. This automatically filled all the holes and crevices where sound leaks could occur.

When drywall came into existence, the large-size boards created substantial gaps between panels. Large gaps occur particularly at corners, along the base, and the wall-ceiling juncture. Joint cement and paper tape were designed to cover these potential flanking paths. However, drywallers quickly found that they could skimp on their efforts and still create a good-looking finished job. As the price war began, the quick, easy, but still aesthetically pleasing drywall job became the norm. Huge cracks and wall penetrations were covered with moldings. They looked great but did not seal potential sound leaks. Probably the most troublesome sound leak occurs at the juncture of the wall with the floor.

Standard drywall practice is to install the panels tight to the ceiling or the upper wallboard panel using a "toe lifter." To accommodate this technique, walls are usually an inch or so higher than the floor-ceiling height. Using standard-length wallboard, this means there will be a ¾-in. or 1-in. wide gap

Figure 5-7 Caulking recommendations at floor-wall intersection from ASTM E497.

along the floor. Covered by standard floor molding, this crack is never noticed. Since sound will travel through any crack, even if the path is circuitous, the floor gap becomes a significant sound leak. Proper acoustical control methods call for this gap to be fully sealed with a "permanently elastic acoustical sealant." Since the gap is out of sight, out of mind, expensive to seal, and rarely inspected, most drywall installers ignore the plans and specifications. Worse yet, no one catches the error until the drywall installer is long gone from the job site, probably having been paid and left the state for other work. Pursuit is impractical, and no one wants the mess of making corrections. It is a classic situation where the noise literally "slips through the cracks" (Figures 5-6 and 5-7).

Similar cracks occur in drywall where pipes penetrate the walls. A simple escutcheon plate covers the holes that are cut oversize to accommodate installation. To make matters worse, pipes are typically installed after the drywall applicator is gone. Plumbers and HVAC installers do not normally consider it their job to install caulking or sealants. Worse yet, they do not realize they have penetrated a sound barrier. They simply cover the hole with a plate that is not airtight or sound-sealed. The result is a dramatic drop in the STC of the barrier due to a major flanking path. By simply sealing these holes, one can most times bring the barrier up to its expected sound barrier performance.

How does one find these sound leaks? The most practical solution is to train the installers and utilize designs and specifications that call for noise control. After the fact, a trained acoustician can tell very quickly. Listening to a sound source in an adjacent room generally makes the sound paths obvious. Merely knowing where to look for a crack is sufficient. Finding a practical solution after the fact is an art that has been fine-tuned by many acousticians and acoustical consultants. One thing is very clear: correcting a flanking noise problem is inevitably 10 times more expensive than prevention at design and installation.

Why use "acoustical sealants" and what makes them different from caulking? "Acoustical sealants" are typically made from a caulking material that remains "permanently elastic" throughout the useful life of the building. While expensive, acoustical sealants made with silicone-base materials stay rubbery and bond tightly to wood and gypsum wallboard. When a sealant dries or becomes hard and brittle, small building movements or expansion and contraction of dissimilar materials will cause them to crack and become ineffective as noise barriers. Sealants that are designed for thermal barriers or to hold glazing in place are usually good acoustical sealants. Since most acoustical sealants are never exposed to weather, they are usually made with less-expensive materials. Major manufacturers of gypsum wallboard market acoustical sealants as an accessory. Typically the sealant is provided in a caulk tube but is available in bulk for pressure-type caulking equipment. A proper acoustical sealant should meet ASTM standards.[5]

Sealants used in laboratory testing are idealized. Sound transmission loss testing in an acoustical laboratory will optimize the role of sealants. One of the most important issues in constructing a sound barrier specimen is to eliminate any possibility of sound leaks. If the object of the test is to determine the sound transmission loss performance of a material or system of construction, by definition, sound leaks are eliminated. Laboratories go to great lengths to assure that there are no flanking paths between the source and receiving room. A particularly vulnerable spot for sound leaks is the juncture between the specimen and the laboratory structure. In most instances the laboratories use an abundance of a soft, pliable, heavy material that looks and acts much like modeling clay. Called duct seal, this material is pushed into place by hand to assure it completely fills the gap and is airtight. With its high density, this material, when properly installed, yields a high assurance of eliminating sound leaks for barriers up to 60 STC. It is the classic example of a good barrier, namely, a heavy, limp mass. It stays elastic indefinitely and can be used over and over provided it does not get contaminated by dust, dirt, and construction debris. This same material is used to seal leaks when conducting sound transmission loss tests according to ASTM E336 in the field.

Building movement tends to open up sound leaks. With the recent earthquakes in southern California, it is likely that any sound barrier there has been considerably compromised. Even the best sealants cannot bridge the gaps generated by the forces involved. It is a given that multifamily dwellings no longer will provide the degree of noise control they did prior to the earthquake. Out of necessity, corrective measures focus on structural and possibly aesthetic repairs rather than sealing sound leaks. (Note: Repairs designed to seal sound leaks at the same time structure and aesthetic repairs are made will be far less expensive than corrective measures for noise control after the fact.)

All building movement will tend to disrupt acoustical sealants. While earthquakes are the most obvious,

---

[5] ASTM C919 Practice for Use of Sealants in Acoustical Applications, and E497, Standard Practice for Installing Sound-Isolating Lightweight Partitions.

an astute noise control expert will look for dried and cracked seals caused by age, dry ambient conditions, and nonuniform expansion and contraction of dissimilar materials due to temperature fluctuations. Warping caused by excess moisture and subsequent drying is also a major cause for acoustical sealant failure. In cold climates such as the midwest and northeast, expansion and contraction is a major cause of cracks. These types of failures are usually noted and repaired quickly on the exterior envelope since cracks jeopardize the thermal envelope. Unfortunately, cracks on interior noise barriers are overlooked. Eventually, they become a major source of complaint by occupants who hear their neighbors.

5.5 LABORATORY VERSUS FIELD TESTING OF TRANSMISSION LOSS

A common misconception is that laboratory tests will always yield a better sound transmission class than the same configuration tested in the field. This is not necessarily true. This incorrect assumption likely stems from the fact that seldom is a sound barrier constructed carefully in the field, and inevitably sound leaks or flanking paths occur in the field that were eliminated in the laboratory. In fact there are many instances where the field result can be higher than the laboratory test result. Theoretically the results of laboratory and field sound test data will be the same for an identically constructed sound barrier. Here is why:

5.5.1 Procedure ASTM E90 Standard Test Method for Laboratory Measurement of Airborne Sound Transmission Loss of Building Partitions[6] is the procedure used in most laboratories to evaluate partitions and floor/ceiling assemblies. E336 Standard Test Method for Measurement of Airborne Sound Insulation in Buildings is intended as a field test version of E90. Both procedures measure the STC of the barrier. When tested in the lab, the result is written STC. When tested in the field the correct terminology is field sound transmission Class (FSTC). E336 also can be used to evaluate the barrier as installed, including any faults, sound leaks or flanking paths. The latter is reported as noise reduction. Please do not confuse STC, FSTC, and NR. Many, including acousticians, utilize these terms interchangeably, further adding to the incorrect assumption that field tests will always be worse than laboratory tests. Note that:

- **NR evaluates the sound barrier in use.** Flanking paths are included.

- **FSTC evaluates the sound barrier in the field.** Flanking paths are eliminated.

- **STC is a laboratory test of the barrier.** Flanking paths are eliminated.

It is imperative for the person who evaluates test reports listing STC, FSTC, and NR values to understand this important difference. Confusion on this difference between laboratory and field results has caused considerable unnecessary pain and/or lawsuits.

5.5.2 Size The size of the test specimen can have a significant effect on sound transmission loss values. In the laboratory, standard specimen sizes are generally 9 by 14 ft for partitions and 14 by 20 ft for

---

[6] ASTM E90 and E336 may be found in American Society for Testing and Materials Volume 04.06. The latest version is 1993. A description of these test procedures may be found in Section 4.6 "Sound Transmission Standards."

floor/ceiling systems. These specimen sizes generally are large enough to show the effects from panel junctures, stiffness, and frequency-dependent effects. In the field, the specimen size is typically mandated by conditions. Given the same size of specimen, there is good correlation between the field and laboratory test results when flanking paths are eliminated. Unfortunately, circumstances rarely can be duplicated between the lab and field. Consequently, it is impossible to equate the two. For example, a field specimen with a resilient face mounting that is 8 ft by 8 ft may not be large enough to allow the resilient mount to function. Likewise, edge constraints may force the face to vibrate differently than a much larger specimen, thereby limiting the STC. By contrast, a very large barrier, say 12 by 30 ft, has been known to provide better STC values than the 9 by 12 ft specimen in the laboratory.

5.5.3 Utilization. How should the field and laboratory values be utilized? Both the field and lab tests may be utilized for a project. If the project is too small to warrant the cost of field analysis to verify performance, it is appropriate for the specifier to call for the sound barrier to have an STC of $\underline{X}$ per ASTM E90 and installation per ASTM E 497[7]. If the project is large enough and sound isolation is a significant issue, it is appropriate to specify the barrier by requiring the initial design to satisfy an STC of $\underline{Y}$ per ASTM E90, installation per E497 and an FSTC of $\underline{Z}$ per ASTM E336 as part of the building acceptance. Both techniques can be used to effectively guarantee noise barrier performance. The latter specification was used by the United States General Services Administration as a performance specification for over 1.2 billion square feet of federal office buildings. Performance specifications are a powerful tool to assure guaranteed acoustical performance on the completed building.[8]

LABORATORY/FIELD TESTING

- Laboratory tests per E90 provide data for direct comparisons of construction systems. Field tests rarely have identical construction details.
- Test results in the field per E336 provide actual results. They may be equivalent to tests on a similar construction per E90 if room and specimen sizes are similar.
- Large size specimens may produce higher ratings because stiffness is different.

---

[7] ASTM E497 Standard Practice for Installing Sound-Isolating Lightweight Partitions, Volume 04.06. This standard gives specific installation instructions for wood stud, steel stud and gypsum wallboard partitions for optimum sound barrier performance.

[8] Acoustical performance specifications require the party bidding on the project to guarantee that they will meet the criteria. If the installed barrier does not measure up at the time field tests are conducted, the system offeror must make the corrections at its own cost. This procedure fulfills a great need for an acoustical guarantee. An offeror quickly realizes it is much less expensive to design the barrier correctly and use care in installation.

5.6 EFFECTIVE SOUND-ABSORBING MATERIALS

Improving the quality and quantity of sound in a source room is a matter of adding efficient sound absorbing materials to that space. As discussed in Section 2.3, "porous absorptive" materials are the most common.

5.6.1 Examples of Sound- Absorbing Materials

Typical examples of porous sound-absorbing material are mineral fibers including glass fiber and rock wool. Open-cell plastic foam materials are also common. These materials are soft and flexible and allow sound to enter them easily. They are considered "porous" since an early test of their sound absorbing efficiency was their "air flow resistance." Materials that resist the passage of air have poor sound-absorption properties. Efficient sound-absorbing materials tend to allow air to flow through them with relative ease. A simple test of how efficient a sound absorber is can be conducted by holding the material tightly against your mouth and blowing. If you cannot easily blow through the material, it is likely to be a poor sound absorber. A good absorber will have little resistance to your breath. Gypsum wallboard, plywood, and hard sheet materials are very poor sound absorbers at normal speech frequencies. Building insulation, open cell foam, blankets, clothing, and soft furry things tend to be good absorbers. The ideal sound absorber is the open sky, since sound directed to the open sky is never reflected.

Prior to the advent of fire ratings and incombustible material requirements in building codes, the most effective sound-absorbing materials were made of wood fiber. Cork was one of the first sound-absorbing materials sold commercially, along with a board made of cane fiber. All of the fibrous materials convert impinging sound energy into heat, which is dissipated.

Other types of absorbers that have been effective include the Helmholtz resonator. Here, sound enters a small orifice into a larger chamber. Sound bounces around inside and cannot easily exit the small orifice. A good example is a soda bottle. While efficient absorbers, a Helmholtz resonator is usually able to absorb only a very narrow frequency band of impinging sound. Another absorber utilizes a thin membrane stretched across a stiff material having small holes. Why this method works is still being argued by acousticians.

5.6.2 Sound-Absorbing Misconceptions

<u>Sound absorbers are not a cure-all solution.</u> Probably the most widely misunderstood aspect of sound-absorbing materials is that many believe that adding them to a space will magically make it acoustically acceptable. Not so! While true that absorbing sound in a room will reduce sound reflections and reduce the reverberation time this may not be the real noise problem. Reducing the reverberation or echoes in a room makes it more comfortable for a person to understand speech. But it will reduce the sound level by only 3 to 8 dB_A. This misconception has prompted the author to invent the "super duper sound sucker." Anyone who buys one of these devices will solve all acoustical problems – so says the literature. Obviously, this can easily become a rip-off of major proportions.[9]

---

[9] Apparently, there is great hope that someone will invent a product that will magically solve all noise problems. Early marketeers of sound-absorbing materials added to the problem by overstating the performance of their product. In any event, there are still those who tout the wonders of their material when in fact, it is acoustically worthless. If your goal is to reduce echoes in a room, then by all means add all the sound-absorbing material you can find. If reducing noise level is your goal, then sound absorption is only one of many tools needed.

Sound absorption efficiency is not the same at all angles of incidence. The industry standard for measuring the sound absorption characteristic of a material is a test procedure known as ASTM C423, *Standard Test Method for Sound Absorption and Sound Absorption Coefficients by the Reverberation Room Method.* This procedure has been carefully structured to create a "diffuse sound field." Specifically, the test chamber is configured and designed to reflect sounds randomly. The object is to produce a sound source that will strike the specimen being tested at all angles of incidence. While this is a noble testing effort, some materials are more efficient at absorbing sound at specific angles of incidence. Some materials are efficient absorbers at angles of incidence that are common in a space, such as normal (perpendicular) to the surface. Others are poor absorbers of normal incident sound but are excellent absorbers of grazing incidence sounds.[10]

If the purpose of your application is to reduce the reverberation time in a large-volume space, then it is appropriate to select a material having a high sound absorption coefficient at all angles of incident sound and at all frequencies that are vital to speech clarity. Thus, the noise reduction coefficient is a correct way to rate the acoustical efficiency of the material.

However, if your purpose is to absorb sounds at specific angles of incidence or at frequencies other than those between 250 and 2000 Hz, then ASTM C423 may not provide the information needed. More detailed analysis of the sound absorption at specific angles of incidence may be required. A typical example of a procedure designed for measuring sound absorption at specific angles of incidence may be found in the family of procedures adopted by ASTM Committee E33 for open plan offices. This family of procedures measures sounds typically found between two work-stations as the sound reflects over a part high barrier off the ceiling or around the barrier off a vertical surface.[11]

Sound absorption measurements using the "impedance tube methods" do not provide a very reliable means of evaluating sound absorption performance.[12]

Painting a sound absorber may be disastrous. Sound-absorbing materials are notorious for getting dirty. They usually have a rough texture, for both aesthetic and acoustical reasons. Consequently, they collect dust, dirt, and stains more easily than other building materials. When these materials are painted with a typical oil or vinyl paint, the thick film paint surface will block the entry of sound to substrate fibers. The result will to be render the material acoustically reflective. A ceiling board with an NRC of 60 may be reduced to an NRC of 20 with one or two coats of vinyl paint. This change in NRC will likely cause considerable

---

[10] This fact was discovered by R. N. Hamme et al. when they were evaluating the effect of sound absorption and speech privacy in open plan offices. In an open plan environment, sound typically strikes a surface between 30 and 60° from normal. They found that some materials having the same NRC had quite different sound absorption characteristics in this environment. For example, a typical mineral tile ceiling board and a glass fiber ceiling board, both with NRC 80 were vastly different in absorbing sounds within the 30 to 60° angle of incidence. It was found that the mineral board gained most of its sound absorption at grazing angles and was almost a reflector at normal incidence. Conversely, the glass fiber had poor absorption at grazing angles and excellent absorption properties at normal incidence. For more details see the *Noise Control Manual*, Chapter 8, and *Planning and Designing the Office Environment*, Chapter 2, 1992, both authored by D. A. Harris and published by VNR/Chapman and Hall. Additional details may be found in a paper presented at the Acoustical Society of America (ASA) titled "A Tribute to Richard N. Hamme – The Father of Open Office Acoustics," May 13, 1992, by D. A. Harris, and the background material for the Public Buildings Service (PBS) research report known as PBS C.3 prepared by Geiger & Hamme Laboratories for the United States General Services Administration (GSA).

[11] See ASTM E1374 Standard Guide for Open Office Acoustics and Applicable ASTM Standards, Section 2, "Referenced Documents," and PBS C.1, C.2, and C.3 (See note 10).

[12] ASTM C 384 Test Method for Impedance and Absorption of Acoustical Materials by the Impedance Tube Method. This procedure was developed primarily as a research screening tool. While it is useful in rank ordering materials, this method cannot be correlated with ASTM C423.

degradation of the acoustical environment.

It is possible to paint ceiling tile and maintain the original, or close to the original, NRC. There is nothing magical about the paint. The key is to spray the paint in very light, watered-down coats. A thick film must be avoided. Brush application will fill the voids and is not recommended. For smudges and small areas of discoloration, use white shoe polish or talc powder applied with a damp sponge. Do not cover with a sound-blocking film.

Materials with an NRC of 70 and higher should not be painted. Stained or discolored materials should be replaced or cleaned. Even a slight film will dramatically reduce the ability of the sound to penetrate and get absorbed.[13]

Surface texture and surface shape rarely improves sound absorption. Considerable research has been conducted by ceiling tile manufacturers to find a combination of aesthetically desirable surface materials that are also good sound absorbers. Many theories abound. Some have merit but most fall miserably short of the marketing statements. About the only rules of thumb that have proven merit are: Thicker is better and more surface area is better. A big proviso is, of course, that the material is exposed (i.e.: without with a coating).

Thickness is the main determinant of absorption efficiency. For highly efficient sound-absorbers with 60 or higher NRCs, such as glass fiber, light-density mineral fiber, and open-cell foam, thickness is the major factor in achieving higher sound absorption coefficients.[14] In fact the thickness effect is so overwhelming that all other potential variations that could affect sound absorption are rendered minimal. Unless there are limiting circumstances such as in aircraft, it is economically advantageous to increase NRC via extra thickness. This effect will become readily apparent on viewing test data for fiberglass materials (Figure 5-8).

Surface shape and texture has little effect on absorption. Open-cell foam producers have touted shape as having excellent acoustical value. The fact is that, unless the shape is very large, surface shape has little or no effect on the same overall thickness. The idea of providing "minianechoic wedges" may have considerable aesthetic appeal but provides no significant acoustical value. This approach has some market impact, however, since open cell foam is nearly 3 times as expensive as glass fiber or mineral wool product of similar thickness. Since foam is easily cut into intricate shapes, it is feasible to split a foam slab in half with wedge-shaped cuts. This process yields double the amount of coverage with essentially the same overall thickness. In effect, the price per square foot of foam becomes more economical.

---

[13] Manufacturers may provide special instructions for cleaning. Specialized contractors that are familiar with painting sound absorbing materials do exist. However, many claim to know how to clean the tile without affecting the sound absorption performance when in fact they use normal painting techniques with disastrous acoustical results. Check with manufacturer of ceiling board for recommendations.

[14] A major study of sound absorption vs. thickness and other attributes was conducted by Owens Corning Fiberglas, Interior Products Operating Division in 1985. It was shown that for fibrous materials, thickness was the prime product attribute (by a factor of 10) that established NRC. Materials evaluated were in the density range of 0.5 lbs per cubic foot (i.e. building insulation) to 7 pounds density, and included ceiling tile. Other attributes such as density, fiber diameter, resin content were also tested. In each case thickness was 10 times more efficient at improving the NRC. While density and fiber diameter did have a measurable affect on NRC the bottom line cost was no match for increased thickness. A similar trend was noted for open cell foam materials. This study also confirmed that film facings thicker than 0.5 mil will dramatically lower sound absorption coefficients, particularly for frequencies above 2,000 Hz. See Appendix 3 for sound absorption coefficients and NRC of various materials.

Rough textures have the effect of opening up the face to allow more surface area for sound absorption to occur. While texture is a practical consideration in design with a desirable aesthetic appeal, there is no clear evidence that a rough texture yields significant improvement of sound absorption. There is one exception. If the surface texture or shape is large enough – that is, the size equates to the wavelength of the sound being considered – then increased absorption values can be expected. An example of this phenomenon can be found in the design of anechoic wedges for an anechoic chamber. Since the wavelength of most sounds found in the speech frequency range are at least several inches and sometimes several feet, this size is a distinct limitation for most architectural applications.[15]

Increased surface texture is sufficient to warrant consideration for ceiling tile materials that have NRCs between 45 and 60. Historically, wood and mineral fiber ceiling boards are ⅝. and ¾-in. thick. This is primarily because of their span characteristics and fire ratings. Naturally, thicker boards are more expensive. In an effort to reduce costs,

Thickness (in.)	Density (lb)	Material	NRC*
1	1	Fiberglass	0.70
2	1	Fiberglass	0.90
3	1	Fiberglass	1.15
4	1	Fiberglass	1.15
1	3	Fiberglass	0.70
2	3	Fiberglass	1.00
3	3	Fiberglass	1.10
4	3	Fiberglass	1.15
1	5	Fiberglass	0.65
2	5	Fiberglass	0.95
3	5	Fiberglass	1.10
4	5	Fiberglass	1.10

*Tests on Type A mounting per ASTM C423

Figure 5-8 Increased thickness improves sound absorption coefficients and NRC.

make a lighter product, and provide new and aesthetically pleasing finishes, manufacturers found they could also increase the NRC slightly by providing texture. The first products were drilled. ¼-in. diameter holes, drilled to a depth of ½-in. and spaced 1-in. on center improved the NRC from 30 to 50. While it is not clear why, one might speculate that the holes allowed the sound to penetrate the painted surface and increased the surface area exposed to the sound wave. It made a good sales story at the time anyway. Designers quickly tired of the uniform holes drilled on a grid. Random holes then appeared with no detrimental acoustical effect. As this technique evolved, it was found that similar effects could be accomplished with a "fissure" punched in with a fissure roll pattern. Unfortunately, the NRC with fissures that were aesthetically acceptable were not good enough to be acoustically competitive. Some manufacturers tried small holes, about the size of needles or tack holes in addition to or in lieu of the fissures. NRCs of 55 to 60 for a ⅝-in. mineral ceiling board became common-place. While designers wanted to get rid of the holes, it became apparent that they were needed to maintain the NRC.

When cast mineral ceiling tile appeared, higher NRCs became possible. However, the phenomenon

---

[15] Miniwedges, sold by some open-cell foam manufacturers, have about the same sound absorption values as the same material of a similar thickness without wedges. These products are a classic case of "They look like they should be acoustic absorbers so they must be good absorbers." This is not necessarily so. However, when the wedges are several feet in dimension, they will show improvement.

occurred primarily because texture allowed the material to trap more grazing incident sound. The higher NRCs were not indicative of adsorption efficiency in the angles of incident sound needed for most applications. In reality, the cast mineral products found a loophole in the testing process.[16] Designers liked the new look of a cast product. Fire ratings were also improved. However, they are very expensive by comparison to the standard mineral fiber board known in the industry as "dumb-dumb-board." As a result marketers began promoting these products for the top-of-the line applications, including open plan offices. Unfortunately, they have poor absorption at the specific angles of incidence required for open plan office configurations. Cast mineral products are ideally suited for natatoriums, auditoriums, entranceways, and single-purpose rooms where control of reverberation time, aesthetics, resistance to high humidity, are prime concerns.

5.6.3 Sound Absorbing Myths
As indicated earlier, sound absorption has taken on an aura of magic in solving noise control problems. However, sound-absorbing materials must be added to a space extensively and spread uniformly around the room to reduce reverberation and echoes. Even the most idealized application of absorbing materials will reduce noise levels by only 10 dB_A. This statement may seem repetitious, but, it is vital to good acoustical design. The following stories and myths will help amplify the misunderstandings.

Myth: doubling the amount of sound-absorbing material will double the noise reduction. While doubling the amount of sound-absorbing material will improve (reduce) the reverberation time, there is no direct-line relationship. In fact the relationship is closer to being logarithmic than linear. In other words, adding a small amount of absorbing material in a highly reverberant space will have a noticeable effect. Each time you double the amount of sabins of absorption, the effect is only half as noticeable. When over 50% of the surface areas in the space are rendered absorptive with a material having an NRC of 80, a doubling of sabins of absorption may only reduce the sound level by 3 dB. A practical maximum noise reduction is 10 to 12 dB.

As an example, a church multipurpose room was found to be "noisy." It had gypsum board walls, ceiling, and a hardwood floor. With a reverberation time near 9 seconds, echoes and noise levels with a cheering crowd were extremely annoying. A manufacturer's representative proposed a sound-absorbing device that could be hung in the center of the room to solve the problem. He claimed that the device, about the size of a garbage can, was "tuned" and had the same absorptive characteristic as a whole ceiling of conventional ceiling tile. It was true, the sabins of absorption did equate a standard painted wood fiber ceiling tile ceiling. When the parishioners purchased their "super duper sound sucker," they found some minor improvement in the acoustical environment but were very disappointed. Someone concluded that maybe this wasn't enough absorption so they bought a second "super duper sound sucker." It seemed only logical that if one unit worked, two would work twice as well. Funny thing, nothing noticeable happened with the second absorber.

Why didn't the second absorber do the job? There are a number of reasons. First, sound-absorbing material must be spread around the space to be effective. Simple geometry will demonstrate that a single point of absorption, no matter how efficient an absorber it is, will absorb only the sounds that reach the absorber. In a highly reverberant room, it may take sounds many reflections to reach the absorber, if at all. Putting in a second highly efficient absorber did double the possibility of sound reaching an absorber.

---

[16] See note 10 above.

However, nearly all of the sound still reflected from acoustically hard and reflective surfaces many times before it reached the absorber. Second, the effectiveness of absorbers is not a linear relationship. As noted above, four or maybe even eight sound-absorbing cans may be required to obtain the same degree of improvement as the first can. Worse yet, while reverberation time decreased, the sound levels in the space were hardly affected for all the same reasons.

The solution for this application would be to apply a sound-absorbing wall and ceiling material that covers in excess of 50% of the wall and ceiling area. Most of the sounds will then be absorbed on the first reflection, thereby providing a far greater impact on reducing both the reverberation time and the noise level.[17]

The density myth. Adding sound-absorbing material to the cavity of a sound barrier has proven to be an effective way to improve the sound transmission loss characteristics of a wall or floor/ceiling system. Partition systems that employ acoustical separation between opposing barrier faces, (e.g., staggered or double studs, resilient channels, or sound deadening board) adding a sound-absorbing material to the barrier cavity has generally improved the sound transmission class by 5 to 7 points. Note that this improvement is not generally possible without acoustically separated, or resiliently mounted faces.

It has heretofore been assumed that the density of cavity wall insulation would have a large effect on the performance of insulation in wall and ceiling constructions. This is not necessarily so. An extensive research program sponsored by Owens Corning Fiberglas demonstrated that density variations over a wide range "have almost no effect on performance."[18] In this test program, the same wall constructions were tested with different densities of insulation. Tests were run with controlled materials under identical conditions at the Fiberglas Sound Laboratory and Riverbank, a leading independent acoustical laboratory.

Three basic wall constructions were tested:

Wall 1 – Standard wood-stud construction, resilient furring channels, a core of insulation and gypsum wallboard faces.

Wall 2 – 2½-in. steel studs with a core of insulation, one layer of gypsum wallboard on one side and two layers of gypsum wallboard on the other side.

Wall 3 – Same as wall 2 except 3⅝-in. steel studs.

---

[17] Reverberation time and noise levels in a large space such as a multipurpose room may be calculated by using the "Worksheet for solving reverberant sound problems" in Appendix 2. Sound-absorbing data is given in Appendix 3.

[18] Owens Corning Publication 1-BL-4589, dated January 1969.

STC values did not vary with densities. The walls with ¾-lb-density Fiberglas building insulation were as good as the walls with building insulation of greater density (e.g., nominal 3-lb-density rock wool building insulation) (Figure 5-9).

Since this study was conducted, there have been instances where high-density insulation did provide higher STC values. These results are unique to the particular construction details utilized and do not erode the findings of this study. This anomaly to a myth promulgated by overzealous marketing campaigns further emphasizes the need for a specifier to carefully investigate the circumstances of a test and apparent minor construction differences.

STC COMPARISON
(Typical partitions with sound absorbing core)

Wall	Cavity Fill Type	
	0.75-lb/ft³ Fiberglas	3 lb/ft³ mineral wool
1	50	50
2	51	51
3	49	50

Figure 5-9 Wall systems with light-density Fiberglas and high-density rock wool core have similar STC ratings.

Myth; sound absorption cannot be tuned. While it is true that most materials exhibit a signature sound absorption characteristic, there are many products that can be tuned to provide specific sound absorption properties at specific frequencies. The most notable example of tuned absorption is the Helmholtz resonator. An excellent example of a Helmholtz resonator is an old-fashioned soda bottle. With a sizable cavity and a small opening, the bottle traps sound that enters the opening; the bottle is therefore an efficient sound absorber. Just as this bottle may be tuned to play a specific note by blowing across the top and varying the volume by filling it with water, sound absorption efficiency may be tuned to a specific frequency.

Other examples of tunable sound absorption characteristics are found in ceiling tile, diaphragmatic action absorbers, and specialty absorbers.

In the case of ceiling tile, the tile designer may vary the fiber diameter size, thickness, hole size, fissure shape, and face texture. While there are no rules of thumb available, results of proprietary manufacturers' research have demonstrated variations in sound absorption values at different frequencies. After considerable trial-and-error testing, some manufacturers have been able to optimize the sound absorption coefficients for patterns and textures that are compatible with design aesthetics, light reflection, weight, and fire resistance.

Similar tuning may be accomplished in open-cell foam sound absorbers by varying the size of the air voids and the holes in the foam cell wall. A wide variation of sound absorption efficiency can be effected by changing thickness and surface undulations (e.g. mini-wedges).

Highly efficient sound absorbers such as independent can absorbers have been specifically tuned to desired frequencies. This makes them very desirable for use in industrial applications where a noise source may have a predominant frequency. However, they are not necessarily appropriate for broad-band noise sources such as speech.

A unique sound absorber that utilizes diaphragmatic action for absorption can be tuned for certain frequency bandwidths, generally those of speech and high frequency noise sources. An example of a diaphragmatic absorber is a perforated sheet such as pegboard with an onion-skin material stretched across the back of the board. As the onionskin material vibrates, it absorbs sound quite efficiently. NRCs of 0.80 have been achieved with this technique. These materials may be tuned to be more efficient absorbers at specific frequencies by changing the number and diameter of the holes in the face material. A particularly interesting product using diaphragmatic action as an absorber is a diffusing light fixture lens. Dropped in a conventional grid system and backed by fluorescent lights in the ceiling plenum, the suspended ceiling provides a uniform luminous ceiling plus sound absorption.[19]

Myth: hard-surface materials are always poor sound absorbers. While it is true that most materials like gypsum wallboard, plywood, hardboard, plaster, and similar hard, stiff materials are poor absorbers in the range of 100 Hz and higher, they do absorb very low frequency sounds. NRC and typical sound absorption tests generally cover the frequency range from 125 Hz to 5000 Hz. Plywood, glass, and gypsum wallboard generally have an NRC of less than 0.01. However, it is a well-known fact that these same materials exhibit good sound absorption at frequencies below 25 Hz. Their flexural characteristics act as a diaphragmatic absorber at these low frequencies. As an example, witness the movement of thin-plate membranes in an earthquake. Only very stiff materials such as concrete and stone walls several feet thick exhibit poor absorption at very low frequencies.

Walls having little or no high-frequency absorption but good low-frequency absorption have been used successfully in designing auditoriums and amphitheaters. They are particularly useful in controlling low-frequency noise produced by musicians with high-tech electronic instruments and many "woofers."[20]

Myth: empty beer bottles are good sound absorbers. Beer bottles are a crude form of a Helmholtz resonator and consequently can be good absorbers. However, Helmholtz resonators are very frequency-sensitive and may not absorb the kinds of sounds needed.

This simple observation was tested early in the days of sound testing. There is an interesting story told by those who remember the early days of sound absorption testing at Riverbank Laboratories. It seems the absorption theories about Helmholtz resonators were upheld after a series of tests on arrays of beer bottles in various sizes as resonator cavities. As the story went the rounds, there was always the question, "What happened to the beer?" All one got from the tester was a smile.

A similar experience relates to the use of beer bottles inside a wall cavity. In the early 1960s the author was teaching a course in sound control construction techniques to a group, of drywall contractors. After an explanation of the effect that sound absorption had in improving the STC of wall systems, a class member had a very intriguing observation. It seems that several months before, he had been on a commercial construction job where a crew was building a steel-stud wall with gypsum wallboard faces. Apparently his

---

[19] Unfortunately, this product is no longer readily available in the marketplace. The plastic material used as the facing did not exhibit fire resistance. Efforts to allow sprinklers behind the material were rebuffed by fire officials and the inventor had insufficient financial resources to fight the required political battle or find an incombustible material that exhibited similar acoustical and light-diffusing properties.

[20] Woofers produce the lowest-frequency sounds in a speaker system. As the electronic sound system industry has progressed, producing low-frequency sounds has become in vogue. Low-frequency sounds below 50 Hz are not heard by the ears. Rather, the sound vibrates the whole body, a sensation that is apparently "neat" to some. Others find it quite obnoxious. In any event, it takes a lot of energy to produce these low-frequency sounds. Likewise, specialized constructions are required to provide sound absorbers at these low frequencies. Providing a low-frequency sound barrier is even more difficult.

superior instructed him to hang empty beer bottles on a wire in the stud cavity. He wanted to know if this was a good acoustical practice. On further investigation it was determined that contractors had, in fact, tested the theory and found that it worked. To this day the writer does not know if this was a put-on or not. But it did become known that the testers drank all the beer. One wonders if the empty bottles are still in the wall.

5.7 MYSTERY: WHAT ARE THE LIMITS OF ACOUSTICAL BARRIERS?

The art of designing architectural sound barriers has engrossed many specialists including the author, his associates, and competitors. Efforts to find a less expensive way to build a wall or other building element has been a major pastime and focus of building materials research engineers. Many texts, technical papers, and manufacturer's literature exist to document some of the findings. Others are proprietary information and exist only in the authors' files. The reader will find much of this material, including information not heretofore published, included in the test data pages of Appendix 3. While specific findings are intriguing to those close to the action, the technical aspects must of necessity be included to make much sense of it all. In the end, the result is lots of boring reading. For this reason data, information, and findings of interest have been summarized for ease in understanding.

A recent paper written by John Kopec, director of Riverbank Laboratories, does an excellent job of addressing the many issues currently evolving in the design of acoustical barriers, be they walls, floors, ceilings, doors, flanking paths or the like. John's paper is reproduced here in its entirety, as he prepared it, to assure that his humor and delicacy in handling highly proprietary information is retained.

PUSHING THE LIMITS OF ACOUSTICAL BARRIER PERFORMANCE
By John W. Kopec, Director of Riverbank Acoustical Laboratories, Geneva IL

During the 1970s, a few corporate representatives advocated that the science of architectural acoustics was dead. They stated that existing documents contained all the information required for an effective noise control computer program. Thus, deriving a solution to most noise related problems was simply a matter of parameter input. Those representatives tried to convince their audience that whatever noise control products were available then, were as good as there would ever be. To provide technical support to the latter, they referenced the mass law principle.

The mass law principle used the weight of a material expressed in square feet as a variable. The weight was plugged into a formula, and the result was predicted transmission loss (TL) values for the material. In general, if the mass of a material was doubled, the TL increased 6 dB.

It is true that the mass law principle has some merit. When it is applied to a single "limp mass" material at the middle (speech) frequencies, minimum transmission loss (TL) values can be reasonably predicted. Also, it can be used to flag barriers with design errors. If a barrier receives TL values less than the predicted mass law values, it is likely that the barrier has a design error.

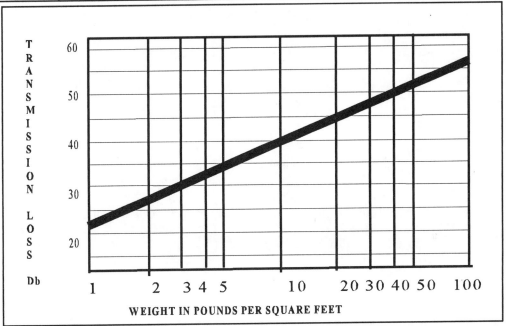

Figure 5-10 Mass law relationship to TL.

However, basing acoustical limits on the mass law principle alone is essentially "Neanderthal acoustics." First, if the mass law principle is applied to composite barriers, other factors are involved. If these other factors are neglected, the predicted TL values can be highly inaccurate. Second, there are different mass law formulas based on different parameters. Different curves were made up from the different formulas in the 1970s. With no mention of other curves, and with no consideration for composite barriers, the dead science advocates used the curve in Figure 5-10 to establish the following acoustical limits:

So Called Acoustical Limits on Doors. No 1¾ inch thick, fully operable metal door would achieve a Sound Transmission Class (STC) rating of 50 or greater. This conclusion was based on the mass law curve which showed that the barrier materials would have to weigh at least 70 lbs/ft^2. Two problems would result. First, few 70 lbs/ft^2 products were available, and second a typical 3070 (3' by 7') door would have to weigh at least 1470 lbs. In addition, noise leaks due to the hardware components made achieving an STC 50 even more difficult. Door knobs, cylindrical locksets, hinges, frame components and the various seals all contributed to reduce a door's acoustical barrier performance.

Although the above acoustical limits were originally stated in regards to metal doors, the STC limit established for wood doors by the manufacturers was that it appears that no operable 1¾ inch thick wood door could achieve an STC of 47. Especially without fixed bottom seals that create tripping hazards. Their hypothesis was also based upon the mass law lbs/ft^2 theory and from comparative old test results.

So Called Acoustical limits on Windows. Unless a standard sized (thick) window was glazed with massive, costly 1 to 2 inch thick solid (monolithic) glass, it would never reach an STC of 50.

So Called Acoustical Limits on Gypsum Board Walls.   Gypsum board walls provided less noise control than block walls because of their limited available mass. Only by doubling or tripling the wall thickness could gypsum board walls achieve comparable STCs, unless of course they contained other more massive materials.

So Called Acoustical Limits on Floors.   Massive, solid floor constructions achieved STCs of 60 and greater. However, the problem was that they had very limited structure borne noise elimination capabilities. Solid floors could not obtain an Impact Insulation Class (IIC) of 55 unless they were covered with a carpet and pad.

The lesson here in regards to the above so called limits is in line with the cliche, never assume anything, or, better yet, never predict acoustical limits involving architectural acoustics. What those dead science advocates didn't realize was the tenacity and purpose of some Riverbank clients.

Opening new Acoustical Door limits.   Three Riverbank clients obtained an STC of 50 and two even higher on a fully operable metal 3070, 1¾ inch thick door. Obviously, the design of a door that exceeded an STC of 50 involved much more than mass law principles. The proof was in the weight of the doors. The doors weighed between 10 and 15 lbs/ft^2. According to the mass law chart, the latter would only receive an STC 40.

Just recently an STC 47 was achieved on a fully operable 1¾ inch thick **wood** door assembly. What makes both the metal and wood door achievements even more remarkable was that the new STC breakthroughs were achieved without the use of fixed bottom seals.

In a sense, the wood door breakthrough was the result of perhaps another kind of acoustical first. Without the involvement of a sell out, merger or take over, two different door manufacturers (competitors), decided to combine their respective expertise and by doing so were able to obtain an STC level previously thought impossible. Actually it goes even a bit further than that. Both manufacturers had obtained their original respective areas of expertise through the usage of their own acoustical consultant. In other words, this STC breakthrough was the result of **two acoustical consultants** and **two manufacturers** combining their expertise in the development of one fully operable wood door/metal frame assembly, fantastic – it's about time.

Often the RAL staff realizes situations where – if manufacturer (A) would combine with manufacturer (B) greater results would occur. Unfortunately because of other clients with like products or proprietary related reasons we can not suggest same. Thus, when one of the above referenced manufacturers informed me of their joint effort test, and scheduled same, I was most elated. I realized that a possible STC breakthrough could occur and as it turned out, did.

Viewing New STC Window Limits.   Two clients exceeded 50 STC with standard thick windows. Even higher results are expected in the future. New developments are ongoing.

Climbing Up and Over the Wall STC limit dilemma.   As for block walls versus gypsum board walls, typical single plane block walls usually achieve STCs from 50 to 58. The STC was primarily dependent upon the density and type of blocks used. As of today five Riverbank clients have either reached or exceeded 60 STC using gypsum board constructions. Also, each of those clients have raised their previously lower STC rated gypsum walls anywhere from 6–10 classes. The above wall tests also had in common the same construction firm.

Walking Over Those Old STC/IIC Floor Limits. One client achieved a 50 STC and IIC without carpet and pad on a floor construction. Also some recent underlayment tests show improved results on some basic wood joist constructions.

In all of the above cases, the mass of the barrier was important. Each client knew how much the mass contributed to the TL ratings, and each client exceeded the predicted minimum mass law TL values.

Independent Efforts Show Seven Common Factors to Success. Although each client was unaware of the others' accomplishments, I was in the unique position of making comparisons. I found that in achieving success, each client utilized seven common factors. They were as follows: composite testing, unbalanced barrier components, floating barrier components, isolated insulation, airspaces, acoustical/thermal breaks (support members), and resilient/acoustical sealants.

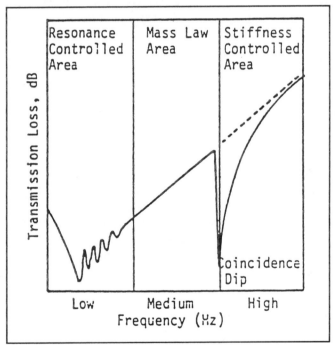

Figure 5-11 Typical TL characteristics of a barrier.

Each factor contributed to raising the TL values of the above referenced noise barriers. Each factor was unique in the sense that it improved certain areas of the TL curve (Figure 5-11). Some factors removed or reduced specific problem areas observed in previous tests. Although each factor showed small increases in STC, when combined major STC increases prevailed.

As far as how specifically each client accomplished the above factors, sorry that's proprietary. However, I will relate to what frequency area of the TL curve each factor addressed.

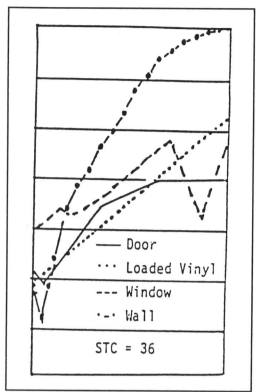

Door

··· Loaded Vinyl

--- Window

·-· Wall

STC = 36

Figure 5-12 Four different barriers, same STC.

1. Composite Testing. Each client conducted a series of preliminary TL tests. Each used their results in conjunction with results from other manufacturers of products considered for use as noise reflectors. Fortunately, each client ignored the misleading single number STC rating and concentrated on the TL values. They all realized that very noticeable differences in a TL curve could exist, and yet the STC could be the same (Figure 5-12). Each conducted tests to find which component improved certain TLs at one or more frequencies. Comparison tests were run on an array of components to establish a pertinent data base. For example, the door manufacturers collected data on their respective components. Those components included frames, jambs, seals, automatic door bottoms, door knobs, locksets, face skins, septum materials, insulation, dampening materials, hinges, and other hardware.

By combining the highest TL components into a composite barrier, each client improved their STC rating by at least two classes. Then, by applying the other six factors, additional increases in STC occurred. Thus, each client made tremendous advances over their original STC rating.

2. Unbalanced Barrier Components. In this case, unbalanced meant the physical differences between each barrier component in the composite barrier. Barrier components were those components that covered the entire specimen area. They consisted of metal, wood, glass, drywall, concrete, ceramic tile, underlayment material, loaded vinyl, plastic, etc. These clients had no two barrier components alike. Sometimes the unbalanced barriers were de-tuned. By de-tuning, the client addressed problems associated with the coincidence dip, diaphragming, flexural wave, and pure tone effects. Unbalancing addressed the lower-mid to high frequencies.

3. Floating Barrier Components. By floating the entire barrier or barrier components, each client obtained improved TL values at the low and lower-mid frequencies. In other words, the door, window, wall, or floor barrier components were isolated from the jamb, frame, structural support components, or box header/joist assembly respectively. Each client accomplished the task of combating low frequency problems with less costly and space saving means than used previously. There were STC increases from 4 to 10 classes. This factor provided equivalent STCs on the same specimens in the field.

4. Isolate Insulation. By mounting their acoustical insulation in a precise manner, each client improved the middle and high TL values significantly. Previously, TL increases due to adding insulation were small

180

when applied to massive type barriers. Also, covering insulation with a thick, impervious skin reduced its sound absorbing qualities considerable. However, the techniques used by these clients isolated both exterior skins form each other. They removed a common homogeneous path, a path in which sound passed from one side of the barrier to the other. Their techniques provided dampening and increased mass/density on one skin. Thus they reduced or removed the coincidence dip. Two clients incorporated another principle. They mounted the insulation in such a way that they created a resilient (mechanical) spring that reduced skin vibrations. This technique was akin to a vibration isolator or automotive shock absorber.

5. Airspace. Airspaces were an effective factor to improve the TL performance of the barriers at all the pertinent test frequencies. The degree of effectiveness depended on the distance from one barrier element to the other. It also depended on the degree of isolation from the support (structural member) components that connected the elements together. With isolated airspaces, the noise had to jump from one barrier component to the other. If each barrier component had a different resonating frequency, it was more difficult for noise to transgress through the barrier. Because more energy was used up radiating through the barrier, there was less to radiate out on the opposite side. Each client incorporated airspaces most effectively.

6. Acoustical/Thermal Breaks (Support Members). Again, removing homogeneous paths (isolating) increased the TL values. Typical supporting membranes were as follows: frames, jambs, edge components, studs, runners, headers, sills, plates, thresholds, joists, sub-girts, and beams. Good acoustical breaks incorporated in support membranes improved the TL at all of the test frequencies. Good engineering was involved. The problem was how to create a structurally sound (no pun intended) barrier with an acoustical break in the components. Again, the clients used different techniques to accomplish the same objective. They isolated one side of the barrier from the other.

7. Resilient/Acoustical Sealants. Besides the obvious significance of sealants for preventing noise leaks, the key word here is resiliency. Each client used resilient (elastic) type acoustical sealants throughout. Resilient sealants were used effectively as insulators, isolators, dampening agents, void fillers, and vibration restrainers. In certain instances the resilient sealants were used to float the barrier or barrier components. All the values improved.

SUMMARY: What is surprising is that although the above seven factors related to higher STC's, some acousticians had difficulty in accepting some of the results because they based their conclusions on past experiences rather than, as I had to tell them, to "Really read our test reports in depth." In some cases a single word was purposely entered into the report just to stress a vital issue. Nevertheless, two acoustical consultants telephoned RAL and in essence both stated – Are you trying to tell me that these gypsum walls that are now being promoted by various wallboard or insulation manufacturers, have essentially the same kind of components and weigh virtually the same as walls tested a few years ago but now are obtaining STC's 6 – 10 classes higher? The answer is YES.

Fortunately both consultants had a copy of a manufacturer's Riverbank report and therefore I wasn't breaking any RAL client proprietary rules by having the consultant read the report back to me. As each consultant read, I would periodically break in and say "Well, what do you think *that* sentence means?" It didn't take each consultant very long to fully comprehend the significance of what was stated and relate to the above seven factors. Each realized that the manufacturers had indeed come up with sound barrier improvements that increased their respective STC's. What is also good news, is that one of the consultants was in a situation where he had to duplicate the higher STC results in the field, and did.

Another issue is field tests vs lab tests. Many believe that you will always obtain anywhere from two to five STC's lower on a field test then what the lab test achieves. Well, I can definitely say that within the past decade, numerous field tests have provided the same or within one STC of the lab results. Of course, that will occur only when all the possible flanking paths in the field are accounted for and most importantly the barrier in question is constructed or installed the same as it was in the lab.

In any case, fellow acousticians agree that the best method for noise control is to eliminate the source. If source elimination is not possible, one must combat the noise by whatever means works. The means available are to either reflect, absorb, insulate, isolate, dampen, seal, cover, or restrain the noise. The Riverbank clients referenced above incorporated each of those means within the seven common factors and thereby were able to achieve breakthroughs in acoustical barrier performance. In doing so, those clients advanced the science of Architectural Acoustics.

CONCLUSION: Riverbank clients made a mockery of what the dead science advocates predicted by proving that the science of architectural acoustics is alive and growing. Those clients as well as others, are developing new and better products for noise control. Not referenced above are recent advances in office screens, theater seats, folding walls/partitions, and absorption products. Many of the advances have led to many new or revised theories in architectural acoustics. If some present tests continue as they are, other theoretical revisions will soon be in order. Which perhaps explains why some large acoustical oriented constructions resulting from computer modeling based on old theories, and data, have thus far proven ineffective.

In any respect, from all indications, the decade leading into the next century will be of major importance to acousticians worldwide. The demand for more quiet will be at an all time peak. Already various national and international government agencies, standard organizations, and trade associations are scrambling to unite the world's acoustical community. Riverbank is already involved in these unification efforts. Undoubtedly various INCE, ASA, ASTM and NCAC members will participate in this much needed endeavor.[21]

Thanks to John for an excellent assessment of sound barrier design! I concur.

5.8 SOUND BARRIER DESIGNS

In addition to the nonproprietary information supplied by John Kopec there are a number of myths, misconceptions, and mysteries about specific sound barrier designs. A few of the more interesting and well-known stories follow.

Myth: sound-deadening board works no matter how it is used. Not necessarily so!
The term sound-deadening board refers to a family of products made from wood and mineral fiber into sheets or boards nominally 4 ft by 8 to 10 ft. Thickness is usually ½ to ¾-in. The process uses a Fordernier

---

[21] From "Pushing the limits of Acoustical Barrier Performance," ASTM Standards Technology Training – Acoustics and Noise Control Course Notes, November 5, 1993, Author; John W. Kopec, director, Riverbank Acoustical Laboratories, Geneva, Ill..

or Oliver machine similar to low-quality paper machines. The early form of the product was a wood fiber board coated with asphalt and used extensively for exterior sheathing. Later forms were made from mineral fibers to achieve incombustible ratings. The boards are flexible and soft, with low strength.

In the 1960s the Insulation Board Institute (IBI) began searching for new product uses for wood fiber insulation board. Someone must have said "Hey, that looks like it has acoustic properties." Since the base board had been used for making wood fiber ceiling tile, this was a logical conclusion. Realizing that partitions have better sound transmission loss characteristics when a sound-absorbing material is included in the core, the industry scrambled to prove the theory. Hundreds of tests by manufacturers and IBI ensued.

Researchers quickly found that sound-deadening board provided significant STC improvement only when the partition had staggered studs or a resilient mounting. It worked well on steel studs since they are inherently resilient. However, **merely nailing sound-deadening board as an underlayment to gypsum wallboard was not acoustically worth the effort.**

Additional research indicated that a resilient effect could be created by adhering the outer layer of gypsum wallboard to the sound deadening board by laminating with strips of joint compound. Tests demonstrated that the location, size, and character of the lamination strips were critical to the improvement of STC. For example, the best STC values were obtained when the adhesive was still wet – obviously not acceptable in practice, and when the adhesive was located midway between the studs. The effect was to make the board act as a "leaf spring" and thereby provide the resilient mounting needed.

Unfortunately, marketing efforts were so effective with this magical new product called "sound-deadening board" that the contractors thought it would solve all their acoustical problems. Advertising failed to emphasis the "small print" about specialized construction details. Since contractors did not want to hear all of the conditions, many buildings were constructed with sound-deadening board nailed in place just like sheathing. STC values were totally lost. Manufacturers did not attempt to correct the wrongs in spite of warnings by their research engineers because it could mean lost business opportunities. This shortsighted and misleading marketing approach was promulgated through efforts by the IBI which inundated the industry with "technical literature" and articles in construction trade journals.

Eventually, wood fiber sound-deadening board got a bad name – not for the acoustical failings, but because of several major fires. It seems the board "punked." A plumber's torch was usually the ignition source, and the board smoldered for days before full blown combustion occurred. While mineral fiber board did not punk, it lost market share when it was found that the same lamination techniques worked with ¼-in. gypsum wallboard.[22]

From a technical standpoint, the evolution and demise of sound-deadening board provided a wealth of information that prompted huge gains in improving the STC of partitions and floor/ceiling systems. It also brought about improvements in the testing procedures. For example, ASTM E90 now has an entire section devoted to the proper description of specimens including cure times for adhesives, plaster, and like materials. One also hopes that marketeers have learned the lessons of overzealous sales literature, claims, etc. Unfortunately, buyouts, takeovers, and downsizing have removed a huge portion of the building research budgets. Acoustical research is near a standstill in major building materials manufacturers' research

---

[22] The author invented a gypsum wallboard version of sound deadening board that provides the same STC, is much cheaper, is easier to cut and handle, is incombustible and has a 1-hour time-design fire rating. Test data on wood fiber, mineral fiber, and gypsum wallboard sound-deadening board products are listed in the partitions test data portion of Appendix 3.

departments. As a result, little new information is forthcoming and technical literature is woefully lacking.[23]

Myth: resilient furring channels are all alike. NOT SO! The venerable resilient furring channel is one of the most successful acoustical products invented. First introduced to the market-place in the early 1960s by USG, Donn, and National Gypsum (Gold Bond), resilient furring channels have not changed in design for over 30 years. They still enjoy nearly the same sales volume they did in the 1970s. Why has the channel been so successful? Because it is simple to understand and use and is reliable. Wall systems with resilient channels were the first to gain the same STC in the field and laboratory.

So what is the acoustical difference between resilient furring channels? For openers, the designs are quite different (Figure 5-13). The original USG RC-1 channel, still offered today, has a single leg with oblong cutouts in the leaf leg and is made of 25-gauge electrogalvanized steel. The Gold Bond resilient furring channel is formed into a "hat" section with two equal legs formed of expanded metal. It also is 25-gauge galvanized steel and has not changed in nearly 30 years. The Donn channel is one-legged with long slits in the spring/leaf leg. It too was 25-gauge galvanized steel. Recent copies of the RC-1 Channel have no cutouts and/or are made from heavier gage metal. The Donn product was withdrawn from the market approximately 10 years ago.

Acoustical performance of the three original designs is very similar. If one was better, the nod went slightly to the Donn design since it had more flexibility.[24] No other resilient channel on the market can measure up to the acoustical performance of the original designs. Most have no

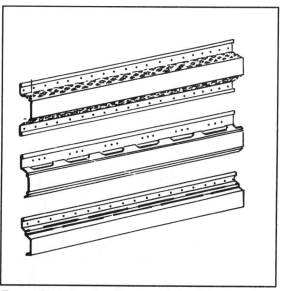

Figure 5-13 Resilient furring channel designs that are acoustically efficient.

acoustical merit at all other than a copycat name. All are the result of rip-off designs and none provide any acoustical value. When challenged by a knowledgeable acoustician, the rip-off suppliers provide bogus test reports or no test data to support their claims. It is a wonder why these suppliers of copycat channels have not been sued for at least false advertising or more serious offenses.

Why don't the copycat channels work? There are many reasons. First, designs for the original resilient furring channels were the result of considerable research and testing in both the laboratory and field. A key to their success is the resiliency or flex characteristics. In the RC-1 channel, the length, shape, and size of

---

[23] The issue of marketing and sales impacts on acoustical materials and systems is covered in a later section in this chapter.

[24] The flexibility of the Donn channel probably lead to its demise since the resulting wall or ceiling sagged in use.

the slots in the web sections are critical to acoustical performance along with the selection of 25-gauge material. Attachment of the channel to the stud or joist must be where the nail holes occur, otherwise the flex or spring action will not perform as tested. Similar, testing went into the design of the Gold Bond channel. In copycat channels, flex action is far too stiff due to heavier gauge steel.

How can a resilient channel system be acoustically negated? The most common form of problem is screws that are too long. If the screw is more than ⅜-in. longer than the gypsum wallboard thickness, there is a risk that the screw will short-circuit the channel. Short circuiting happens when a screw that attaches the gypsum wallboard also penetrates the stud, joist, or framing. Channel spacing is also critical. Most tests were conducted with the channels spaced 24-in. on-center (OC) and laid perpendicular to the stud or joist. Any other spacing or attachment than that shown in the test report will negate the acoustical performance.[25]

5.9 FLOOR/CEILING SYSTEMS AND DESIGN

A floor/ceiling system has two major acoustical attributes that must be considered together in selecting the appropriate design for your project. Not only must the floor/ceiling block airborne noise, just like the partition, but it must also control impact noise.

Control of airborne sound transmission through a floor/ceiling system is essentially the same process as for a partition assembly. As described above, the object is to seal the cracks, make the assembly as massive as practical, separate the floor side and ceiling side with resilient mounts, add a sound-absorbing material to the core, and employ techniques similar to those discussed earlier. About the only difference in designing a good STC for a floor/ceiling assembly is the fact that the floor must also serve its prime function, namely, too support the loads of the floor. We acousticians tend to spend little time worrying about the STC and airborne sound transmission for floor/ceilings simply because they are inherently good at the onset. This is due to the massive construction required to support floor loads. Ratings for STCs on floor/ceilings typically start at 45 STC for even a simple floor/ceiling. Sophisticated systems will require resilient channels, two plies of gypsum wallboard, and a thick blanket of sound/thermal insulation. Ratings of 60 STC are common and satisfy most criteria.

By contrast, impact noise becomes a vital concern to the acoustician, particularly in residential construction. In North America, the most common construction techniques utilize wood joists, plywood subfloor, and gypsum wallboard ceilings. This lightweight system sounds and acts much like a drum head. Impact noise, basically from footfall sounds, is easily transmitted to the adjoining apartment or living space below. In Europe, other developed countries, and in typical commercial construction, floors are masonry and massive by comparison. Massive floors such as a poured concrete floor are relatively good at containing impact/footfall sounds. One other major item is working in our favor. Carpet and pad has become an industry standard for residential construction in North America. When used, carpet and pad will easily reduce impact sounds - at the source - to the point where footfall sounds are not intrusive to neighbors. It is only when hard-

---

[25] Note of caution: Only Gold Bond is still making and marketing the original Gold Bond resilient furring channel. USG sold its interest in manufacturing the RC-1 channel, thereby leaving the door open for look- alike products that are even named RC-1. Be sure you have the real thing before utilizing. Apparently the major manufacturers have abdicated their role as industry policeman thereby encouraging copycat rip-off products. See Section 7.4.

surface floors such as vinyl flooring, ceramic tile, and hardwood are exposed to impacts that impact noise becomes a major concern. With the trend toward more use of hard-surface floor materials by designers, the acoustical design community has been burdened with a noisy problem. No longer is carpet and pad always acceptable to designers, builders, owners, and building occupants.

Misconception: wood floor systems are always noisy. New developments have identified an interesting product using constrained-layer damping technology developed for metals used in industrial noise control. The author has been fortunate to participate in this development.

Present solutions require top-quality carpet and pad and/or expensive and massive floor systems. A floor system has been developed that allows the use of hard-surface flooring materials such as vinyl, hardwood, or ceramic tile to be applied over a plywood subfloor. The key component is a plywood subfloor material having a viscoelastic core. Utilizing the principles of constrained-layer damping, the floor is quiet, only moderately more expensive than conventional plywood, and requires no special installation techniques. Damped plywood is ideal for quieting residential and commercial buildings including single- and multifamily dwellings, stores, offices, schools, clinics, and modular buildings. Initial evaluations on modular commercial structures show low noise levels in the occupied space. Recent tests of wood floor/ceiling systems for both airborne and impact noise show excellent results. A paper that presents several versions of the damped plywood applied to lightweight wood and steel joist floor systems is presented in Section 7.5. This paper describes the situation, test development, and solution.

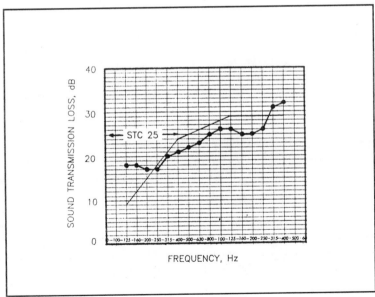

Figure 5-14 Example of ASTM E413 STC reference contour fitted to transmission loss data (STC 25).

5.10 MYSTERY OF THE STC TRAP

5.10.1 Background:
The construction industry dutifully adopted sound transmission class , as determined by ASTM E413 <u>Classification for Rating Sound Insulation,</u>[26] as a universal single-number rating system. STC is the single-number rating of choice for identifying the sound transmission loss characteristics of a multitude of sound barriers including walls, part high barriers, floor/ceiling systems, exterior facades, sound enclosures, roof/ceilings, windows, doors, other building components, and composite barriers containing one or more individual components. STC is applied to both laboratory measurements, and adaptations are used for field data. ASTM E413 may be used calculate the sound transmission class (STC) (Figure 5-14), field STC (FSTC), noise isolation class (NIC) and normalized NIC (NNIC) for barriers tested per ASTM procedures E90, E336, E596,and the new ASTM ceiling attenuation procedure. In spite of the many cautions to the contrary by acousticians, and the ASTM documents themselves, thousands of construction systems are rated by STC. Most manufacturers now only provide the STC for their barriers.

5.10.2 The Trap
As with any single-number rating system, there will always be exceptions. We all learned, some of us the hard way, that averages and single-number rating systems such as STC create many problems. As an example, consider the poor guy who was told the stream he was about to cross averaged only 1 foot deep. Imagine his surprise when he fell in a hole 12 ft deep and nearly drowned. To the uninitiated, STC appears to be an easy -to-use technique for selecting the barrier of choice for buildings. This nice single-number rating scale provides designers, architects, builders, and estimators with a simple system to rate the acoustical barrier attributes. The higher the STC the better. Considering the other issues for selecting a barrier such as code acceptance, structural integrity, fire resistance, material availability, and cost, STC is indeed a welcome simplification by the design community. Fortunately, it works – most of the time. **However,** acoustical testing has identified several types of barriers having STC ratings that are much lower than they deserve when one considers the sound transmission loss values over the entire test frequency spectrum between 125 and 5000 Hz.

5.10.3 Which Barrier Would You Choose?
Two 50 STC barriers are shown in Figure 5-15. While they are far different in their sound attenuation characteristics, both have the same STC rating. Barrier A has excellent sound transmission loss characteristics at all frequencies except 125 Hz.   The STC is established by the 8-dB rule. ASTM E413, paragraph 5.4.1, states that "The maximum deficiency at any one frequency does not exceed 8 dB." By contrast, barrier B has sound transmission loss characteristics that closely follow the standard STC curve over the entire frequency spectrum. The STC 50 for barrier B is governed by the 32-dB rule. ASTM E413, paragraph 5.4.1 states that "The sum of the deficiencies is less than or equal to 32 dB." If you were selecting the barrier and all other conditions were acceptable, which barrier would you choose? A prudent person would select barrier A.

---

[26] E413 and the following listed documents are found in American Society of Testing and Materials <u>Annual Book of Standards</u>, Volume 4.06. A description of these standards may be found in Section 4.6.

Barrier A provides excellent sound attenuation at all frequencies that are important to the reduction of speech and normal airborne sounds between a source and receiver in apartments, condominiums, offices, and hotels. That is precisely the situation that STC was designed to address. To emphasize this point, ASTM E413, paragraph 4.2, states "These single-number ratings correlate in a general way with the subjective impressions of sound transmission for speech, radio, television, and similar sources of noise in offices and buildings. This classification method is not appropriate for sound sources with spectra significantly different from those sources listed above. Such sources include machinery, industrial processes, bowling alleys, power transformers, musical instruments, many music systems and transportation noises such as motor vehicles, aircraft and trains. For these sources, **accurate assessment of sound transmission requires a detailed analysis in frequency bands**." (Bold provided by author.) In other words, STC is appropriate only as an initial screening device. Final selection of the barrier should be based on analyzing the entire frequency spectrum and comparing it with the anticipated type of noise source.

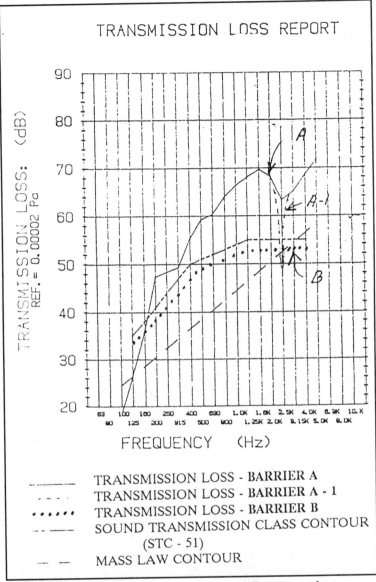

TRANSMISSION LOSS - BARRIER A
TRANSMISSION LOSS - BARRIER A - 1
TRANSMISSION LOSS - BARRIER B
SOUND TRANSMISSION CLASS CONTOUR
(STC - 51)
MASS LAW CONTOUR

Figure 5-15 Two STC 51 barriers can have different sound transmission loss curves. Barrier A is limited by the 8-dB rule at 125 Hz. Barrier B is limited by the 32-dB rule.

In most instances this will be speech and sounds predominantly centered in the 500- to 4000-Hz frequency range. Barrier A would be ideal for this situation. If the major sound source contained primarily sound at 125 Hz, then barrier B might be the preferred barrier. Since STC was originally conceived, and is still intended, to rate the effectiveness of a barrier in the speech frequency range, it should not be used to rate low-frequency conditions. Barrier A is discriminated against because of the arbitrary 8-dB rule.

5.10.4 Other Circumstances?

Yes, other acousticians have argued for many years that the 8-dB rule is not needed and arbitrarily penalizes certain types of barriers unnecessarily. Consider the case of a barrier having a curve similar to barrier A. The dotted line (shown as A-1 in Figure 5-15) indicates a deep spike at the one-third octave frequency band of 3150 Hz. It has been demonstrated by others that this spike will not be subjectively adverse and will not be likely to be noticed by the listener. Considerable data exists to support this contention. In the case of the 125-Hz deficiency of barrier A, some will argue that we are encountering more low-frequency noise in our environment, therefore the 8-dB rule provides a "hurdle." There are two faults in this argument. First, the acoustical community has not yet found ways to accurately measure low-frequency noise with enough consistency to be sure of the values. Secondly, the STC technique carefully points out that it was not intended for this purpose. If we are to retain the 8-dB rule, we must then rewrite the E413 parameters.

5.10.5 What Is Proposed?

Those who have a vote on ASTM Committee E33 are requested to consider the case presented and express a no vote on ballots for E413 and recommend dropping paragraph 5.4.2. Those who are in the position of specifying barriers are cautioned that they should review the entire frequency spectrum of barriers with similar STC numbers. If the decision is based solely on STC, the specifiers should be forewarned that they may be ruling out perfectly acceptable and even better sound barriers.

To those who will say "We always did it this way," be aware that you are standing in the way of progress. You are automatically discriminating against new and improved systems of construction. In many ways designing and constructing the "built environment" is like the auto industry. Quoting Lee Iacocca, "In this business you must lead, follow or get out of the way." Let's remove an arbitrary obstacle by deleting the "8-dB rule" from ASTM E413 and evaluate barriers on their performance over the entire frequency range.

Chapter 6 ENVIRONMENTAL NOISE CONTROL

6.1 INTRODUCTION AND GENERAL CONSIDERATIONS

In recent years, we have become more acutely aware of the quality of the environment in which we live. Noise pollution is one factor affecting the environmental quality to which we are all exposed. As a result, government regulations have been instituted for the purpose of controlling the ever-increasing noise pollution problem. Noise is clearly a major pollutant in our technology-driven society. While it may not kill you outright, like a chemical spill, it has been proven to be a significant health hazard.[1]

A key thought to keep in mind whenever considering environmental noise is that noise is always defined as unwanted sound. What may be unwanted to one individual may be highly desirable to another. And that role can reverse depending on specific circumstances. An example is an outdoor rock concert. If you are an "aficionado" or "rock hound," you may love the performance while in the audience, but may find that sound totally obnoxious and unwanted if you are a nearby resident trying to sleep or have "quiet time."

The extent of noise pollution in our environment varies widely. The degree of noise at a particular location depends on the density of population and the proximity of noisy equipment, highways, railroads, air traffic, and industry. Individual noise sources have a noise level of their own. Each noise-producing element has its own noise signature of frequency spectrum, loudness, time duration, and aesthetic impact. For example, the squeal of a rail car around a curve will produce a sound frequency near or even above the upper level of normal hearing. By contrast, the low-frequency rumble of the heavy train may produce frequencies below normal hearing and impact the body like an earthquake. Most of the sounds we hear fall within the normal hearing range of 50 to 20,000 Hz. Loudness may range from the rustle of leaves to the roar of a jet airplane. Time duration may be the split second crack of a bullet or bomb or the continuous roar of a generator. Aesthetic impact may be the gentle sounds of an orchestra playing soothing music to the ruckus of a major confrontation. All of these elements must be evaluated when considering environmental noise. In most instances the ear of the beholder determines when a sound is an environmental hazard. Criteria for mitigation of a noise hazard are typically geared to noise that is obnoxious and unwanted by most of the potential recipients or clearly will create a health hazard.

Federal, state, and local environmental noise criteria and standards have been instituted to deal with exterior noise exposure.[2] Criteria have been established for a multitude of noise situations. Noise requirements may be found in land use planning documents, building codes, noise ordinances, and municipal codes. Land use planning documents generally require a notice of mitigated determination of nonsignificance and/or an Environmental Impact Statement (EIS). The section in an EIS that deals with noise is generally

---

[1]   See Section 7.10 in which panelists review the allegations that acoustical materials are a hazard to health; noisy workplaces cause hearing loss, stress, and loss of life; and poor workplace acoustics cause significant loss of productivity in both the office and factory. The panel consisted of experts from ASTM Committee E33 – Environmental Acoustics, manufacturers of acoustical materials, and concerned users.

[2]   See Chapter 4 for detailed descriptions and examples.

known as the "Noise Element."[3] Building codes mostly deal with construction practices required for noise control, depending on the existing or predicted overall community noise levels. Building codes may also contain construction equipment noise criteria. Federal, state, city, and community noise ordinances all contain specific noise criteria. Some may be a simple nuisance ordinance while others may be comprehensive.[4]

6.2 COMMUNITY NOISE DESCRIPTORS

A plethora of community noise descriptors have evolved in recent years. Each was conceived to describe a particular noise signature. The following noise descriptors may be found in typical community noise ordinances or regulations for noise criteria:

A-Weighted Sound Level (dB$_A$)

$L_a = 10 \log (P_{a2}/P_{o2})$ dB$_A$

L_a in dB$_A$ is the most common single-number rating system for measuring the loudness of a noise. It may be read directly on most sound level meters by selecting the designated scale. Most meters make the conversion automatically. It is obtained by applying the A-weighted frequency response of a sound level meter to the measured noise, where P_a is the A-weighted sound pressure and P_o is the reference sound pressure of $20P_a$. This quantity is easily measurable with a simple sound level meter and nearly always appears on sophisticated sound level meters. L_a in dB$_A$ is often used as a basis of noise regulations and criteria.[5] ASTM Procedure E1014, Guide for Measurement of Outdoor A-Weighted Sound Levels, provides guidelines for measurement of dB$_A$ sound levels outdoors at specified locations along particular site boundaries, using a general-purpose sound level meter.[6]

Equivalent Sound Level (L_{eq})

$L_{eq} = 10 \log 10^{La/10}$ dB

Also known as the equivalent-continuous sound level or average sound level, this descriptor is based upon the measured time-varying A-weighted sound level L_a for a given period of time. It is calculated to have a constant value which is equal to the mean sound energy of the measured time-varying sound. In practice L_{eq} is obtained by measuring the A-weighted sound level at short time periods over a specific time frame using a sound level meter that displays L_{eq}. While the value is given in units of dB it should always state the time

---

[3] See Chapter 4 for an example of an EIS "Noise Element."

[4] See Chapter 4 for detailed explanations of federal, state and local noise ordinances including examples of a village noise ordinance and an extensive noise ordinance excerpted from the City of Los Angeles Municipal Code.

[5] See Figures 1-3, 1-6, and 1-7, the relation of common sounds to various rating scales. See Appendix 2 for procedures to add decibels and determine dB$_A$. See Chapter 4 for regulations using dB$_A$

[6] ASTM designation E1014 may be found in the ASTM Annual Book of Standards, Section 4, Volume 4.06, Thermal Insulation; Environmental Acoustics.

period. It is a technique to indicate the distribution of overall sound levels in a community during the measurement period. Wide swings of sound levels and impact sounds will produce values that are misleading.

Day-Night Average Sound Level (L_{dn})

$L_{dn} = L_{eq}$ over a 24-hour period, with a 10-dB nighttime penalty (10:00 p.m. to 7:00 a.m.)

The day-night sound level is based upon the measured time-varying A-weighted sound level and is simply a modification to the L_{eq} whereby the time period is defined as 24 hours and a weighting factor of 10 is applied to the measured nighttime noise levels between 10:00 p.m. and 7:00 a.m. to account for increased community sensitivity to noise during those hours. This quantity is presently in use by the U. S. Department of Housing and Urban Development (HUD) for environmental noise regulations. Developed by the U. S. Environmental Protection Agency (EPA), L_{dn} is the "standard metric for determining the cumulative exposure of individuals to noise" and has been adopted by the Federal Aviation Administration (FAA) and the U. S. Air Force to describe noise near airports.[7]

Community Noise Equivalent Level (CNEL)

CNEL = L_{dn} with an additional evening penalty factor of 3

The CNEL is based upon the measured time varying A-weighted sound level and is simply a modification to the L_{dn} in that an additional weighting factor of 3 is applied to the measured evening noise levels between 7:00 p.m. and 10:00 p.m. to account for increased community sensitivity to noise during those hours. CNEL is presently in use by the state of California, among others, for environmental noise regulations. For example the city of Palmdale, Calif., requires the CNEL not exceed 65 dB$_A$ outside and 45 dB$_A$ inside for all new single-family dwellings.[8] Mitigation measures are required if predicted levels do not meet the criteria. To satisfy outside levels, increased distance and sound walls may be employed. Inside levels may be met by utilizing exterior envelope construction with sufficient sound transmission loss characteristics.

Perceived Noise Level (L_{pn})

The perceived noise level is a measure of the "noisiness" of a sound (e.g., as from an aircraft). It is calculated from measured octave or one-third octave band sound level data and is based on standardized properties of the human hearing sensitivity as determined by psychoacoustic methods. For typical noise spectra, $L_{pn} = L_a + 13$ dB.

Effective Perceived Noise Level (L_{epn})

$L_{epn} = L_{pn} + C + D$

The effective perceived noise level is a modification of L_{pn} that allows for adjustments for the presence of a prominent pure tone (C), and which takes into account the event duration (D).

---

[7] Day-night average sound level contours have been generated for most major airports in North America. Copies are usually available from the airport administrative offices which usually have one or more persons specifically assigned responsibility for noise assessment and control. An example of an FAR Part 150 Noise Exposure Map is contained in Section 7.9.

[8] See Section 7.6.

Composite Noise Rating (aircraft) (CNR)

$CNR = L_{epn\,(max)} + N + K$

The composite noise rating is used exclusively for aircraft noise and is a modification of L_{pn} that allows for adjustments based upon the number of events (N) and time of day (K). The CNR was originally adopted by the FAA to describe noise in the vicinity of airports and therefore, CNR contours are available for many airports. CNR is being replaced by NEF.

Noise Exposure Forecast (aircraft) (NEF)

$NEF = L_{epn} + N + K$

The noise exposure forecast is used exclusively for aircraft noise and is a modification of L_{epn} that allows for adjustments based upon the number of events (N) and time of day (K). Older CNR contours have mostly been converted into NEF, the preferred aircraft noise descriptor.

Since the L_{dn}, CNEL, CNR, and NEF all relate to measurements of sound energy, and all incorporate similar day-night weighting schemes, the following approximations can usually be made for comparison purposes: L_{dn} = CNEL = NEF minus 35 = CNR minus 35.

The intent for development of these community noise descriptors was to provide a method for the prediction of community response to environmental noise based upon measurable quantities.

6.3 ENVIRONMENTAL NOISE STANDARDS

6.3.1 Federal Noise Regulations

On the federal level, noise regulations have been written and are enforced by:[9]

- Environmental Protection Agency (EPA)
- Department of Transportation (DOT) including;
    - Federal Aviation Administration (FAA) [Federal Aviation Regulations (FAR)]
    - Federal Highway Administration (FHWA)
    - Federal Rail Administration (FRA)
- Air Force Regulations
- Department of Labor - Occupational and Safety Health Administration (OSHA)
- Department of Health, Education, and Welfare (HEW)
    - Housing and Urban Development (HUD)
    (HUD Circular 1390.2, Part 51 provides minimum national standards applicable to HUD programs to protect citizens from excessive noise in their communities and residences.)
    - Federal Housing Authority (FHA)
    (FHA Minimum Property Standards establish minimum acceptable noise criteria and construction practices within and between living units. All construction guaranteed by an FHA loan must comply. Veterans Administration Minimum Property Standards are similar to FHA.)

---

[9] Noise control criteria for each agency is presented and discussed in Chapter 4.

6.3.2 State Noise Regulations

At the state level, noise criteria are typically contained in state building codes and land use regulations. Most states have adopted one of the following model building codes:[10]

■   National Building Code (NBC) [administered by Building Officials and Code Admin. Int., Inc. (BOCA)]

■   Standard Building Code (SBC) (administered by the Southern Building Code Congress International)

■   Uniform Building Code (UBC) [administered by the International Congress of Building Officials (ICBO)]

Each state has adopted some or all of one of the model codes and has altered, added, or deleted sections to meet its needs.   Some have no noise criteria, while several states, most notably California and Washington,[11] have adopted extensive noise control regulations.

California Noise Insulation Standard, Title 25, requires noise insulation from exterior sources to ensure that the interior intrusion does not exceed CNEL 45 dB. Residential areas exposed to exterior noise levels in excess of CNEL 60 dB require an acoustical analysis showing that the structure has been designed to limit noise intrusion to below the allowable limit.

Washington State Revised Code, Chapters 70 and 173, requires that the Environmental Designation for Noise Abatement (EDNA) not exceed the required level anywhere on the receiving property. Values of EDNA shall not exceed the following:

Class A (residential area)	$\leq$ 55 dB$_A$
Class B (commercial area)	$\leq$ 57 dB$_A$
Class C (industrial area)	$\leq$ 60 dB$_A$

Notes: Class A (residential) limits shall be reduced by 10 dB$_A$ between the hours of 10:00 p.m. and 7:00 a.m. At any hour of the day or night the applicable noise limitations may be exceeded for any receiving property by no more than: 5 dB$_A$ for a total of 15 minutes in any one-hour period; or 10 dB$_A$ for a total of 5 minutes in any one-hour period; or 15 dB$_A$ for a total of 1.5 minutes in any 1-hour period.

6.3.3 Local Noise Regulations

On a local basis, noise regulations may be part of the county or city ordinances or codes. Each jurisdiction has chosen the location for noise requirements that best fits its format. Generally the noise requirements are established by the agency Planning Department. However, many may be found in the building code while others are under the jurisdiction of the police or sheriff. Where the municipality is well organized, guidelines for environmental noise compatible land used will have a "Noise Element" as part of the general plan. Where an Environmental Impact Statement is required, there should be a "Noise Element" included in the EIS that spells out existing noise levels, predicted noise, and mitigation measures that are required for compliance.

---

[10] See Chapter 4 for detailed listing and related information.

[11] See Chapter 4 for noise provisions of the Washington State Revised Code, Chapters 70.107 and 173-60.

6.3.4 Which Regulations Apply?

It should become clear that all of the foregoing noise regulations may apply to a particular project. Unfortunately, it is not possible for a lay person to determine which one will be required for a specific project. Suffice to say, it behooves the owner and developer of a property to review all the federal, state, county, city and municipality noise regulations that may apply to the project. Listing the criteria for each is a time-consuming and complex job. In the end, of all the noise criteria that may apply, you the owner and builder must satisfy the most stringent criteria of all the performance requirements. To ignore this apparent bureaucratic nightmare can be very expensive. It behooves an owner/developer to acquire the expertise of a local acoustical consultant. If you do not satisfy the requirements, the expense for reconstruction may be astronomical.

6.4 METHODOLOGY FOR MEETING CRITERIA

6.4.1 Establishing the Noise Source

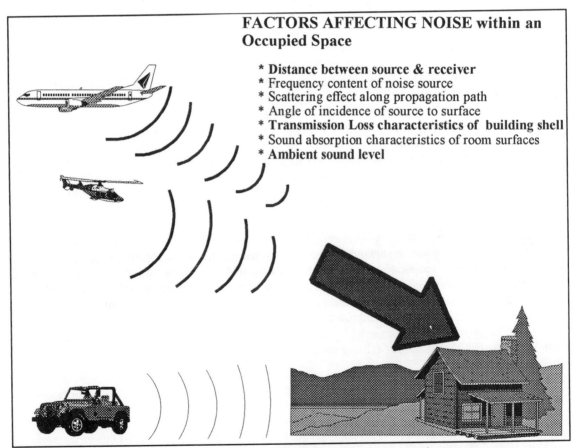

FACTORS AFFECTING NOISE within an Occupied Space

* **Distance between source & receiver**
* Frequency content of noise source
* Scattering effect along propagation path
* Angle of incidence of source to surface
* **Transmission Loss characteristics of building shell**
* Sound absorption characteristics of room surfaces
* **Ambient sound level**

Figure 6-1 Factors affecting noise level in an occupied space from an external source.

The noise insulation provided by a building shell is dependent on the characteristics of the noise source, including loudness, frequency, duration, and angle of incidence, the transmission characteristics of the structure; and the sound absorption characteristics within the receiving room (See Figure 6-1). For this reason, the first step in applying the pertinent noise regulations is to establish the sound source characteristics.

First, identify the source by measuring the noise output level in dB, frequency content (e.g., loudness at one third octave intervals between 50 and 20,000 Hz), duration of sound (e.g., a specific time period or an average loudness such as L_{dn} or CNEL), angle of incidence (e.g., grazing, random or focused between 30 and 60°), line or point source (e.g., a moving vehicle or a fixed source.)

Source identification can best be accomplished by performing a site survey in person, if the site is easily accessible. The next best choice, particularly if the site is not local, is to obtain a site plan which includes adjacent streets and site orientation, plus a local area map. Given this information, it is then possible to ascertain the probable predominant noise sources, whether they be aircraft from nearby airports, vehicle traffic on nearby highways or railways, or local roadway traffic. Less obvious will be stationary sources such as adjacent penthouse mechanical equipment or nearby industrial process equipment. If there is indication that industrial sources exist, a site survey is advisable.

Noise output data acquisition can be accomplished in either of two ways. The most convenient method is to request data from the proper authority.

For aircraft, request local airport authority or regional FAA office to provide CNR, NEF, L_{dn}, or CNEL contour plots. Every major airport in North America will have this information. Unfortunately, some may stonewall requests for noise information and yet others may provide out-of-date information. Be persistent, this information is a matter of public record.

For highway & railway, request the local Department of Transportation (DOT) office to provide L_{dn}, $L\%$ (A-weighted noise levels exceeded a given percentage of time), L_{eq}, or CNEL contour plots. To double-check the values, request traffic count data which can be used to calculate the noise exposure.[12] State and Federal highway officials generally have this data for all interstate and local freeways. Finding the party that has the information may be a daunting task. Again, persistence will eventually prevail. Railroad officials likewise are required to have the information. However, obtaining the data may be even more daunting since there are meager public funds supporting the rail system.

For local roadways, request local authorities to provide L_{dn} or CNEL contour plots. They are usually found in the community noise element of the general plan along with local DOT traffic count data. In other communities, this information is generally not available unless an Environmental Impact Study has been conducted. A properly conducted EIS must have a section titled "Noise Element" that contains present and predicted noise types, levels, and character along with noise mitigation requirements.

The second method of data acquisition is by direct measurement which entails an on-site noise survey., especially if local roadway traffic is the primary noise source, or if stationary sources are involved.[13]

---

[12] Many papers have been written on this subject and appear in acoustical journals such as Journal of the Acoustical Society of America (ASA), Sound and Vibration, and acoustical texts.

[13] ASTM E1014, Guide for Measurement of Outdoor A-Weighted Sound Levels, and E1503, Standard Test Method of Conducting Outdoor Sound Measurements Using a Digital Statistical Analysis System (ASTM Standards Volume 4.06), provide detailed techniques for measuring community noise levels.

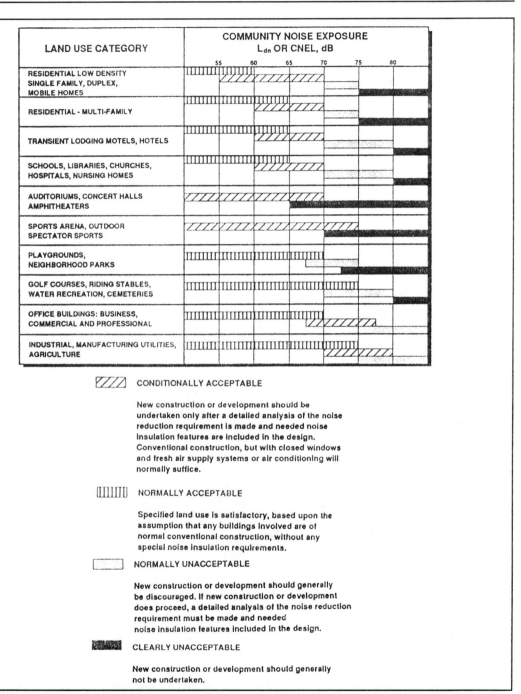

Figure 6-2 Typical community land use compatibility.

198

If the foregoing information is not available and it is not economically feasible to conduct an acoustical analysis of the site, an estimate may be better than nothing. Figure 6-2 provides a listing of typical community noise exposures for various land use categories. Note that levels considered conditionally acceptable will not satisfy everyone. Normally acceptable ranges may satisfy the majority of the sound receivers, normally unacceptable ranges will prompt considerable public outrage, and clearly unacceptable values are usually considered a health hazard. This chart generally reflects the rationale used by government in selecting noise criteria.

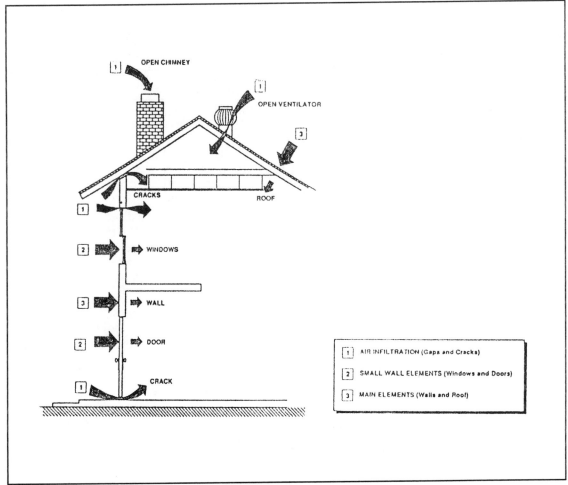

Figure 6-3 Typical noise intrusion paths of a building envelope.

6.4.2 Determining Required Building Shell Attenuation

The noise attenuation (also termed <u>noise reduction</u>) characteristics of an acoustical envelope or building shell are dependent on the collective sound transmission loss characteristics of the wall, roof, doors, windows, and any penetrations such as vents. To determine the acoustical criteria required for the building envelope, one must also establish the acceptable sound level inside the building and, of course, the exterior sound level. In simple terms we can say that the noise reduction (NR) is equal to the exterior sound level minus the interior sound level (that is, NR = L_{outside} minus L_{inside}).

Using a typical inside sound level of 45 dB_A, as required by HUD and many building codes, and an outside level of 80 dB_A, we can readily see that the exterior envelope must have a noise reduction of 35 dB_A. If the building is considered a luxury-type residence or if the receiving room is a bedroom, it may be best to select an inside level of 35 dB_A or lower. With the same outside sound level, the building envelope must then demonstrate an NR $\geq$ 45. As the exterior level increases, the NR of the building envelope must increase by a like amount. A more detailed analysis of a typical building facade may be derived by using the following expression:

$$L_{\text{pint}} = L_{\text{pext}} - TL + 10 \log (S/A) + ADJ$$

Where

L_{pint} is the design criterion level, or predicted average sound pressure level within the test room.

L_{pext} is the sound pressure level measured or predicted at proposed location of building facade.

S is the total exposed exterior surface area of the proposed room of interest.

A is the total room sound absorption in sabins for the room of interest.

TL is the measured sound transmission loss of proposed construction or the TL design criterion.

ADJ is an adjustment factor which takes into consideration certain characteristics of the sound source. In general, for aircraft traverses or sufficiently long lines of vehicle traffic, the sound field incident on the building facade is a reasonable approximation to the reverberant field condition in which TL values are measured. For this case, the term ADJ = 6 dB.

6.4.3 Noise Barrier Design

Selecting an appropriate noise barrier design for a building facade having various wall, window, and door designs and combinations can become a complex process. A simplification of the above equation allows the use of sound transmission class (STC) ratings for transmission loss (TL) and A-weighted values such as L_{dn}, CNEL, etc., for sound levels (L). The revised equation then becomes:

$$L_{\text{int}} = L_{\text{out}} \text{ minus STC} + 10 \log (S/A) + ADJ$$

Note that this expression is only an approximation. It is intended primarily as a tool to assess various design combinations. Direct comparison of the acoustical performance coupled with an economic assessment of construction details can be an effective technique in selecting appropriate systems for final judgment. Note that engineering calculations for compliance with code requirements may necessitate the services of a qualified acoustician[14] analyzing field or laboratory test data.[15]

In addition to the 6-dB adjustment listed above, two additional elements, G and F make allowance for use of STC:

$$ADJ = 6 + G + F$$

where:

Factor G (in dB) is an adjustment for the geometrical arrangement of the noise source relative to the proposed building facade. The inherent sound transmission loss of a building component, of which STC is a single-number representation, is dependent on the angle of sound-incidence. Since STC is determined with random incidence sound, the following adjustments to G should be made for situations where incident sound has a predominant angle of incidence (e.g., a single-source sound or focused sound approaching a facade from a specific angle). If the angle of incidence is random, no adjustment is needed. A $30°$ to $60°$ angle requires ADJ of 2 and an angle greater than $60°$ requires an ADJ of 5.

Factor F (in dB) is an adjustment for the frequency spectrum characteristics of the exterior noise source. STC was originally developed to classify sound insulation properties of interior building partitions against sounds such as speech, radio, and television. For this reason, adjustments to the STC must be made to account for noise spectra that differ markedly, such as aircraft and motor vehicles. F should be determined by adding the following correction value:

Jet aircraft less than 500 ft from observer	= 0
Train wheel/rail noise	= 2
Road traffic with negligible trucks	= 4
Jet aircraft over 3000 ft from observer	= 5
Road traffic more than 10% trucks	= 6
Diesel-electric locomotives	= 6

The factor $10 \log (S/A)$ can be approximated by multiplying the room floor area by 0.3 if the room has highly sound reflective floors, walls, and ceilings. For a room with carpet, drapes, and soft furniture multiply the room floor area by 0.9.

The foregoing design equations can be use in either of two ways. First, if a preliminary exterior wall design exists, the interior noise level can be predicted for the given design conditions or compared to the interior noise criteria for code compliance. Second, if a preliminary design does not exist, the minimum required STC that satisfies the code requirement can be determined without prior knowledge of any construction details. The STC value utilized in this method is the composite value for all building components including walls, doors, and windows, which are exposed to the noise source.

The first method is probably the most useful, since the designer will usually have a particular type of

---

[14] Qualified acousticians generally are members of the National Council of Acoustical Consultants (NCAC). The NCAC Directory lists all members and provides a detailed statement of their particular area of expertise. Copies of the Directory may be obtained by contacting NCAC, 66 Morris Ave., Suite 1B, Springfield, NJ 07081-5859, telephone (201) 564-5859. Similar assistance is available by calling the Noise Control Association (NCA) at (360) 437-0814.

[15] STC and sound attenuation data for a multitude of systems of construction and situations are given in Appendix 3. To determine a composite value for the sound level of several noise sources or the STC for a barrier containing several components such as a wall, door, windows, and penetrations, see the acoustical design worksheets in Appendix 2.

construction in mind for a given building project, and the question is simply whether that particular design will work acoustically. Note that a composite STC must be determined for the exterior acoustical envelope.[16]

A composite STC of a nonhomogeneous wall can be estimated. For a typical wall with an area of 126 ft^2 that contains a window or door having an area of 26 ft^2 the composite STC is as follows:

STC OF A COMPOSITE WALL SYSTEM

Wall	Glazing	STC of wall	STC of window	Composite STC
Wood siding	Single strength	39	28	35
Wood siding	1-in. insul. glass	39	34	38
Brick veneer	Single strength	56	28	35
Brick veneer	1-in. insul. glass	56	34	39

6.5 EXTERIOR ENVELOPE DESIGNS

Design of the exterior acoustical envelope is a function of determining the required sound attenuation or noise reduction for the composite. This process may be established by using existing code criteria as applicable or by using the techniques and formulas listed in the foregoing sections. Since most acoustical data from the laboratory is given in terms of the STC, this section presents typical STC data for exterior walls, windows, and doors that are often used in residential and commercial construction.

Comparison of the individual element noise isolation performance in terms of STC will indicate which element is the weak link. This component can either be changed in size or in type to reduce the sound transmission. Ideally, each separate component could be designed to contribute equally to the interior noise intrusion for the optimized design. Many builders will quickly dismiss this design approach as not economical and some may insist it is not a viable option. In spite of these obstacles, it has been consistently demonstrated that solutions that bring the weak link into compliance with the performance of the other barrier elements will yield the greatest decibels-per-dollar. An excellent example is the contractor or code official who insists that an air vent be installed in a roof or outside wall, invariably requiring that it be located with the shortest ducting, no lining, and the exhaust directly in line with the noise source. There is no barrier that can compensate for a direct hole. One must plug this acoustical hole.

6.5.1 Exterior Walls

STC values for exterior wall systems are unfortunately sparse. Only recently have designers found a need for this information and manufacturers have not yet been challenged sufficiently to provide STC ratings for their systems of construction. While some commercial building systems have been tested for their acoustical performance, the information in Table 6-1 is based on a series of tests conducted in a private

---

[16] See Appendix 2 for worksheet on determining a composite STC.

laboratory.[17]

TABLE 6-1 Sound Transmission Class of Typical Exterior Wall Systems*

EXTERIOR FINISH	CAVITY INSULATION	RESILIENT CHANNEL	STC
Wood siding**	None	None	37
	3½-in. Fiberglass	None	39
	Al. foil type 2P reflective	None	39
	3-in. rock wool	None	38
	None	Yes	43
	3½-in. Fiberglass	Yes	47
Stucco***	None	Yes	49
	3½-in. Fiberglass	None	46
	3½-in. Fiberglass	Yes	57
Brick Veneer****	3½-in. Fiberglass	None	56
	None	Yes	54
	3½-in. Fiberglass	Yes	58

   * 2 x 4 wood studs 16-in. on center (OC), ½" gypsum wallboard interior face.
   ** ½-in. by 10-in. redwood nailed to studs over ½-in. wood fiberboard sheathing.
   *** 15-lb building paper, 1-in. wire mesh nailed to studs, three coats (7/8-in.) stucco 7 to 9 lb/ft^2.
   **** ¾-in. wood fiberboard sheathing, standard 3½-in. face brick spaced ½-in. with metal ties
      nailed to studs. Dry weight. 41 lb/ft^2.

Note that the increased mass of stucco and brick raise the STC by 10 and 20 points respectively over lightweight wood construction. Also note that insulation in the cavity has little acoustical benefit unless the interior face is attached with resilient furring channels[18].

6.5.2 Exterior Doors

Doors are a major weak link in the noise reduction characteristics of the acoustical envelope in most residential buildings. While the door itself can be made to have a reasonably good STC, the perimeter seal fosters acoustical leaks. In northern climates where thermal leakage is of concern, seals tend to also provide better acoustical performance. And where there is a storm door, the combination may provide STC ratings nearly equivalent to a standard wood frame exterior wall. In selecting a door for acoustical purposes, it is vital to consider the entire door assembly, including the door frame, seals, threshold, and latch system. In selecting a door from manufacturers' literature, it is prudent to require that test data be in full compliance with ASTM

---

[17] This test data is based on full-scale tests done in a certified NVLAP acoustical testing laboratory. Full test reports are not available because the manufacturer closed the facility during downsizing.

[18] Additional information about resilient furring channels may be found in Chapter 7.

E90 Paragraph A1.8, Swinging Doors.[19] Reductions of 20 STC are likely if seals are worn or out of adjustment.

Door manufacturers are just beginning to understand the need for acoustically efficient doors. Industrial-type doors with an STC up to 60 are now available. While many utilize materials and hardware that is not presently acceptable to residential buildings, ongoing research indicates you will soon see practical and aesthetically pleasing exterior door systems that have an STC of 50 or greater. STC 40 and 45 are presently being offered by several manufacturers. Since this market is so new, it behooves the designer to shop manufacturers for the best decibels-per-dollar. Caution: Ignorance and outright cheating on ratings have been noted. If in doubt, ask for a copy of the sound test report. Check to be sure the test was conducted according to ASTM E90 within the last several years by an NVLAP-approved laboratory or field-tested by a member of NCAC. Call the tester to verify the test is valid and represents a practical application.

Representative STC ratings of door systems are given in Table 6-2.

TABLE 6.2 STC Ratings of Exterior Doors[20]

DOOR DESCRIPTION	ACOUSTICAL SEAL	STC (normal closed position)
1¾-in. flush solid core wood	Brass weather seal	27
1¾-in. flush solid core wood	Plastic weather seal	27
Same as above plus aluminum storm door	Plastic weather seal	34
1¾-in. flush **hollow core** wood	Brass weather seal	20
Wood french door	Brass weather seal	26
Fiberglass reinforced panel door	Plastic weather seal	25
Steel door, conventional commercial flush	Magnetic	28
Proprietary acoustical doors*	Patented designs and special hardware	35 – 50

* Several major manufacturers of residential doors are promoting high-STC exterior doors.

6.5.3 Windows

Window manufacturers have awakened to the fact that sound attenuation is a major criterion for the selection of windows. Like doors, windows are a weak link in the sound transmission loss characteristics of the exterior envelope of a building. Fortunately, the efforts required to improve the thermal characteristics of windows also produces a positive effect on the sound transmission loss attributes and the STC. As with doors, seals are the key to good STC values. Even standard window units can be dramatically better at attenuating sound with air-tight seals. In most instances the seal must be operable. Consequently, a seal that will maintain the attribute of air-tightness after being opened and closed hundreds of times is the key to acoustical performance. While some extreme noise problems may be solved with fixed windows, this solution

---

[19] Reference <u>ASTM Annual Book of Standards</u>, Volume 4.06, <u>Thermal Insulation, Environmental Acoustics</u>. E90 requires details of the door, frame, clearances, and seals, and if the door is operable, that it be opened and closed five times before testing.

[20] Data from NVLAP test laboratory. Full test reports are not available.

becomes expensive since air quality within the structure must now be controlled with an efficient air conditioning system. <u>Caution:</u> As with doors, ignorance and outright cheating on ratings have been noted. If in doubt, ask for a copy of the sound test report. Check to be sure the test was conducted according to ASTM E90 within the last several years by an NVLAP-approved laboratory or field tested by a member of NCAC. Call the tester to verify the test is valid and represents a practical application. If the unit is intended to be operable, look for a statement in the test report that the unit was opened and closed at least five times prior to the acoustical test. Unfortunately, many tests are conducted with tightly sealed units yet the report may imply they are operable.

Representative STC ratings of window systems are given in Table 6-3.

TABLE 6-3 STC Ratings of Window Systems[21]

Frame material	Type window	Glazing (1)	Sealed STC (2)	Locked STC (3)	Unlocked STC
Wood	Double hung	Single strength (SS)	29		23
		SS– divided lights	29		
		Double strength (DS)	29		
		DS – divided lights	30		
		Insulated - 7/16-in.	28	26	22
	Picture	SS – divided lights	28		
		Double strength	29		
		Insulated – 1-in.	34		
Wood-plastic	Double hung	Single strength	29	26	26
		Insulated – 3/8-in.	26	26	25
	Awning	Double strength	30	27	
		Insulated – 3/8-in	28	24	
	Fixed Casement	Double strength	31		
	Operable Casement	Double strength	30	22	
	Sliding glass door	Laminated – 3/16-in.	31	26	26
Aluminum	Sliding glass	Single strength	28	24	
	Operable casement	Double strength	31	21	17
	Single hung	Insulated – 7/16-in	30	27	25
	Jalousie	1/4-in.	26	20	
Test frame	Single pane	1/2-in. laminated	34		
Test frame	Double pane	1/2-in laminated	45-55*		

(1)   Single strength nominal thickness is 3/32-in., nominal weight of 1.3 lb./ft².
       Double strength nominal thickness is 1/8-in., nominal weight of 1.63 lb./ft².
       3/8-in. insulating glass is composed of two layers of 3/32-in., glass with 3/16-in. air space, nominal

[21]   Data from a NVLAP test laboratory. Full test reports not available.

weight of 2.6 lb./ft^2.

7/16-in. insulating glass is composed of two layers of 1/8-in. glass with 3/16-in. air space, nominal weight of. 3.3 lb./ft^2.

1-in. insulating glass is composed of two layers of 3/16-in. glass, with 5/8-in. air space, nominal weight of 5.2 lb./ft^2.

1/4-in. louvers in jalousie window, nominal weight of 3.3 lb./ft^2.

(2)   Sealed STC means entire window unit was caulked with duct seal, a heavy clay-like material.

(3)   Applies to operable windows mechanically closed without separate lock.

*   Depending on air gap and perimeter absorption. See manufacturers' literature.

Specific STC and sound attenuation data at one-third octave intervals is generally available from manufacturers of windows. This information is particularly useful in designing for specific noise sources such as nearby airports, highways, railways, and factories. Each noise source has a different frequency content. Specific sound transmission loss characteristics should be selected to provide optimum noise reduction for the frequencies that predominate in the noise source. STC values are helpful in making initial screening analysis and when the noise source is broadband in character. Aircraft and vehicle noise is generally broadband in character. Industrial noise may have specific frequencies that predominate. Low -frequency noise, below 125 Hz, requires special analysis.

6.5.4 Roofs and Building Envelope Sound Attenuation

The industry has precious little acoustical test information on roof-ceiling structures. Some field tests have been conducted for commercial buildings and schools. To date, there is only one source for the sound transmission loss characteristics of light frame residences.

While the author was employed at Owens Corning Fiberglas, we had the good fortune to be able to conduct an acoustical analysis of three identical, single-story, typical construction, three bedroom ranch houses constructed with different levels of thermal insulation. House A contained R-38 Fiberglas attic insulation and R-19 wall and floor insulation. House B had R-19 attic insulation and R-11 wall and floor insulation. House C did not have any building insulation. The three structures were constructed side by side with identical construction materials and techniques at the Owens Corning Fiberglas Research Center in Granville, Ohio. All the houses were carefully instrumented for thermal research efforts that had been ongoing for nearly a year.

The acoustical tests[22] consisted of instrumenting the houses inside and outside, with multiple microphone

---

[22]   The tests were very formal and have been reported at various acoustical industry gatherings. On an informal note, the first series of flyovers caused quite a stir among local residents. It seemed the site was a favorite soaring location for a local infamous bird known as a turkey buzzard because of the thermal created by the test house roofs. The name turkey buzzard is very appropriate because the birds look like and are the same size as a turkey. They have a wing span of over 6 ft when soaring. A dozen soaring turkey buzzards make quite an impressive sight. Our first test was intended as an information gathering event to determine if the instrumentation was appropriate. With no air-to-ground communication, imagine the fright when we noticed the turkey buzzards were at the precise altitude of the plane. The jet approached with lights on, gear down, full flaps, and full throttle, right at the birds. We were convinced that the birds would be ingested into the engines, a flame out would occur and the plane would crash into the offices of the OCF Research Center management building. Just when we were all about to run, we noticed that the birds had vanished. With a big sigh of relief we continued with the tests. The moment we curtailed operations, the birds were back soaring. That evening over drinks with the pilots we relayed the buzzard story and

pickups fed to a computerized source designed for instant comparison. Data was stored on tape recordings for detailed analysis in the laboratory to determine sound attenuation and STC. Two separate series of tests were performed. One used a steady-state source consisting of a commercial-test-quality loudspeaker placed on the ground facing the house. The second source was the corporate Learjet which was assigned to make several passes over the test houses at set altitudes while flying with full flaps, gear down, and full throttle. Since the Learjet was one of the loudest jets flying at the time, we were assured of having sufficient loudness to obtain accurate sound level readings inside the structures. The results of our tests are summarized in Tables 6-4 and 6-5.

TABLE 6-4 Steady State Test Results (Ground speakers) (STC)

	House A	House B	House C	
House insulation	R-38/R-19	R-19/R-11	None	/
Living room	35	33	30	
Front bedroom	40	39	37	
Corner bedroom	37	36	32	
Rear bedroom	36	35	32	

Note in Table 6-4 that the STC value from House C, without insulation, and House B with standard insulation, improved by 2 to 4 points. The STC from House C to House A increased by 3 to 5. These increases are somewhat higher than expected from ASTM E90 tests conducted in the laboratory. Note that the steady state tests are essentially testing the horizontal noise, which basically impacts the vertical surfaces or exterior walls of the structure. Roof attenuation is a minor sound path in this test.

For the airplane flyover tests summarized in Table 6-5, noise reduction values were identified and reported as noise isolation class (NIC), which is essentially a field determination of sound transmission class (FSTC). The STC and NIC, while different in some technical respects, serve as an acceptable comparison for this situation. A direct comparison is not physically convenient since it would be quite an effort to create actual flyover sounds in a laboratory environment.

Note that the flyover results are, in reality, a determination of the roof/ceiling structure. Very little sound from the flyover will impact on the vertical wall surface of the one-story structures chosen for this test.

With House C (no insulation) as a reference, the NIC improvements in house B (R-19 insulation in the roof and R-11 insulation in the sidewalls) range from 1 to 5 and in House A they range from 3 to 7. We also saw a trend of increasing NIC with lower altitudes. This is believed to be due to different angles of incidence and the influence of the roof structure. Since all other construction details and materials were constant, the effect of insulation on NIC can be considered significant. Based on other findings, it would be reasonable to expect that resilient mounts plus added mass would be effective in improving the STC or NIC of the exterior envelope. Of course, an upgraded STC door and window unit would have to be utilized. Like the weak link

---

asked if they knew about the birds. The chief pilot said, yes, it had occurred to him so he had turned on his radar. After considerable evaluation, we concluded that the birds disappeared the precise moment the radar was activated and then returned when it was shut down. A formal paper on the event was presented to FAA and appeared in several pilot publications.

in a chain, all penetrations must maintain the STC of the composite acoustical envelope.

TABLE 6-5 Aircraft Flyover Results [Noise Isolation Class (NIC)]
(Sound source at 250, 500 and 1000 ft altitude)

	House A	House B	House C
House insulation	R-38/R-19	R-19/R-11	None
Living room			
1000 ft altitude	36	34	34
500	37	35	34
250	36	34	34
Rear bedroom			
1000 ft	39	37	33
500	40	38	33
250	41	39	34
Kitchen			
1000 ft	--	33	30
500	--	32	29
250	--	33	31

6.5.5 Vents and Other Acoustical Envelope Penetrations

Holes in the exterior acoustical envelope of a residence are probably the single most important aspect of quieting the interior occupied space. Construction of a 40 or 50 STC acoustical envelope can generally be accomplished with conventional construction materials. However, major flanking paths occur by installing exhaust vents, dryer vents, electrical outlets, and any other penetration of the acoustical envelope. These holes may seem innocuous but end up being the weak link in the acoustical barrier. Even a small hole can reduce the composite STC to a point that it is rendered useless as a noise barrier.[23]

A practical technique for upgrading air vents include lining the inside of the air duct with sound-absorbing material and providing many turns where sound may be absorbed within the duct. Generally, a 1-in. thickness of sound-absorbing material placed on the inside of the duct for a length of over 10 ft will attenuate the sound sufficiently to maintain a 50 STC.[24]

---

[23] See Figure 5-6. Note that a 50 STC barrier with a 1.4-in.² hole will be reduced to 20 STC.

[24] Duct attenuation values depend on the size of the duct, the thickness of the sound-absorbing material, and the type of material. Duct attenuation test data for several sizes and types of duct materials are given in Appendix 3, along with a technique to calculate the degree of attenuation that may be achieved with different designs.

Back-to-back electrical outlets should be avoided. Plumbing vents must be carefully sealed. In extreme cases, plumbing vents and chimneys may need to be encased in a gypsum wallboard container filled with building insulation. The new lightweight fireplaces present an interesting acoustical challenge. A tight fireplace door will not only reduce huge energy loss up the chimney but will also reduce noise intrusions. Penetrations for pipes etc. must be carefully sealed with a permanently-elastic weather resistant sealant. Silicone sealants may be required. Sealants should also be utilized on all sill plates, corners, and wherever abutting materials or similar interfaces occur.[25] If vents are installed to vent attic spaces to prevent moisture buildup, be sure to locate the vents away from the noise source.

6.5.6 Building Orientation [26]

If the opportunity exists in the design process, it is wise to locate the building envelope so that direct sound waves will impact on the building facade at an oblique angle. This will cause sounds to be more readily reflected thereby lessening the amount of sound that will be available to penetrate the building envelope. However, this technique will not reduce the need to be sure all sound leaks are sealed. If the building has a preponderance of windows and doors, it is wise to locate them on the side away from the sound source. In northern climates, door entrances and vestibules may be utilized as partial barriers to create an acoustical shadow. Likewise, fences and like structures may be used to provide an acoustical shadow for a window or door that is left open.

Place the quiet rooms away from the noise source. For example, if there is considerable street noise, place quiet sleeping rooms in the back of the living unit. Where flyover noise is of concern, consider placing the sleeping rooms on the lower level of the unit where sounds will be attenuated by the upstairs rooms and the additional floor/ceiling system.

A simple practical analysis of the site where potential noise sources are identified and a ray diagram of the sound paths along with careful building orientation can sometimes completely eliminate noise complaints later. A classic example is the location of a patio that is next to a neighbor's air-conditioner unit. Simply locating the patio around the corner or in an acoustical shadow is many times sufficient to make the patio an acoustically desirable place to cool off on a hot summer evening. This simple planning process will yield decibels per dollar far in excess of any other sound attenuation techniques. As an aid in determining when a noise problem may exist, consider the data in Table 6-6.[27]

---

[25] Use the same principles to isolate and seal for sound leaks that are utilized for thermal isolation. Every effort should be taken to assure the entire envelope is as air-tight as a balloon. Even the smallest hole will leak sound and likewise reduce the efficiency of the thermal envelope.

[26] See Section 1.2 for more extensive discussion.

[27] For more details see Chapter 4, specifically local noise ordinances.

TABLE 6-6 Exterior Noise Acceptability Standards

Site suitability	Day-night ave. sound level (L_{dn})	Special approvals and requirements
Acceptable	Not exceeding 65 dB	Few or none
Normally unacceptable	Exceeding 65 dB but not 75	Special approvals (1), Environmental review (1), Attenuation (2)
Unacceptable	Above 75 dB	Special approvals (3), Environmental Review (3), Attenuation (4)

(Notes to Table 6-6)

(1) All projects located in a normally unacceptable noise zone require an Environmental Impact Statement (EIS) containing a "Noise Element" if located in an underdeveloped area, or where HUD encourages establishment of incompatible land use. Recommended noise mitigation efforts should be implemented by the stated authorities.

(2) Additional noise attenuation measures are required beyond those which are normally associated with buildings as commonly constructed in the area. Additional attenuation of 5 dB is generally required for sites above L_{dn} 65 but not exceeding 70 dB. Additional attenuation of 10 dB may be required for sites above L_{dn} 70 but not exceeding 75 dB.

(3) Unacceptable noise zones require that an EIS be prepared and that mitigation measures be implemented prior to approval to proceed with construction. The authorities responsible for implementation and enforcement may be federal, state, county, or local depending on the particular circumstances. Generally, an environmental clearance is required prior to commencing construction and monitoring during and after completion to assure compliance.

(4) Noise mitigation measures must generally be formally approved by the appropriate authority on a case-by-case basis.

6.6 THE COST OF COMMUNITY NOISE

Many studies have addressed the physical impact of noise in the community.[28] A few have attempted to place a dollar value on the cost of noise.[29] There appears to be a general consensus that a direct relationship exists between noise exposure and the percentage community annoyance (See Figure 6-4). In a typical community few persons seem to be bothered by a noise until it causes them to raise their voice to be heard – typically 65 dB$_A$. Over 50% of the community residents become highly annoyed when background noise levels approach 80 dB$_A$. These levels of annoyance have been the basis for many noise regulations.

While these tolerance levels for noise are well documented, the reader is cautioned that noise relates to

---

[28] Papers appear regularly in the Journal of the Acoustical Society of America, papers and journals for the Institute of Noise Control Engineers, Sound and Vibration Magazine, Noise Barrier Journal, , and other industry technical journals.

[29] Notable examples of investigations of the cost of noise include: The Tyranny of Noise, Robert Alex Baron, Harper & Row, 1971; Road Traffic Noise, Alexandre, Barde, Lamure, Langdon, Applied Science Publishers, 1975; The Economic Impact of Noise, U.S. Environmental Protection Agency, 1971.

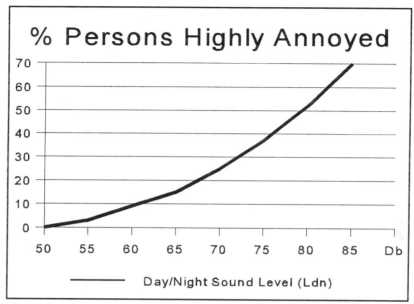

Figure 6-4 Noise exposure versus percent community annoyance.
(T. J. Schultz.)

our environmental situation. Walking about on a city street is quite a different situation from trying to sleep in a tent in the wilderness. We expect the city to generate noise. And when we are walking or riding we are active and generating our own sounds. By contrast, if we are resting in a quiet country setting, we expect low background noise levels. Even more importantly, we perceive noise as a function of the signal to noise ratio. If background noise levels are very low, as experienced in the country, an intrusive noise at 50 dB will become "highly annoying." Unfortunately, most annoyance studies reflect an urban acoustical environment where background sounds are already near 60 dB.

If your project is located where the background noise is low, such as a retirement community next to a national forest, the level of acoustical mitigation must be far more extensive than usual to assure peace and quiet. For this reason, most researchers and economists who have investigated the cost of noise annoyance have been reluctant to identify a decibel level with a dollar level.[30] While the economists are unwilling to place real dollar numbers on the issue, there is no doubt that the dollar impact due to noise is extensive.

In the book Road Traffic Noise the authors devote an entire chapter to "The Social Cost of Noise." Their observations and conclusions indicate that builders, architects, environmentalists, and planners should give great consideration to noise when evaluating the financial aspects of a residential development. Cited studies demonstrated that noise could "lead to a 20–30% depreciation in house value" and that "residents living in

---

[30] The writer coined the term decibel- per-dollar" with another meaning in mind. Here the decibel level referenced is sound level in the community or the background level. When the author coined the term he meant it to be applied to decibel reduction such as found in the terms noise reduction, sound attenuation, or sound transmission loss.

an area with peak noise levels of 80 – 85 dB$_A$ declared that, if they were to repurchase, they would never choose a house near an expressway." While there are few valid studies to cite, it is quite obvious that building a house next to an obnoxious noise source such as a freeway, airport, rock-crushing quarry, or the like will significantly reduce the property value. Ignoring this simple fact has lead to notable financial debacles.[31]

In the book The Tyranny of Noise, author R. A. Baron has a chapter titled "The Price in Dollars." Mr. Baron cites numerous projects and situations where huge dollars were associated with noise complaints. While many are lawsuits, the fact remains that noise causes stress, loss of sleep, accidents because someone did not hear a warning, and a significant degradation of normal physiological fitness. Several quotes from Mr. Baron are especially significant:

"As we continue probing the new concept of environmental quality we will discover that the total cost of excessive noise is something society cannot afford."

"The dollar cost of noise in some of its aspects is vague, hard to put one's finger on, although certainly real enough. But the loss in real estate value is plain for all to see."

Acoustician Lew Goodfriend hit where it hurts when he observed that the geographically extensive impact of aviation and highway noise affects the entire social and economic structure of the city. "Cities pay, in the deterioration of their neighborhoods." This could be serious if it results in middle and upper class families abandoning neighborhoods, quietly and without fanfare, leaving them to the next lower economic level for whom the neighborhood, noisy or not, is a step upward.

An obvious question that is raised by all the foregoing comments is: How does one find quiet? By leaving the city? If we do, we tend to bring our noise woes with us. A case in point:

A couple planning retirement found an apparent acoustical virgin. Bounded by Puget Sound on two sides and massive national forest lands nearly covering the other sides, Port Ludlow, Washington, has lived up to its billing as a quiet "village in the woods by the bay." Sound level readings on a state-of-the-art sound level meter, measured on the front walk, consistently show the ambient or background sound level is below the threshold of the equipment. The levels are so low that the squawk of a sea gull or screech of an eagle can be heard over a mile away from the source. When the breeze flows through the multitude of old growth firs and pines, the background sound is a delightful mask for the occasional high-flying airplane or passing auto.

Protecting this virgin acoustical environment is becoming an intriguing challenge. Because of this delightful setting, developers see Port Ludlow as an ideal location to make gobs of money. Quite understandably, they want to build more, build faster, and place more living units on a smaller plot. And of course, they want to provide all the amenities of the city such as more access by motor vehicles, more roads, higher speeds, access by aircraft, faster boats for water transport, air conditioners, stereos, etc., etc., etc. Existing residents realize that they will soon loose the amenities that created the line "a quiet village in the woods by the bay." What is the solution? Well, we could say: "We are here, you can't come," but that is not

---

[31] An excellent example is the Los Angeles Airport. As the airport expanded, noise levels increased to the point that a considerable number of residences eventually were impacted by noise in excess of 80 L_{dn}. To avoid lawsuits, the City of Los Angeles first condemned a several-mile sector of prime beach-front property located between the end of the runways and the ocean. The purchase price, while unpublished, was exorbitant. As the airport expanded, additional residences were condemned because of excessive noise levels. In an unbelievable sequence of events, a builder eventually repurchased a chunk of land near the end of the runway and constructed a high-rise, high-rent, luxury residential complex. Within days of moving in, residents abandoned their beautiful suites because of the high noise levels. One wonders what intelligence prevailed. This facility now stands empty, at great expense to the taxpayers, builders and financiers, as a monument to acoustical stupidity. Examples such as this abound across North America. Not all are airport noise, many are complexes built next to freeways, railroads, and industrial noise sources. I am appalled at the waste this generates.

politically correct. Is there another way to preserve our peace and tranquility? Possibly. Let's legislate it. How? Simple – write a noise ordinance.[32] But then what happens? Local politicians were elected because they want to protect their "property rights." Some of the property rights advocates insist that a noise ordinance intrudes on their right to do what they want on their property. All of a sudden, a pure environmental issue is now a political issue. Result? The jury is not yet in on the subject here in Port Ludlow, but if one looks around our county to find similar experiences, you will find an acoustically sad result. We tend to bring noise with us. It builds until there is no feasible way to recreate the virgin acoustical environment. Then it masks our efforts to contain additional noise. Soon we have the same noise level of the city along with stress, danger, and loss of the tranquility that we treasure as humans.

What is the answer to this dilemma? Unless we are willing to say "Stop development before it begins," the result is inevitable. About all we can do is slow down the process with legislation.

The bottom line to the cost of community noise is huge. A figure that equates the cost of all construction is not unreasonable. Maintaining peace and quiet will cost a lot. Technically there are many low-cost solutions. Most of them are addressed in this text. The decibel-per-dollar for these technical solutions is amazingly small. If addressed at the design stage, the cost can be less than 5% of the project cost. Unfortunately, because we are human, the real cost of quiet will go to pay for retrofit and litigation expenses. What a waste. With costs like this looming on the horizon, wouldn't it be prudent to incorporate good acoustical design at the beginning of the project? Will we do so? I doubt it. Does anyone have a better way?

---

[32] Just such an ordinance has been written and is reproduced in Chapter 4.

TABLE 6-7 Environmental Noise Control Checklist

Noise Source	Descriptor	Information sources	Mitigation measures
Airplanes	dB_A, L_{dn}, NEF	FAA, Air Force, HUD, EPA, local ordinances	Upgrade acoustical envelope relocate
Highways	dB_A, L_{dn}, CNEL	FHWA, HEW, OSHA, state/local code	As above + sound walls
Railroads	dB_A, L_{dn}, CNEL	FRA	As above + sound walls
Industrial	dB_A, L_{dn}, CNEL	OSHA, HUD, building codes, facility owner/management	Request noise quieting measures at source, upgrade acoustical envelope, relocate, sound walls
Residential/ Commercial	dB_A, L_{dn}	Owner, building codes, state and local ordinances	Request/enforce noise quieting measures at source, upgrade acoustical envelope, relocate, add sound walls
Concerts, race tracks	dB_A, L_{dn}	Owner, building codes, state and local ordinances	Request/enforce noise quieting measures at source, upgrade acoustical envelope, relocate, add sound walls.

Chapter 7 CASE STUDIES

The only formula or format for quieting a residence is common sense. Methodical application of the principals outlined in Chapter 1, supplemented by the information supplied in the balance of the chapters plus the appendixes may be of assistance in organizing the problem into parts that may be a bit more palatable. Nevertheless, applying these principles in practice nearly always creates a quandary. For this reason, reviewing a project that is of a similar nature can be helpful in analyzing your specific problem. These case studies are designed with that thought in mind.

Each noise control problem does have one commonality: it can be broken into an analysis of the **source, path, and receiver** (SPR). Clearly, quieting the source is the most economically effective means to resolve the problem. Only after all potential options to quiet the source are exhausted should one consider treating the receiver. Only after all options for both the source and receiver are exhausted should one consider mitigation efforts along the noise path. Unfortunately, it always seems that we end up dealing with the path of noise since viable solutions at the source and receiver have social limits. Therefore, you will find that most of the case studies address potential noise mitigation along the path. The path is usually airborne sound, although others such as structureborne sound are addressed.

Previous chapters and the test data in the appendix are the basis for these case studies. No effort has been made to organize or coordinate them except that footnotes indicate where pertinent information and test data may be found. Each case study stands on its own merit. Many of the studies are papers that are reproduced here in their entirety.

To help you select the study that best describes your situation, a brief descriptor follows:

- Quieting a pool filter/heater unit   7.1

- Quiet design of a single-family residence   7.2

- Quiet design of a multifamily residential complex (condos, apartments, etc.)   7.3

- 30 years of resilient furring channels   7.4

- Quiet lightweight floor systems   7.5

- Noise assessment and noise control recommendation for a housing development   7.6

- Greek Theater concert noise analysis and mitigation measures   7.7

- Residential sound insulation program, Dayton International Airport   7.8

- EIS "Noise Element" for airport   7.9

- Acoustical environment – a health hazard?   7.10

7.1 QUIETING A POOL FILTER/HEATER

This situation is common in many residential communities across the country. Noise from equipment required to filter and heat the water in a swimming pool is intrusive to neighbors. The noise source is primarily pumps and filters that produce a broadband sound level of approximately 80 dB$_A$ within a few feet of the source. This noise source is similar to an air conditioner, compressor unit for a heat pump, and many types of residential support equipment. **An enclosure for the pumps and filters reduced the sound level by over 30 dB$_A$.**

PROBLEM

The client of Santa Monica, Calif., has a neighbor with a backyard pool. The noise of the filter pump and pool heater disrupts his sleep and will not allow normal conversation his back porch, patio, and garden.

Lots in Santa Monica are quite large for an urban area. Typical of developments in the late 1930s the lots are long and narrow, allowing considerable space in the back-yard for pool, patio, garden, and separate garage. The space between the houses is about 10 ft. Pool support equipment consisting of a large pump, a heater, and associated storage tanks is mounted on a concrete base located on the side of the neighbor's garage nearly on the lot line. A 6-ft-high board fence follows the property line, providing some visual screening but little sound attenuation. All houses in the vicinity are typical California style – one story, stucco with tile roofs. When constructed, the homes in the area sold at a very modest price. In 1990 the community is rapidly becoming upscale and expensive. The client purchased the home when it was nearly new. A new neighbor has added many amenities including the in-ground pool.

The client and his wife consider the noise from the pool equipment to be unbearable. It keeps them from getting a good night's sleep, and when they utilize the patio and porch, which is often in Santa Monica, they cannot have a normal conversation. They must raise their voices to be heard at normal conversation spacings when the pumps are operational. The noise invariably occurs when the neighbor is inside seeking a cool spot to watch his favorite program. The client requested the assistance of an acoustical consultant to provide an analysis and recommendations to mitigate the incessant intrusive noise.

ANALYSIS

A visit to the site and discussion with the owner established that:

- The noise level is clearly intrusive, and broadband in character and is approximately 80 dB$_A$. Noise is relatively uniform throughout the entire backyard area of the client property including the porch, patio, and garden.

- The noise is intermittent. Pumps that circulate the pool water for filtering and temperature control operate for several hours at a time, 24 hours a day. Downtime is less than 50%.

- The noise source is the pumps, resonating tanks, and associated piping. A 6-ft-high wood fence does attenuate sounds by approximately 5 dB$_A$ in the fence shadow, but is ineffective in reducing noise on the client's property. High-frequency sound is transmitted through the cracks between the boards,

216

the boards have insufficient mass to reduce low frequencies, and the entire frequency band is reflected off the stucco walls and underside of the garage eaves. Equipment is located in a corner thereby reflecting considerable sound up, out, and over the fence toward typical receiver locations.

- Background sound levels are surprisingly low for Santa Monica, since it is surrounded by city, but typical of a quiet suburb. Daytime levels average about 50 dB$_A$. Nighttime levels were not measured but may be 5 dB$_A$ lower. There are no major airports, highways, freeways, railroads, or industrial noise producers within a ½-mile radius. Traffic is almost entirely local autos and only a few vehicles per hour.

OPTIONS TO BE CONSIDERED (SPR)

Options for consideration are given in order of economic importance (i.e., decibels-per-dollar). Note that all feasible options for treatment of the source should be addressed fully before addressing the options at the receiver. Only when all source and receiver options are exhausted should one consider quieting along the path.

Source options:
1. Request neighbor to turn off or otherwise quiet the equipment.
2. File a complaint noting a violation of the noise ordinance to local authorities.
3. Replace existing pumps, valves, etc. with quiet pumps and related equipment, including mufflers and other similar noise-suppressing devices.
4. Install noisy equipment and pipes on resilient mounts.
5. Surround noisy equipment with a sound attenuation enclosure (also a path solution, depending on your point of view).
6. Relocate noisy equipment where it will not intrude on neighbors.
7. Install active noise control.

Receiver options
1. Move.
2. Wear ear attenuators.
3. Curtail use of porch, patio, and garden when noise source is operational.
4. Add masking sound to bedroom and living areas.

Path options
1. Extend noise path distance.
2. Construct a sound wall at property line.
3. Enclose the porch and patio utilizing good sound attenuation practices.
4. Upgrade the sound attenuation attributes of the acoustical envelope (i.e., the sidewalls, windows, doors, roof, chimney, etc.).

OPTION ANALYSIS

The following section provides background information and limitations to the utilization of the options described above.

Source options:

1. Clearly the most effective technique for quieting is to turn off the source. In this instance, the neighbor is new, and considered nonapproachable and therefore this option was dismissed as not feasible. The new neighbor may not have known that the equipment noise was intruding on the client because the sound was effectively blocked from their hearing by a garage. The noise was out of sight, and out of hearing, therefore not perceived as a problem. Had the recipients of the noise (the client) been more persistent, they may have learned that the neighbor traveled extensively and was not in residence nearly 50% of the time. A reasonable neighbor may have willingly, (1) installed a timer on the filter equipment so that it operated only when the noise would be less intrusive, such as during the hours of 9:00 a.m. to 3:00 p.m.; (2) turned down the heat, a significant energy saver, to reduce substantially the time of circulation pump operation; and (3) placed a damping material on the resonating pipes and tanks. The cost of the foregoing mitigation measures would be less than the cost of a brief visit by an acoustical consultant. If there was any resistance by the noise-producing neighbor, an offer to pay for the new equipment could have been made; payment would have been far less expensive than all other options.

2. Complaining to the Santa Monica Officials in charge of the noise ordinance is a viable and cost-effective option. This option was dismissed out of hand since the complainant felt he would be considered negatively by both the new neighbor and others. Also expressed was a lack of confidence in and/or fear of becoming embroiled in the bureaucratic process. This option was considered unacceptable by the complainant.

   In reality, Santa Monica has a noise ordinance that specifically addresses this type of noise complaint. Had the complainant gone to the City Building Inspection Department, he would have found that a local inspector would have come to make sound level measurements at the property line, found them to be out of compliance with the ordinance, and issued a notice to comply to the owner of the noise-producing property. The inspector may have determined that the noise was also a significant annoyance to several other neighbors. Formal notification of a violation of the noise ordinance would have been made, requiring the owner of the property producing the noise to mitigate the noise to a level that was in compliance within a reasonable period. If compliance was not achieved, the building inspector has the authority to enforce the ordinance with a fine or make alterations at the expense of the noise-producing owner. In effect, the same issues that were covered in option 1 may have ensued without any formal involvement by the complainant.

3. Purchase quiet equipment. Best accomplished at the onset or as a replacement, this approach is one of the most practical solutions available. The additional cost for manufacturing quiet pumps is likely less than 5% of the total cost of the equipment. Unfortunately, some manufacturers may charge more than their costs because of market anomalies.

Since noise is usually generated by undersized equipment running at high speed, a "lower and slower" mentality design effort can produce very effective decibels per dollar. And, if the equipment changes are not feasible, minor system design changes that incorporate vibration mounts, damping, mufflers, or similar devices are very effective in quieting the original equipment. Shop manufacturers and press them to supply quiet equipment. Noise levels can be reduced by 30 dB at a price addition that is likely to be less than the tax for the item. In addition, quiet equipment will be likely to run longer, consume less power, and require less maintenance. The bottom line: Buy Quiet. Many manufacturers will have noise ratings but few marketeers elect to publish the information. Technical literature sometimes contain noise ratings.

4. All mechanical equipment and pipes should be installed on resilient mounts. Much of the noise problem is associated with inevitable vibrations that are transmitted to pipes, panels,and containers that act as an amplifier. A resilient mounting may be all that is needed for some pump systems. Mounting design has been reduced to a formula these days, and there is no excuse for installing motors or pumps without the proper resilient mounting. There are spring and elastomeric mountings plus flexible connectors for pipes. A resonating panel, pipe, or container is easily identified by placing your hand on the item. If hand pressure reduces the noise, then a damping compound, placed at the node point or the like, will also be very effective in mitigating the noise at the source.

5. Surround the source with an enclosure. This solution is actually a form of path control. However, since the enclosure is best applied close to the noise source, it is usually considered a source solution. This solution was chosen as being the most practical for this case.

   Enclosure design appears simple on the surface. Merely enclose the noise source in a box made of heavy, dense material. A sheet of plywood usually has sufficient noise reduction.[1] Line the inside with highly sound absorbing material such as glass or mineral fiber building insulation or boards.[2] The major problems in designing the enclosure are aesthetics, weather resistance, durability, access for maintenance, and last but surely not least, allowing for air cooling.

6. Relocate the noisy equipment to a spot where it will not intrude on the neighbors. While simple to say, this is a very difficult solution. There are many factors involved in siting the pump system. Long runs of piping are not only expensive, but the line losses for both temperature and energy utilization preclude distant locations. Furthermore, it takes a doubling of distance to effect a 6-dB reduction. It may not be possible to find a site on most city lots where the effect of distance will garner much noise reduction. Furthermore, the noise may now also adversely affect other neighbors. Only on a large estate is this technique successful. In this situation, relocation was out of the question for many reasons.

---

[1] See Appendix 3 for noise reduction and sound transmission class ratings of various materials. A design methodology and example are also provided in Appendix 2.

[2] See Appendix 3 for sound absorption values and Noise Reduction Coefficients (NRC) for various types of sound-absorbing materials. See Appendix 2 - Design Guide Worksheets for techniques in calculating their acoustical effectiveness.

219

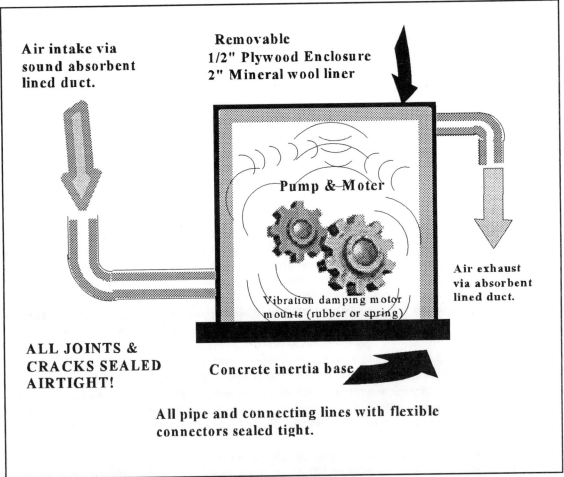

Figure 7-1 *Design of acoustical enclosure for pool filter/heater system.*

In the case at hand, the enclosure design adopted consisted of ¾-in. thick marine-grade plywood, fabricated with tight joints fully sealed with acoustical sealant, lined on the interior with 4-in. thick fiberglass 4-lb-density board (duct board). Alternative barrier materials are cement board, stucco, and masonry. For greater attenuation, you could use typical wall construction for the enclosure, with structural members faced both sides with gypsum board sheathing, a resilient attachment like resilient furring channels, and a core of building insulation.[3] If the interior is subject to fiberglass erosion due to high speed air motion or mechanical rubbing, you could cover with an open-weave mesh or perforated metal. The enclosure provides air circulation by incorporating a chimney-like

---

[3] See case studies on resilient furring channels (Section 7.4) and quiet lightweight floor systems (Section 7.5) for additional ideas.

stack on both the intake and exhaust. The stack is approximately 5 ft long, has several turns, and is fully lined with sound-absorbing material. The lid is removable, with all edges sealed tightly with resilient foam weather stripping. Bolting the lid in place assures a tight seal. Effective noise reduction of 30 to 35 dB_A can be achieved with this design (See Figure 7-1).

Receiver options:

1. Move. If you can't stand the noise, go where it is quiet. This may seem ludicrous. However, physically, each time you double the distance from a noise source it is reduced by approximately 6 dB. A move down the block is a viable but probably not a sociable or practical consideration. Such was the case in this instance, but the option must not be overlooked.

2. Wearing ear attenuators can be a very effective technique to reduce sound levels. Typical industrial earmuffs are rated to provide a noise reduction of up to 30 dB. Earplugs are also very effective but more uncomfortable to wear. Of course, they also attenuate speech.

   Ear attenuation is usually only viable as a temporary measure. In industrial noise control parlance, the Occupational Safety and Health Administration (OSHA) regulations[4] require Temporary Measures to be replaced by Engineering or Management Controls within a reasonable time period. Few would accept earplugs or earmuffs as a reasonable solution to the noise problem in this case study. It is simply not a viable, aesthetically or socially acceptable solution. Obviously, this approach was rejected for this project but should not be overlooked.

3. Curtailing use of the porch, patio, or garden while the noise source is operational has a fancy name in OSHA regulations. It is termed "management controls." While not acceptable for this situation (and ludicrous to many), this approach is technically feasible.

4. Adding a masking sound inside the house has some merit in this situation. It would be unacceptable outside since any sound at or above 80 dB_A is an annoyance in and of itself. Yes, it would mask the noise source but the solution is worse than the problem. Inside the house, the noise source will likely be attenuated by the existing acoustical envelope to a level at or below 50 dB_A. A broadband masking sound at 55 dB_A would render the noise source as inaudible. Furthermore, the masking sound will be acceptable to most people. Considered a viable and inexpensive solution to the sleeping portion of the problem, masking inside the house was recommended. Masking units can be obtained for less than $100 for a single desktop version; more sophisticated units, designed for the open plan office, with multiple speakers are available.

---

[4] See Chapter 4 for additional details.

Path options:

1. Extending the length of the noise path has merit if the property size will allow. The size of the lots in Santa Monica is less than ¼ acre, so there is insufficient distance available to make a significant noise reduction. A reduction of nearly 30 dB$_A$ is needed to render the noise source acceptable. There just isn't enough distance available to utilize this approach.

2. Construction of a sound wall or acoustical barrier between the source and receiver is a viable solution. The barrier must be long enough and high enough to provide an effective acoustical shadow. Placing the barrier near the source will extend the size of the acoustical shadow.[5] Note that the barrier will only block direct sounds. In this instance, the sound is reflecting off adjacent stucco walls and the underside of masonry eaves. These surfaces must be rendered highly sound-absorbent for the barrier to be acoustically effective. Analysis using the barrier worksheet in Appendix 2 shows that a solid fence 6 ft high will reduce the sound level of the source by 3 to 6 dB at the typical receiver location. While this solution is acoustically viable, it is not sufficient to be acceptable and therefore not practical for this case.

3. Enclosing the porch and patio with windows that are sealed in place will attenuate the sound by 20 to 30 dB$_A$. While very effective acoustically, this solution is unacceptable since the cost of enclosure is considerable and defeats the purpose of the outside living area concept. In addition, enclosure may have little effect on the bedroom sound levels. This solution was rejected as aesthetically unacceptable.

4. Upgrading the acoustical envelope of the house, including new windows, doors, etc., may solve the noise problem in the bedroom. However, it will have no effect on the outdoor living space. Since Santa Monica is very temperate all year long, there is no thermal advantage to be gained by adding thermally and acoustically efficient windows and doors. And, by closing up the acoustical envelope, one must add an air-conditioning system. (Note: This may be a practical solution in a northern climate.) Considering the cost of the potential upgrades, this solution is considered economically unacceptable.

ACOUSTICAL GOAL = 30 dB$_A$ REDUCTION

In order to provide an acceptable acoustical environment on the patio, porch, garden, and in the bedroom, a source level of 80 dB$_A$ must be reduced to a level at or below the ambient sound level, in this case determined to be 50 dB$_A$. This means the noise must be mitigated by reducing the source and/or attenuating along the path by at least 30 dB$_A$.[6]

---

[5] See Appendix 2 for an example of how to determine the barrier characteristics including the height, and location with respect to source and receiver, etc.

[6] NR = L_S - L_R where NR is noise reduction, L_S is the loudness at the source, and L_R is loudness at the receiver, all expressed in decibels (dB).

THE BOTTOM LINE

Source options 2 and 3 (i.e., installing resilient mounts and an acoustical enclosure around the sound source) were chosen by the client. The recommended enclosure shown in Figure 7-1 was constructed by the client with the approval of the owner of the equipment. Postevaluations determined that the enclosure was acoustically successful.

The consultant was pleased that a solution was found. However, it is disappointing to know that thousands of dollars were spent to quiet a noise source after the fact when the original manufacturer of the equipment could have provided the same degree of quiet, plus improved efficiency and durability, with an expense in the range of $50. Unless consumers make an issue of noise, manufacturers will continue to overlook their responsibility to keep our acoustical environment pollution-free.

7.2 QUIET DESIGN FOR A SINGLE-FAMILY RESIDENCE

This case study covers the acoustical design of a new house for Dave and Nancy Harris on a lot in Carmarthen Woods, Granville, Ohio, in 1983. The home of Denison University, Granville is a quaint old village located in the hills 20 miles east of Columbus, Ohio.

The goal was is to identify potential acoustical concerns and provide design solutions that can be implemented as part of the design and specifications. The ultimate goal is to achieve a design that, when built according to specifications within the budget, will provide an optimum acoustical environment that meets the desires and criteria of the owners. **This case study outlines a series of design elements that were incorporated into the construction drawings and specifications. This home is quiet even though normal appliances were utilized. The added cost was minimal and did not affect budget and/or bottom line.**

BACKGROUND INFORMATION

The following information, concerns, and criteria were obtained from a review of the preliminary plans and discussion with the owners.

Ambient acoustical condition: The construction site is surrounded by woods and farms. Lot sizes are generally 1 acre or larger. There is no impact of noise from highway, railroad, airport, or industry. An occasional seasonal noise impact may be noted from typical farm equipment on adjacent property. Private aircraft flyovers occur several times a day but not on a regularly scheduled basis. Access roads carry few vehicles. Ambient background sound is essentially nil. It varies primarily with the weather conditions. During a quiet evening, background sound is below the threshold of measurement for most sound level meters. At other times the weather conditions will dictate background sound levels, such as wind in the trees, farm animals, and local birds, including the pileated woodpecker. Because of the large lot sizes and the preponderance of trees, intruding neighbor noise is expected to be minimal. No local noise ordinances are known to exist.

Noise Sources: Pulse air furnace, dishwasher, garbage disposal, clothes washer and dryer, plumbing noise, power tools, autos in garage, air-conditioning unit, woodpeckers, neighbors.

Room acoustics: Goals include: privacy between living and sleeping areas, convenient distribution of a quality sound system designed for classical and contemporary music (i.e., stereo, CD player, tuner, tape player over multiple speakers located throughout the home.), speech intelligibility in entertainment spaces, quiet/privacy in sleeping areas, desire to hear exterior wildlife sounds while resting, and quiet floors and stairways (e.g., low impact noise).

Background noise levels: The background noise level is considered to be nil. In acoustical terms, the design level is 25 dB_A or less.

ACOUSTICAL ANALYSIS AND RECOMMENDATIONS

Noise sources:

Furnace: A significant noise source in this house is the pulse air furnace. This type of furnace has a unique noise element. Operation is similar to an internal combustion engine in that it has a pulsating heating mechanism that sounds like a Volkswagen. The furnace is highly energy efficient. It has no flue but has an air intake and outlet made of plastic pipe. Warm air is conveyed via a standard forced air fan to conventional air distribution ducts. This type of furnace can produce noise levels in excess of 60 dB_A, which will be very intrusive in this quiet environment. It must have significant noise quieting to be acceptable. Any extra benefits such as thermal efficiency will be considered a plus.

Proposed solution:

1. Specify that the furnace manufacturer and installer provide a muffler package for the intake and outlet pipes. If not available, utilize design proposed by acoustical consultant consisting of a pipe within a pipe (e.g., 2-in. diameter main line with a 4-in. diameter surround) with sound-absorbing material in the void space (e.g., 1-in. fiberglass pipe covering). Inside pipe shall have ¼-in. diameter holes approximately 2-in.(OC) to allow sound to be absorbed in fiberglass liner.[7]

2. Locate the furnace under the first-level stairway in a room that is capable of a 50 STC.

3. Surrounding walls and ceiling system shall be composed of 2 plies of gypsum wallboard mounted on resilient furring channels with R-19 fiberglass insulation in the core. All joints with floor, walls, and ceilings shall be sealed airtight with acoustical sealant.

4. Interior of room shall be lined with 2-in. thick 4-lb-density fiberglass board to absorb sound.

5. The door shall be solid-core exterior type with an airtight weather seal and door-activated threshold. Door hardware shall be of a type that will draw door tight into weather seal.

6. All pipe and electrical penetrations shall be made airtight with "Duct Seal," a permanently elastic clay-like material.

7. Air intakes and air distribution network shall be encapsulated with ½-in. gypsum board or acoustically equivalent wrap on the outside for at least 8 ft after it exits the furnace room.

8. All air intake, return, and distribution ducts shall have an interior fiberglass lining at least 1-in. thick

---

[7] A concerted effort was made to entice the manufacturer to provide a quiet version of the furnace. The manufacturer advised the consultant that "We are working on that," but did not have an approved system for market. The alternative solution was constructed, proved successful, and may now be available. Unfortunately, this type furnace did not have the durability expected and may now be removed from the market. Details of design are given here since they were implemented in the house design and are applicable to other noisy furnace designs.

throughout the entire house. They shall be 1-in. fiberglass duct liner board, 1-in. thick fiberglass flex ducts, or equivalent. Open-cell foam is not acceptable because it burns.

9. Furnace shall rest on resilient mounts or a 3-in. thickness of 8-lb-density fiberglass board.

10. All plumbing pipes shall be installed on resilient mounts and a resilient coupling shall be made next to unit connection to avoid structural vibration transfer.

11. Plastic pipe intake and outlet lines shall be routed in a special chase wall where the normal chimney would have been located. This chase shall be lined with fiberglass building insulation. Exit pipes shall extend above the roof line pointed upward to the open sky.

(Note: The foregoing efforts were implemented in the completed house. No furnace noise intruded in any indoor or outdoor living space during furnace operation.)

Dishwasher: The washer specified was selected because it was the quietest (55 dB$_A$ at 4 ft) available at the time of purchase. The owners reviewed the operation of at least six popular brands that also met their criteria for dishwashers. Installation specifications require the dishwasher to be installed on a resilient pad in a cabinet that is lined with 1-in. thick fiberglass insulation to absorb some of the sound. Kitchen location was selected for efficiency, convenience, and low noise impact on the dining room and other living spaces. Specify plumbing hookups with flexible couplings and resilient mounts to isolate the sound of electronic valves from being transmitted and amplified by the structure.[8]

Garbage disposal: The garbage disposal unit specified was selected because it was the quietest one available at the time of purchase. The owners reviewed the operation of several popular brands that also met their criteria for disposing garbage. Installation specifications require the disposal to be installed on a resilient ring to the sink in a cabinet that is lined with 1-in. thick fiberglass insulation to absorb some of the sound. Specifications required plumbing hookups with flexible couplings and resilient mounts to prevent vibrations from being transmitted and amplified by the structure or cabinets.

Clothes washer and dryer: Washer and dryer are located on the lower level, where noise will not intrude into living spaces. Specifications call for all plumbing lines to be isolated with a flexible coupling and resilient mounts. Dryer exhaust is located where sound and steam will not intrude on exterior living areas.

Plumbing noise: All plumbing lines shall be copper, installed on resilient pad mounts. This will reduce the likelihood of noise being transmitted via the structure and dampen any "singing pipes." All shutoff valves, particularly those serving automatic washers of any type, shall have an air pocket chamber located in the near vicinity of the upstream pipe. This will eliminate pipe banging. All hookups, both supply and drain, to automatic dishwashers, clothes washers, and disposals shall have a flexible connector.

Power tools: The workroom where power tools may be operating was located on the ground floor, off the garage. The entire room was finished with gypsum wallboard on resilient furring channels. All spaces

---

[8] Dishwashers and other kitchen appliances have become quieter in the 1990s as a result of manufacturer efforts to market quiet equipment. Many models have a noise rating of 50 dB$_A$ which is reasonably quiet and suitable for most kitchens. A new ASTM procedure, E1574-95, Standard Test Method for Measurement of Sound in Residential Spaces, should be required in current specifications. Some kitchen equipment suppliers will provide additional noise-quieting packages on request.

between floor joists and stud walls were filled with fiberglass building insulation rated R-11 or higher.. Doors accessing the space were solid-core exterior type with a full weather seal and a cam-type latch set. All penetrations and cracks were sealed with acoustical sealant. Additional sound-absorbing material composed of 1-in. thick glass-cloth-faced fiberglass ceiling board was installed on the inside of the room ceiling and upper wall surface to absorb sound. All electrical outlets in the ceiling were surface-mounted to avoid penetrations. All workbenches were free standing to avoid structural transfer of noise and vibrations.

Garage: The two-car garage is located on the lower level, under the kitchen and dining area. Sleeping areas are located well away from this potential noise source with the master bedroom on the opposite end of the house and other bedrooms on the second level. To assure the noise from the garage door opener did not intrude on the living space, an opener was selected to be quiet and was installed on resilient mounts. All walls, ceilings, and doors separating a garage from living space must be 1-hour fire rated, so ⅝-in. Type X gypsum wallboard was specified. A fully weather-sealed solid-core exterior door to the living space was specified. Since the garage was not heated, all walls and ceilings contained at least R-11 fiberglass building insulation between joists and studs.

Air-conditioning unit: The compressor and exterior portion of the air-conditioning unit was carefully located under the kitchen deck so that the noise it produced would not intrude to other outdoor living areas. This location also offers the opportunity for the unit to be carefully enclosed if the noise proved intrusive.[9] Present day specifications should require the unit to not exceed 60 dB_A at 10 ft or an equivalent noise rating. All service lines should be installed with a flexible coupling and resilient mounts.

Woodpeckers: The presence of pileated woodpeckers is of special concern. For the reader that does not know about pileated woodpeckers, the bird is about the size of a crow, is an endangered species, and will produce a very loud "rat-a-tat-tat" that can be heard from an amazing distance in the woods as it goes after bugs in wood. The rapid but intermittent noise may be music to a wildlife enthusiast but will definitely disturb sleep. Estimates are that the sound level could exceed 70 dB_A at 30 ft. The owners were concerned since the exterior of the house is entirely cedar boards. A woodpecker that went after the siding could very well pass over the line from "a wonderful wildlife experience" to "annoyance." Another choice of exterior material, such as masonry, would easily reduce the noise possibility. However, the owners were not willing to make this dramatic aesthetic change. As it turns out, an experimental sheathing material composed of 2-in. thick 4-lb-density fiberglass board was applied to the entire exterior of the house under the cedar board siding. The fiberglass board acted as a damping material, thereby localizing the impact sounds and providing additional sound transmission loss attributes to the exterior envelope.[10]

Neighbors: Noise from neighbors was considered minimal. The large lots, over 1 acre, meant that

---

[9] See a more detailed description of how to quiet equipment of this type in Section 7.1.

[10] A test of the foregoing installation and premise was contemplated but did not occur. Unfortunately, the owner moved away before there was a woodpecker attack. However, years later, a friend passed along the story that a local did encounter a woodpecker attack on the house. Apparently some wood repair was required but the noise was not a problem. The writer can only speculate as to the effectiveness of the proposed design.

distance between living units was over 150 ft. In addition, the lots on either side required driveways and garages to be facing away from the lot in question. About the only noise that may have been intrusive would be the neighbors air-conditioning unit. Since the subject house was the first constructed in the community, it was not possible to anticipate this problem. It was the owners' intent to contact the neighbor's builder and provide recommendations for quiet as soon as a building contract was in the offing for the adjacent property.

Room acoustics:

Design for good room acoustics in a single-family residence have been overlooked by most designers, builders, and owners. Noise has generally been considered as unwanted sounds that intrude upon a residence from outside sources. This concern is addressed at length in Chapter 6, and has prompted the adoption of noise ordinances.[11] The following issues deal with noise transmission between living areas and sound distribution and sound quality within individual rooms within the single-family dwelling.

Privacy between living/sleeping areas: The key to providing quiet for resting is to locate the bedrooms away from potential noise sources. A residence floor plan that isolates the bedrooms is the most successful technique. In this project, the master bedroom is on the other end of the house, away from the kitchen, washer/dryer, garage, furnace, and other potential noise sources. Guest bedrooms are located upstairs and except for noise coming through the floor/ceiling are quite isolated from noise sources. The master bedroom bath suite is separated by a door to the dressing room plus another door into the bathing area.

It was recommended at the onset that all sleeping areas be separated from other living areas by buffer zones and/or partitions having an STC of 45 or greater. Specific solutions are as follows:

1. Walls between master bedroom and living room as well as guest bedrooms and living room are constructed of wood studs, resilient furring channels, a core of fiberglass building insulation and conventional ½-in. gypsum wallboard.

2. All back-to-back outlets and penetrations were eliminated and all junctures, cracks, and potential sound leaks caulked airtight with acoustical sealant. All bedroom doors are solid-core type with acoustical seals.

3. Guest bedrooms are separated by the bathroom, a full-length closet and a hallway.

4. To assure noise does not penetrate up through the floor/ceiling from the kitchen and dining area into the guest rooms, the first floor ceiling has resilient furring channels with a core of fiberglass building insulation.

5. All supply and return air ducts are lined on the inside with 1-in. thick fiberglass board to absorb sound as it may travel down the ducts. All rooms are ducted separately to avoid a direct sound line between adjacent rooms.

6. The floor-ceiling between the downstairs recreation room and the master bedroom is constructed of resilient furring channels with a core of fiberglass building insulation.

7. Separate fireplace flues were installed to service the back-to-back fireplaces between the living room and master bedroom.

8. Plumbing was eliminated in the wall separating the master bedroom from the master bath suite. A common plumbing wall exists between the master bath and the powder room.

---

[11] See Chapters 2 and 4 for additional information.

Sound system design: A state-of-the-art stereo system was the central part of the acoustical design. Since speakers were desired in all rooms, the entire house was prewired for speaker cable outlets, TV antenna, and phone outlets. Base equipment and all controls were located on an easy-to-reach shelf, part way up the front stairs along with extra storage for tapes, records, music, and related books. This cabinet has a cooling duct into the chimney flue. Speaker locations were chosen with two major considerations: (1) no speaker would be mounted on a wall separating a living space from a sleeping space and (2) speakers were located to provide maximum stereo separation plus a mix of sound in the most likely receiving spot. This meant that the living room speakers were mounted on the second level balcony. No additional sound absorption was added, since there would be adequate sound absorption from drapes, carpets, and furniture.[12]

Design for speech intelligibility: This project has such a low ambient sound level and sufficient sound absorption from the carpet, drapes, and furniture that no additional provisions for speech intelligibility were necessary. If the decor had been hard-wood or ceramic tile floors, blinds with no curtains, and furniture with little padding, some additional sound absorption in the form of acoustical wall and/or ceiling panels would have been recommended to reduce the reverberation time (i.e., echoes).

Hearing the wildlife: A review of the exterior design will reveal that there are many exterior doors and windows. Not only do they bring natural light into the interior of the house, but nearly all units are operable. When open, these doors and windows will bring the sound of the outdoors to the interior space. This also brings with it natural ventilation. All doors and windows are thermal insulating glass with efficient weather seals, thereby allowing the sounds of storms to be excluded when in the closed position.

Impact sounds: All stairways should be mounted so that they transfer their load to the floor in use rather than an adjacent wall. In this case, all stairways are covered with carpet and pad and considered adequate for eliminating undue impact noise. The kitchen and bath floors were specified to be a foam-backed vinyl that effectively reduces impact sounds.[13]

FIELD ANALYSIS AND EVALUATION

Nearly all of the foregoing concepts and recommendations were implemented. Unfortunately, the author was unable to make a detailed analysis of the acoustical attributes because of a corporate transfer. The house was essentially complete and ready for occupancy when the last analysis ensued. All the efforts appeared to be very successful. The furnace noise was nearly nonexistent. Other concerns were likewise acceptable.

---

[12] Current stereo equipment incorporates a much broader range of frequencies, essentially no sound distortion or background sound and reverberant qualities compared to that which existed in 1982. A semireverberant quality of the room covered up poor-quality sound reproduction on older equipment If the current equipment were to be utilized in this project, it might be appropriate to render all surfaces in the entire listening space acoustically absorbent. This will allow full adjustment and utilization of the equipment. This extra cost is unwarranted unless the electronic reproduction system is state-of-the-art or recording-studio quality.

[13] Unfortunately, the manufacturer of the foam-backed vinyl has ceased making this product. A new product available today would be preferred. Called constrained-layer damped plywood, this product replaces conventional subfloor material. It is excellent at reducing the impact noise and improving the perceived solidity of any floor system including vinyl, ceramic, or hardwood flooring. Additional details are given in Section 7.5.

The following figures show the house plans discussed in various sections above.

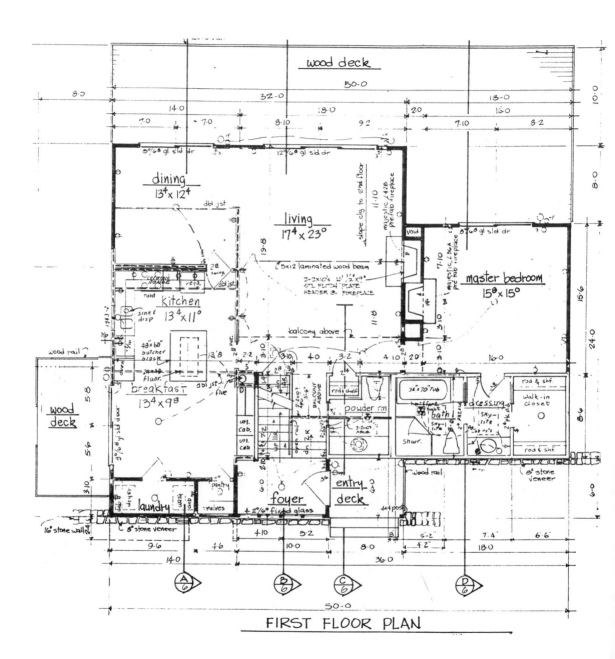

FIRST FLOOR PLAN

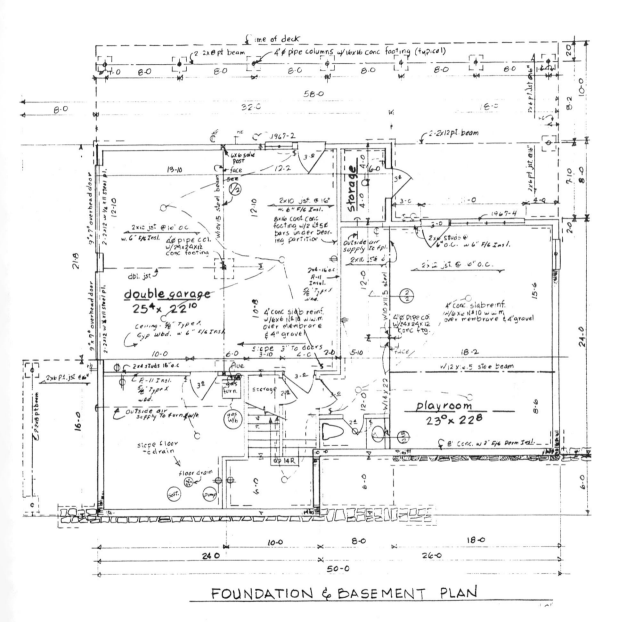

FOUNDATION & BASEMENT PLAN

231

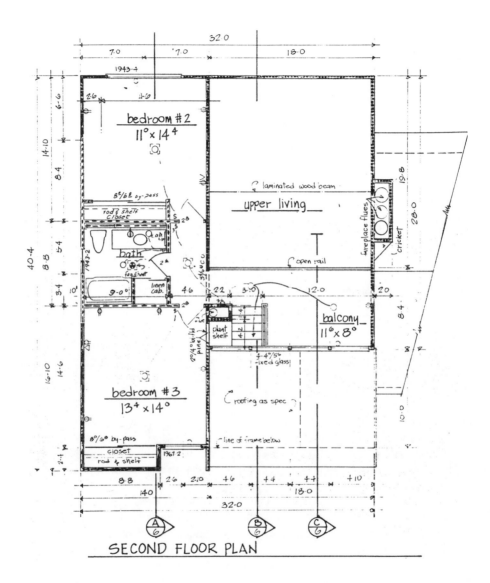

SECOND FLOOR PLAN

232

7.3 **QUIET DESIGN OF A MULTIFAMILY RESIDENTIAL COMPLEX**

This is an acoustical analysis of an existing complex of buildings, each with six luxury condominiums. The site contains eight buildings, all with essentially the same layout. Each building has two side-by-side living units on each of three floors. (Plans are given at the end of the study.) The complex overlooks the Port Ludlow Marina, nestled in a quiet cove of Ludlow Bay in Port Ludlow, Wash. The promotional literature describes the setting as "a scene that takes your breath away. You see the famous Bay, with birds, sailboats, and changing tides. You see tall and green trees by the thousands – no hundred thousands! You see the beach, one for you and one across the Bay." Part of a resort and retirement community of largely single-family dwellings, Port Ludlow is a planned community with a projected build-out for approximately 4000 persons and many northwest amenities including a 27-hole golf course, yacht harbor, tennis courts, parks, trails, condos, resort apartments, an inn, restaurants, and associated meeting andclub facilities. **Retrofit recommendations are provided with predicted acoustical benefit.**

PROBLEM IDENTIFICATION

Review and analyze the site and plans, identify potential acoustical concerns, and provide recommended design solutions. The goal is to provide potential owners with an optimum acoustical environment commensurate with the luxury-type accommodations and ambiance of Port Ludlow.

BACKGROUND INFORMATION

Information gained by an on-site review and analysis of the existing plans indicated:

Ambient acoustical condition: The site is bounded on the back, or view side, by a large grassy area sloping down to Port Ludlow Bay. Immediately off-shore is a typical Pacific Northwest log dump where logs are sized and deposited in the saltwater for soaking and collection for towing by tug to a mill.[14] The log dump has recently been deactivated. On the uphill side, the site has parking spaces scattered among old-growth spruce, fir, and cedar trees along with many native plants. Beyond the parking and green-belt area is the main highway in Port Ludlow, consisting of a winding two-lane road with a speed limit of 35 to 40 mi/hr. The complex is served by one entrance.

During the off season, ambient sound conditions are very quiet, with a background level of under 25 dB_A, a common occurrence. During high-traffic periods, the background level due to the highway may reach 55 to 60 dB_A. In addition to the highway, the site is impacted by occasional floatplanes that take off and land on the bay, power boats cruising at less than 5 knots, dinghy outboards and, until recently, a tiny but noisy tug that maneuvered logs into place on the log jam for pickup occasionally by a large oceangoing tugboat. On peak tourist season weekends, Ludlow Bay is a popular destination for hundreds of cruising boats and associated parties.

---

[14] Port Ludlow used to be the site of a mill that processed logs into timbers and wood chips. The mill closed in the early 1900s and many of the buildings were demolished or moved. In the late 1970s Pope and Talbot conceived and eventually constructed Port Ludlow as a planned community on the site of the old mill, surrounding residences, and commercial buildings.

Most of the permanent residents are active retirees. During the winter months the units are less than 50% occupied. There are few children except on special occasions. The units being evaluated were constructed approximately 15 years ago. There is no local noise ordinance. However, construction was governed by the Uniform Building Code and the state of Washington noise criteria. Some effort was made to conform, since vertical separation between units is double-wall construction.

Noise sources: Community noise sources include boats, airplanes, auto's, construction equipment, tugboats, and parties. Interior noise sources include neighbors, dishwashers, furnaces, washer/dryers, garbage disposals, vacuum cleaners, TV and stereo, etc.

Room acoustics: No effort has been taken to control room acoustics. Most units are carpeted, have drapes, and have gypsum wallboard walls and ceilings. Most kitchens and baths have vinyl floor covering. Exterior stairways and decks are wood, sometimes covered with indoor/outdoor carpeting. Rooms have reasonable reverberation time for speech if fully furnished. Reverberation time is long enough to be annoying for speech if furnished sparsely or with hardwood floors, hard furniture and blinds on windows.

Background noise levels: During the off season or a quiet time, the lowest background level is 25 dB$_A$. At high season, background levels may reach 50 dB$_A$. Design parameters are based on a background level of 40 dB$_A$.

ACOUSTICAL ANALYSIS AND RECOMMENDATIONS

Noise sources:

Boats: Port Ludlow Bay has a speed limit of 5 knots maximum. Most boats that enter the harbor are coming in to dock, anchor, or slowly cruise. As a result, motor noise is not intrusive and rarely exceeds the ambient noise level. Occasionally, a dinghy or jet ski will be intrusive. However, their noise is intrusive only when they are exceeding the 5-knot speed limit. No special measures other than law enforcement are required.

Float planes: Landings and takeoffs from Ludlow Bay are commonplace. While there are no scheduled flights, charter flights may reach 20 on a high-activity day in the summer. Flights may be as few as once a week during the off season. A typical landing occurs within several hundred yards of the subject site. However, the light planes are throttled way back on landing and are not a noise intruder. Takeoff is quite a different situation. Noise levels of 65 to 70 dB$_A$ for a period of less than 1 minute are common at the site.

Unfortunately, not much can be done about the noise from floatplanes. They are governed by FAA rules. FAA officials have made it clear that unless airplanes cause community noise levels to exceed the 80 L_{dn} criteria, no action can be taken.[15] Fortunately, most residents are fascinated by the float planes and are willing to accommodate the noise they produce. As a result of the low frequency of flights, the short duration of the noise and the acceptance by residents, no noise mitigation efforts have been implemented nor are anticipated. If the noise does become a nuisance, a noise ordinance has been written for Port Ludlow that will require

---

[15] L$_{dn}$ is day/night loudness level. See Section 4.2.2 for additional details.

compliance. The simple solution, if the ordinance is enacted, is to restrict takeoffs to the far reaches of the bay, a mild inconvenience to pilots.

Other planes are an occasional problem, since Port Ludlow is under the flight path of scheduled and charter flights to other sites on North Puget Sound. At present all of these flights utilize light or commuter aircraft which are not significant noise intruders. Whidbey Island Naval Air Station is some 20 miles away, and an occasional A-6 or similar craft will create a considerable noise intrusion. Of particular interest are the Navy submarine hunters that fly weekend warrior training missions at mast height over Admiralty Inlet with return runs at tree top height. These four-engine prop planes are a major noise intrusion. However, the intrusions are few and not considered a serious problem. Likewise, Port Ludlow is not under any normal commercial flight paths. As a result, no noise mitigation measures are being considered.[16]

Highway noise: The main highway that parallels the site property is a two-lane road with speed limits of 30 to 40 mi/h. During the tourist season, traffic noise is a potential noise intruder. Peak noise levels of 65 dB$_A$ at the building facade are commonplace. A long-term average has been established at 55 dB$_A$ for design purposes. Fortunately, all but local delivery vehicles and autos are now banned from this road, thereby reducing the noise intrusion considerably. Prior to this ban, heavy vehicles with "jack brakes" were a significant noise intrusion.

For several reasons, highway noise is not considered intrusive to these living units. First and foremost is the fact that the living units are oriented toward the water. The side exposed to traffic noise is almost void of windows. Those windows that do occur on this side are double-pane insulating glass. The building facade is estimated to have a noise reduction capability of 35 dB$_A$. Thus, a noise level of 65 is reduced by 35 as it impacts the exterior envelope, yielding an interior noise level of 30 dBA, well below levels considered acceptable for sleeping. Master bedrooms generally overlook the view, leaving only bathrooms, kitchens, extra bedrooms, offices, and utility areas exposed to highway noise. This buffer technique works very well. In addition, the speed limit, extensive greenbelt area, and downward-sloping site further mitigate potential highway noise.

The foregoing techniques are considered adequate and no additional measures are considered necessary. Should highway noise become a problem that is not adequately handled by speed limits, a noise barrier erected next to the highway is recommended. A 5 to 8-ft high barrier with a sound transmission class (STC) rating of 20 or greater would produce a considerable acoustical shadow, since the living units are well below the grade of the road.[17] Specific vehicles that produce a noise intrusion may not meet the criteria in the proposed Port Ludlow Noise Ordinance.[18]

Construction equipment noise: Port Ludlow always seems to be under construction, with some form of

---

[16] Should the proposed Port Ludlow Noise Ordinance be adopted, the naval aircraft, Boeing test flights, and some commercial overflights will not comply with the ordinance criteria. A jurisdictional confrontation may ensue. For full text of the Port Ludlow Noise Ordinance see Section 4.3.

[17] See Appendix 2 for design guide worksheets covering the technique for estimating sound attenuation from an acoustical barrier. Note that the barrier placed close to the source produces the largest acoustical shadow.

[18] See Section 4.2 for details of the Port Ludlow Noise Ordinance. Also see Section 4.2.1 for additional explanations of highway noise.

equipment for digging, banging, sawing, power generations, etc. that produces an intruding noise. Fortunately, most contractors are aware they produce noise and make a sincere effort to implement noise mitigation on their vehicles and equipment. Mufflers are now very efficient and relatively economical. If the noise is intrusive, direct communication with the owners will likely cause them to reduce the noise to a palatable situation. If they cannot adequately quiet the equipment, most will limit their use to normal work hours of 8:00 a.m. to 5:00 p.m. For those who still do not comply, the state of Washington has an appropriate ordinance and criteria to enforce compliance. The proposed Port Ludlow Noise Ordinance has a specific section and criteria for noise from construction equipment.[19] Noise levels at the property line in excess of 65 dB_A will generally be out of compliance.

Tugboats: It is rare that large tugboats create a noise that is intrusive. Their engines are well below the water line and large motors tend to operate at low rpm. The small tug used to pull logs around the log boom was noisy because the operator removed the mufflers. This operation has now ceased. Most operators will comply readily when challenged. If the noise persists, it may not be in compliance with local noise ordinances. No mitigation measures are required.

Parties: Boater parties are notoriously noisy and can become a noise intrusion. Port Ludlow residents have come to expect parties on the major holiday events and no complaints of significance have been voiced. Usually, an emissary to the party voicing concern is sufficient to quiet the event to manageable proportions. If the noise persists, local law enforcement is handled by the county sheriff who is overtaxed and may be of little help. When and if the Port Ludlow Noise Ordinance is enacted, specific rules apply to quieting a loud party. Except for several days a year, this noise source is nonexistent. No additional mitigation measures are required or anticipated.

Interior noise sources:

Neighbors: Most building codes require that the demising wall or floor/ceiling system that separates living units in multifamily dwellings have a sound transmission class rating suitable for the situation.[20] For example, the FHA criteria for living units in a low background noise environment, such as found in this project, requires the demising barrier (e.g. wall and/or floors) to be 55 STC and 55 IIC.[21]

While there may not have been any specific criteria applicable at the time the case study units were designed and constructed, the units do have horizontal separation composed of a double wood stud wall with a core of building insulation and gypsum wallboard facings. This assembly when tested in a laboratory per ASTM E90 generally exhibits a 50 to 54 STC. The field installation is estimated to have an STC of 45 to 50. Confirmed by interviews with occupants, the horizontal separation between neighbors appears to be nearly

---

[19] See Sections 4.3.4 and 4.2.4 for Occupational Safety and Health Acminustration (OSHA) criteria.

[20] Federal Housing Administration and Vetrans Administration (FHA/VA) Minimum Property Standards, all the major model building codes and many local building codes have specific criteria for airborne and impact noise ratings such as STC and impact isolation class (IIC) for demising barriers between living units in multifamily construction. See Chapter 4 for specifics.

[21] See Chapter 4, Tables I - III for specific airborne and impact noise criteria.

adequate. Since most units are unoccupied much of the year, the source is eliminated. When occupied, neighbors are occasionally disturbed by a loud TV, stereo, or conversation. It is recommended that the demising wall be upgraded with an additional layer of gypsum wallboard on each face when economically feasible.

Vertical separation (e.g., floor/ceiling system) between living units is inadequate. The floor/ceiling assembly between over and under living units is conventional wood joists with plywood subfloors and gypsum wallboard ceilings. There is a good likelihood that the core also contains building insulation. While no specific tests have been conducted, it is very evident that the STC is less than 45 and the IIC for carpeted areas is close to 55, but far less for noncarpeted areas. Consequently, neighbors who happen to be in the lower units have considerable neighbor noise intrusion from both airborne and impact noise. Added to the concern is the fact that the unit layouts are not "stacked." Kitchens occur above master bedrooms in several instances.[22]

Had these deficiencies been noted at the time of design and construction, simple and inexpensive upgrades utilizing resilient furring channels, building insulation cores, and foamed-backed vinyl or carpeted flooring should have been utilized. Construction today, should include damped plywood underlayment[23] and floor/ceiling designs having a minimum of 55 STC and 55 IIC.[24] The upgrade of these units will be expensive. It is recommended that all floor covering be replaced with a thick, high-quality carpet and pad, including kitchens and bathrooms located above bedrooms and like noise sensitive areas, and all ceilings have an added layer of gypsum wallboard installed over resilient furring channels.

Dishwashers: Noisy dishwashers from the time period the units were constructed are notorious. Not only do they intrude within the occupied space, but neighbor dishwasher noise can be very disturbing. It is recommended that a noisy dishwasher be replaced by a new quiet model and that it be installed with resilient mounts, flexible joints for all pipe connections and that the interior of the cabinet be lined with a 1-in.-thick layer of sound-absorbing material such as fiberglass or open-cell foam.[25] Noise levels should not exceed 50 dB_A for installed units. Ask manufacturer for noise rating.

Furnace: The units in this complex are served by electric-forced air heat, although some have been upgraded in recent years with heat pumps. Noise from the furnace fans is relatively low, since furnace units are located in an outside access furnace room and long duct lines and numerous pipe bends tend to attenuate the sound of fan noise within the occupied space. Also, the furnaces are isolated from the living units because of their location in a separate room. Noise may be experienced from the outside compressor units for the heat pumps. The pump and motor noise may exceed 65 dB_A within several feet of these units. If they are judiciously placed, that noise may not be intrusive. However, a few units are located in the immediate vicinity

---

[22] Chapters 1 and 2 contain design suggestions.

[23] See case study in Section 7.5.

[24] Appendix A contains test data for many acceptable floor/ceiling assemblies.

[25] Fiberglass is noncombustible. Most open-cell foam materials, while they are easier to handle and have better sound-absorbing properties, are flammable. Foams should not be installed in buildings requiring noncombustible-rated materials such as high-rise and commercial structures.

of outdoor living areas such as decks and patios. These units should be moved and/or quieted.[26] For those furnace units where fan noise is intrusive, remove the metal ducts and install fiberglass-lined distribution ducts for at least 10 ft out from the furnace on both the intake and distribution side. It is assumed that each furnace is individually ducted to each living unit thereby eliminating any cross talk between neighbors via ducts. Since no complaints of this type have been forthcoming, it is assumed this is the case.

Washer/dryer: In each layout, the washer and dryer are located in the central part of the plan. A typical unit will produce noise levels between 60 to 70 dB$_A$, which is considered intrusive. Manufacturers have done little to quiet these machines in recent years. It is recommended that all the machines be mounted on a resilient mount, flexible couplings installed in all water lines, folding closed doors be replaced with solid doors with thermal seals including drop closures, and available interior wall and ceilings of the closet be lined with a sound-absorbing material such as acoustical ceiling board. If the closet is not vented, an exterior vent should be installed to avoid excess heat buildup.[27]

Garbage disposal: Quiet garbage disposals are now available. It is recommended that new equipment with a low noise rating be selected that is installed on resilient mounts and that flexible pipe connectors be installed in all intake and outlet lines.

Vacuum cleaners: A typical vacuum cleaner will produce a near field sound level of 80 dB$_A$. Until manufacturers realize that quieter equipment will sell, it will be difficult to purchase quiet equipment. Central vacuums reduce the noise in the occupied space. However, the base equipment is still loud and must be carefully located to minimize the noise impact both within the living unit and the neighbors unit. The best noise mitigation procedure is to limit using vacuum cleaners to the hours between 9:00 a.m. and 7:00 p.m.

TV and stereo: Electronically produced sound can become intrusive in two ways. Airborne sound levels may be made excessive, either to satisfy someone with a hearing loss, or to "hear and feel the sound" of a favorite tape or compact disc. Structureborne noise may be transmitted by speakers installed on a resonating panel that is activating in either the source or receiving area. For airborne sound, turning down the level is by far the most effective. If a simple request to a neighbor to turn it down is not accepted, then the barrier, be it a demising wall or floor/ceiling, must be upgraded. To improve the STC of the barrier, add another layer of gypsum wallboard over resilient furring channels and be sure all flanking paths are blocked effectively.[28] To mitigate structureborne sounds, be sure the speakers are located away from a demising wall or floor/ceiling. Instead of resting the speaker cabinets on the floor, place them on a shelf resting on a resilient isolation pad. If the owner of state-of-the-art electronic music equipment lives next door, purchase the owner a set of quality earphones and demonstrate their superior clarity of reproduction. Since our younger society grew up with loud music, it is difficult to force them to turn down the volume. Their culture has created a

---

[26] See case study in Section 7.1 for tips on how to quiet exterior noise sources such as pumps and motors.

[27] Details may be found in Section 3.4

[28] See Section 1.1 and Sections 2.4 to 2.9.

whole generation with significant hearing loss. Earphones will hopefully allow listeners to enjoy the music their way without adversely impacting their neighbors. If that technique doesn't help, turn to enforcement of noise ordinances. Of all the noise problems we encounter, this is probably the most difficult, since it is not a technical but a social problem.

Pianos, musical instruments and musicians: Whole books have been written on how to isolate musical instruments used for home practice from disturbed neighbors. The most obvious technique is to impose limits on when the practice sessions occur. If this is not feasible, then the source room must be isolated from the neighbors by placing the instrument on a resilient pad if it is a piano or the like, upgrading the sound barrier performance of the acoustical envelope, and then adding sound-absorbing material to the source room. A very effective technique for children learning to play the piano is to purchase an electronic piano with earphones. In short, if the project requires isolation of a musical instrument, all the techniques for mitigating sound at the source and along the path must be employed. Each circumstance requires detailed analysis by an acoustician for effective and practical results.

Room acoustics:

Most residential settings require little attention to room acoustics, and this is true for this project. Carpet and pad, draperies, and soft furniture are usually sufficient to reduce the reverberation time to under 2 seconds, considered acceptable for speech clarity in most residential spaces. For very large rooms having hard surface floors, lots of uncovered glass, and metal or all-wood furniture, reverberation time may cause echoes and reduce speech intelligibility. While not the case in this instance, the addition of sound-absorbing acoustical ceiling panels and/or wall panels may be required.[29]

Stairs, entries, and decks: This project has a special noise problem with exterior stairs, hallways, and decks. The classic northwest design utilizes exterior wood stairs, bridges and open hallways. In this particular case, the floor systems are all wood and they are structurally tied to the living units. As people walk these routes, footfalls are commonly heard in the adjacent living spaces. A simple design change could have avoided this occurrence. By isolating the structure of the stairs, bridges, and hallways from the living units, structurally transmitted noise would have been eliminated.

Retrofit solutions are not so simple and may be expensive. An initial solution would be to cover all these spaces with a thick and soft outdoor carpet such as Astroturf. If this is not acceptable, a redesign of the structures utilizing techniques that isolate the supporting structural members from the existing units is required. New pilings and/or resilient mounts may be necessary along with new structural calculations and engineering designs to satisfy building code criteria.

---

[29] Sound-absorbing materials and ratings plus a worksheet for calculating the reverberation time are provided in Appendix A.

SUMMARY OF RECOMMENDATIONS:

1. Enact the proposed Port Ludlow Noise Ordinances so that if an unacceptable noise occurs, there is legal recourse to quiet the source.

2. Floor/ceilings should be upgraded to STC 55 and IIC 55.

3. Install quiet equipment and/or appliances.

4. Install outdoor carpeting on all exterior stairs, walkways, halls, and decks.

5. Construct a 5 ft-high noise barrier along the adjacent highway.

6. Educate residents on the value of thinking quiet in their activities (i.e., turn down the TV and stereo, don't yell at midnight, leave dancing to the dance hall, etc.). Respect the quiet environment of Port Ludlow and **revel in the sounds of the Pacific northwest wilderness – a place of natural acoustical beauty.**

FLOOR PLAN - LEVEL THREE

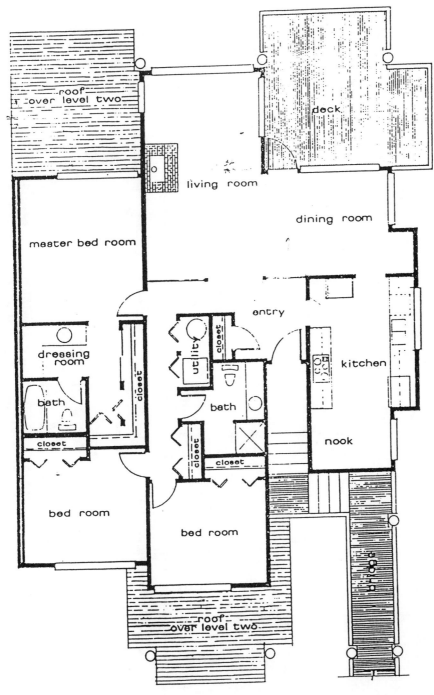

FLOOR PLAN - LEVEL TWO

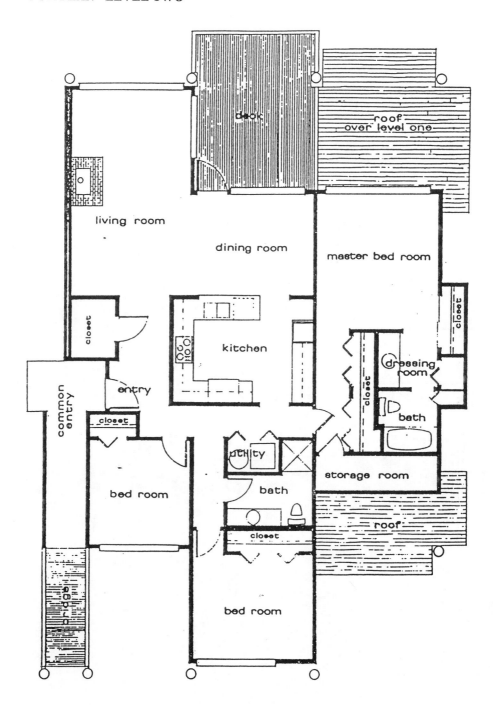

FLOOR PLAN - LEVEL ONE

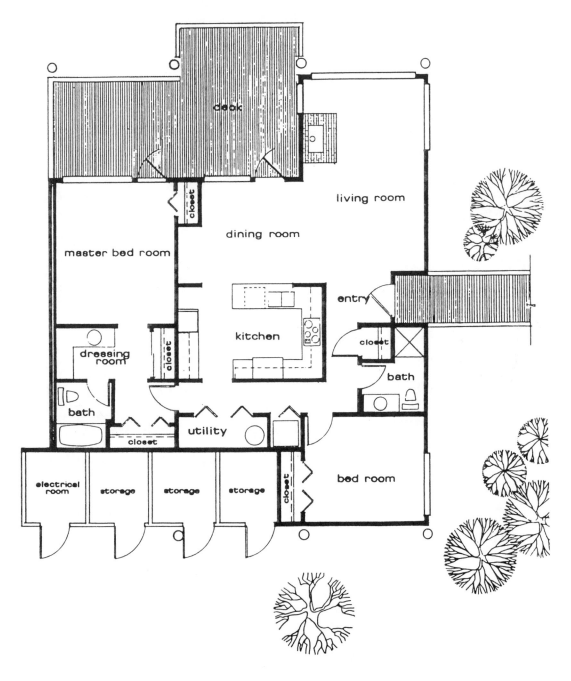

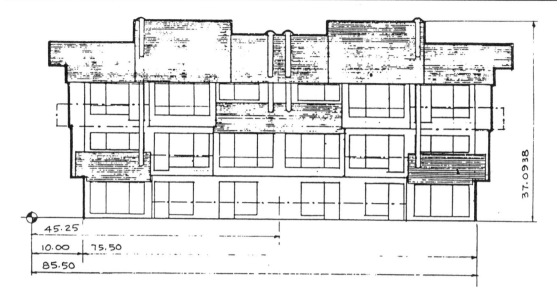

SOUTH ELEVATION

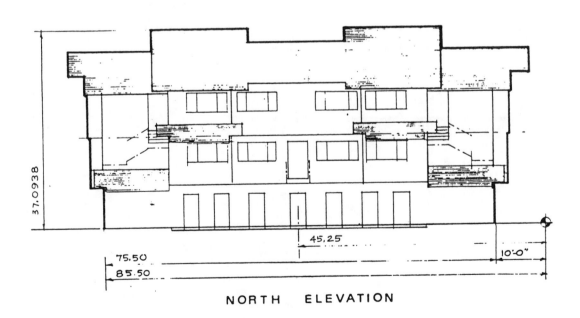

NORTH ELEVATION

244

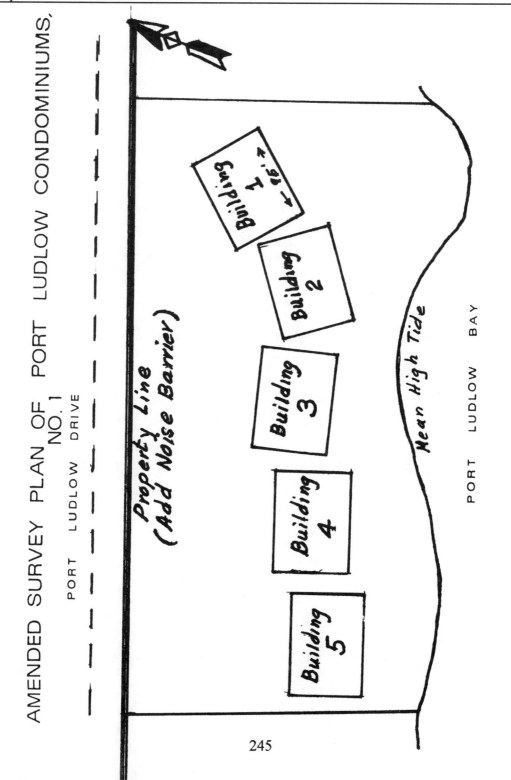

AMENDED SURVEY PLAN OF PORT LUDLOW CONDOMINIUMS, NO.1

PORT LUDLOW DRIVE

Property Line
(Add Noise Barrier)

Building 1

Building 2

Building 3

Building 4

Building 5

Mean High Tide

PORT LUDLOW BAY

7.4 **30 YEARS of RESILIENT FURRING CHANNELS** AND STILL GOING

Resilient furring channels have become one of the most successful acoustical products invented. First introduced to the market-place in the early 1960s by United States Gypsum Co. (USG), Donn Products, and National Gypsum Company (Gold Bond), resilient furring channels have not changed in design for over 30 years. They still enjoy nearly the same sales volume they did in the 1970s. Billions of feet have been installed to provide a sound barrier between living units in multifamily dwellings across North America. Why has the channel been so successful? Because it is simple to understand, is easy to use, and has proven performance. **This study describes the acoustical attributes and application nuances of many wall and floor/ceiling systems that utilize resilient channels. Results of field and lab tests indicate that resilient furring channels provide a user-friendly technique for improving the sound transmission loss characteristics on light wood frame structures.**

Over 30 years ago, the author, then a young engineer at National Gypsum Co. Research Center, received an assignment to find a way to compete with the USG RC-1 channel. My response was "What is a resilient channel?" As the project evolved, I thought this was just a marketing gimmick. I wasn't convinced that it would provide better sound transmission class (STC) and impact isolation class (IIC) for wood-stud walls and wood-joist floor/ceiling systems. But methodical research of the technology proved otherwise. Extensive ongoing research with resilient clips demonstrated that adding resilient connectors did upgrade the acoustical attributes of both walls and floor/ceiling assemblies. And, by adding a sound absorbing material, like building insulation, to system cavities the STC and IIC are made even better.

Drywall steel studs had just been invented. Extensive research and development effort had produced efficient self-drilling screws and screw guns. Since earlier efforts with resilient clips and furring strips produced nail pops, it was a natural progression to wed the two technologies, namely, screw attachment to a resilient member. Steel-stud research demonstrated significant improvements in STC ratings because the steel studs were inherently resilient. Resilient channel design then was reduced to finding a correct balance of strength and resilience for wood studs and joists. Acoustical, fire, and structural testing ensued to demonstrate that our design did provide superior acoustical performance while maintaining the load-bearing requirements typically found in residential construction.

A negative aspect of both resilient channels and drywall steel studs was that resulting walls felt "spongy." Structural engineers and code authorities were skeptical. Some rejected the lower structural stiffness. But calculations by structural steel engineers coupled with a strong marketing effort by steel-stud manufacturers eventually eroded the old strength and stiffness criteria for walls and ceilings. A major barrier to the market acceptance of resilient channels was breached.

Inventing a product to compete with the RC-1 channel proved challenging. USG had a patent and a family of acoustical tests with efficient STC ratings. The RC-1 channel utilized a simple roll-forming process with a minimum amount of material that yielded low production costs. Complicating the situation was another channel introduced by Donn Products. After months of experimentation, the Gold Bond resilient furring channel emerged. With expanded metal legs providing the correct degree of resilient attachment, the channel could be made on existing roll-forming equipment and required even less material than the RC-1 and Donn channels. Extensive acoustical, fire, and structural testing ensued along with extensive field trials. The result: a competitive channel and U. S. Patent 3,333,379.

What is the acoustical difference between resilient furring channels? For openers, the designs are quite different (See Figure 7.2). The original USG RC-1 channel has a single leg with oblong cutouts in the leaf leg and is made of 25 gauge electrogalvanized steel. Decoupling the wallboard face from the wood stud or joist is accomplished with a dual-spring technique. A one-leg attachment provides a cantilevered spring while the span between the framing members allows flexure like that of a leaf spring.

The Gold Bond resilient furring channel is formed into a "hat" section with two equal legs formed of expanded metal. Base metal is also 25-gauge galvanized steel. Resiliency is achieved via the expanded metal legs. As the metal is expanded, it provides the ideal spring constant for optimum STC ratings while retaining sufficient strength and stiffness to satisfy the structural and fire integrity of the wall or floor/ceiling assembly. Both legs may be affixed to the wood member without

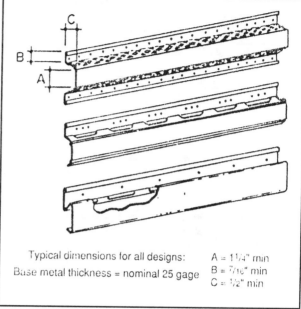

Typical dimensions for all designs: A = 1¼" min
Base metal thickness = nominal 25 gage   B = 7/16" min
                                          C = ½" min

Figure 7.2 Resilient furring channel designs that are acoustically efficient.

negating the acoustical spring effect. This technique reduces potential squeaks, provides a stable base for positive screw penetration, and assures a positive attachment for structural integrity and dependable time-design fire ratings.

The Donn channel had one leg with long slits in the spring/leaf leg. It too was 25-gauge galvanized steel. Long slits in the attachment leg provided the ultimate in acoustical separation between the gypsum wallboard and wood studs or joists. Unfortunately, the design did not have sufficient structural capacity. When attached to the bottom of ceiling joists, many ceilings sagged and did not have competitive fire ratings. Acoustical performance of the three original designs is similar. They all had very competitive STC ratings on both wall and floor/ceiling systems. If one was better acoustically, the nod went slightly to the Donn design since it had more flexibility.[30]

Typical examples of partition and floor/ceiling systems (see Figure 7-3) demonstrate that resilient channels improve the acoustical performance of a standard wood stud and gypsum wallboard partition by approximately 5 STC. Note that channels are usually placed on only one face. Applying channels on both faces of the wall only increases costs since nearly the same STC is achieved by installing channels only one side. Unbalanced wall systems provide wall faces with dissimilar stiffness. The result is an acoustical cancellation effect.

---

[30] The flexibility of the Donn channel probably led to its being withdrawn from the market over 15 years ago. Several ceilings sagged in use and the design did not achieve practical fire ratings. It also required more material, which increased production costs.

RESILIENT FURRING CHANNEL PERFORMANCE

SYSTEM CROSS SECTION	SYSTEM ELEMENTS	STC/IIC*
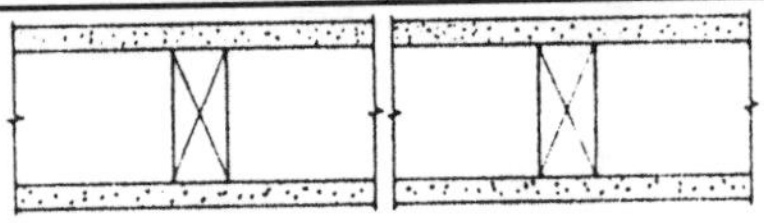	• 2 x 4 wood studs 16-in. OC • ⅝-in. Type X gypsum board **Wall system 1**	35
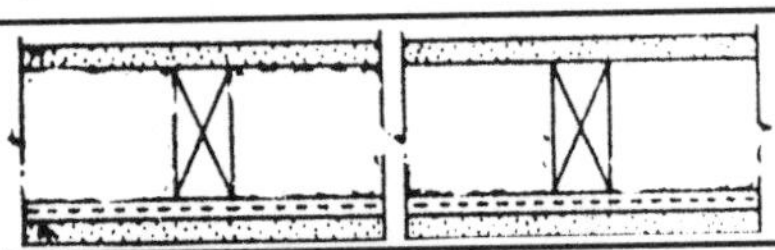	• Wall system 1 above + • Resilient furring channels 24-in. OC	40
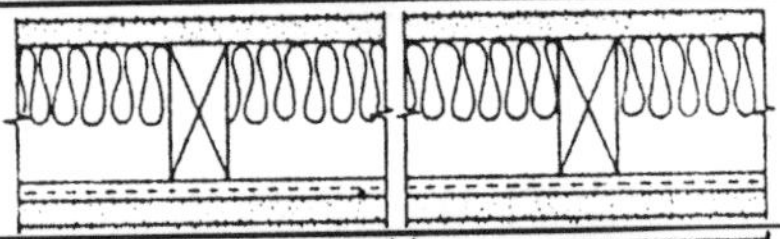	• Wall system 1 above + • Resilient furring channels • 1½-in. glass fiber insulation	45
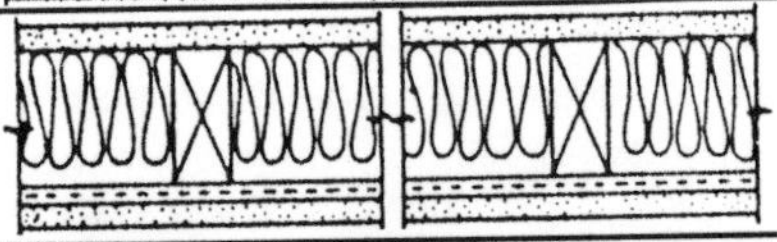	• Wall system 1 above + • Resilient furring channels • 3-in., 3-lb-density mineral fiber	50
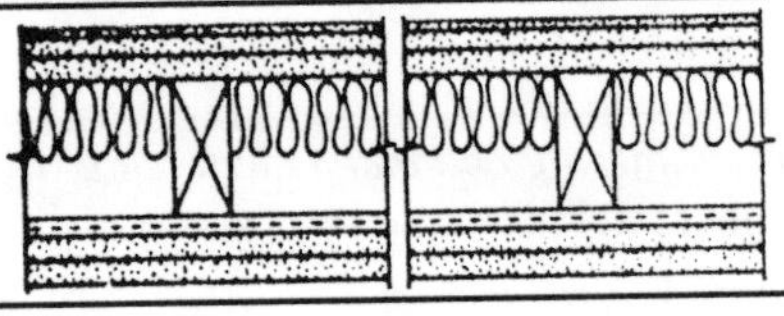	• 2 x 4 wood studs 16-in. OC • 2-ply ⅝-in. Type X gypsum board • Resilient furring channels • 3-in. mineral fiber insulation	55
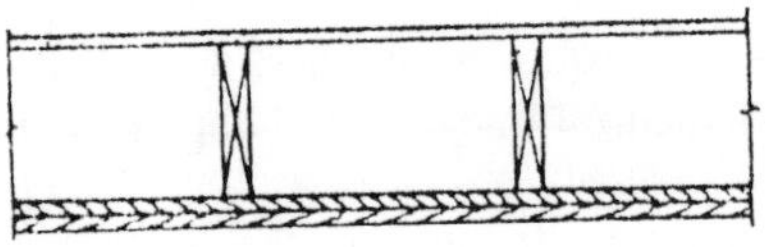	• 2 x 10 wood joists 16-in. OC • 1-in. (2-ply ½-in.) plywood floor • ⅝" Type X gypsum board ceiling **Floor system A**	35/32
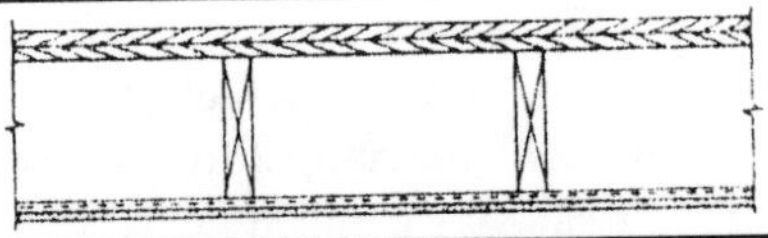	• Floor system A above + • Resilient furring channels 24-in. OC	45/39 IIC = 67 with carpet and pad
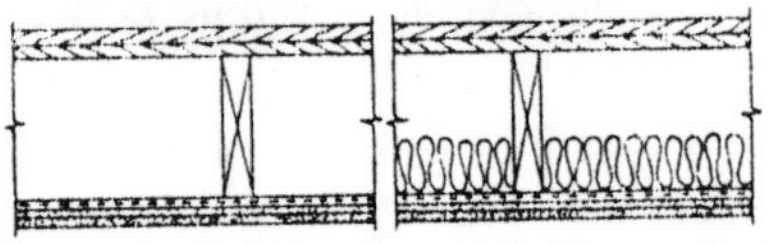	• Floor system A above + • Resilient furring channels • 3½-in. glass fiber insulation	55/68 IIC = 68 with carpet and pad

*   STC:- Sound transmission class tested per ASTM E90.

IIC: Impact isolation class tested per ASTM E492.

Notes: • ½-in. and ⅝-in. wallboard produce similar ratings.

• Type X wallboard usually will provide a 1-hour. time-design fire rating.

• Gypsum and lightweight concrete floors have similar ratings to systems shown.

• Channels are 24-in. OC; gypsum wallboard screws are 12-in. OC.

Figure 7-3 Acoustical ratings for typical systems utilizing resilient furring channels.

Adding a sound-absorbing blanket, such as glass or mineral fiber building insulation, improves the STC and IIC of the acoustical system by another 5 points. As a result, a wall with both resilient channels and a blanket in the core will achieve an STC 45. This is approximately a 10 STC improvement over a standard wood-stud gypsum wallboard partition system. Similar results are obtained on wood floor/ceiling systems having resilient furring channels and a core of building insulation.

APPLICATION CONSIDERATIONS

Application of resilient furring channels may seem simple and relatively foolproof. However, there are many nuances that may not be readily apparent to the architect, contractor, applicator, or specifier that require considerable attention during construction. The following items should be carefully considered and properly addressed by specification and then assured by site inspection and/or acoustical testing of the installed system.

- **Acoustical flanking paths** – A wall or floor will block sound only if all the potential flanking paths are eliminated. Key flanking paths include doors, windows, back-to-back cabinets, electrical outlets, HVAC ducts, and penetrations for pipes. These paths must be acoustically eliminated and/or have an STC rating greater than the barrier rating (e.g., STC 45 or higher depending on system chosen).[31]

- **Acoustical sealant** – Permanently elastic sealants must be utilized at all interfaces between wall or floor/ceiling assemblies and the abutting surface. A particular location that requires great care in sealing is the juncture between a resiliently mounted wall or ceiling surface and an exterior wall or a floor. If the sealant hardens, eventually it will acoustically tie the elements together. Lower STC and IIC ratings are inevitable because the resilient mounting has been short-circuited.

- **Stud or joist spacing** – Most load-bearing structural systems require a spacing of 16 in. OC. The resilient channel spring constant was designed for this spacing. Stud or joist spacings less than 16 in. OC will reduce the resilience of the wall or ceiling surface, thereby lowering the resulting sound transmission loss characteristic of the assembly. While greater spacings may provide greater resilience, one runs a considerable risk of negating the fire and structural ratings. Walls will also become soft and squishy, and ceilings may sag.

- **Spacing of resilient channels** – Most systems tested utilize resilient channel spacings of 24 in. OC. Closer spacing will likely lower the STC and IIC ratings. While wider spacings may provide an acoustical benefit, most fire and structural ratings are no longer valid.

- **Attachment of resilient channels** – Attachment to wood studs or joists will vary depending on the channel selected and the system details used in the fire test. Both the RC-1 and Gold Bond channels can be attached with screws or nails. Self-drilling wood screws are preferred because they reduce the possibility of squeaks. This is particularly a concern for ceilings since the system will tend to flex due to walking and furniture or equipment loads. With only one attachment leg, the RC-1 channel

---

[31] For a more detailed discussion see *Noise Control Manual*, D. A. Harris, Van Nostrand Reinhold, 1992 or Chapters 2 and 5 in this book..

requires one fastener per stud or joist. By contrast, the Gold Bond channel may be attached on both legs at each stud or joist with no alteration of the acoustical rating. Typical application calls for one fastener at each stud or joist with a staggered pattern (e.g., alternate leg at adjacent stud/joist). Note that some systems may require more fasteners or a specific kind of fastener to achieve the fire rating. The RC-1 channel can be installed with the leg up or down on a wall. It is recommended that manufacturers' literature be consulted on this issue. Some fire ratings may be affected. Acoustically it may not make any difference. Placing the attachment leg on the bottom will assure the channel is functioning as intended so the weight of the wallboard will pull the board away from the stud.

■ **Attachment of gypsum wallboard to channels** – The key items here are screw length and location and direction of wallboard. All resilient channels provide a gap of approximately ½-in. between the back of the wallboard and the stud or joist. **Great care must be taken to select screws that are long enough to assure penetration and engagement of the channel yet not so long that they penetrate the wood stud or joist.** STC and IIC ratings will be seriously compromised if the channels are short-circuited by screws that are too long. Unfortunately, this is a major cause of field failures. On the other hand, resilient channels may tend to bend away and not allow screw penetration or proper holding strength.

Screw spacing is critical to the acoustical, fire, and structural ratings. Typical screw spacings are 12 in. OC along the channels. Closer spacings may be required for fire ratings but will generally compromise the sound rating. Some sound ratings are achieved by spacing the screws 16 in. OC so they occur midway between the stud or joist spacing. This technique, while acoustically effective, will negate fire ratings in nearly every instance. Unfortunately, this discrepancy is easily overlooked and manufacturers' literature does not provide sufficient warning of this possibility. System failures and sizable lawsuits are the inevitable result of this oversight.

Wallboard should be installed perpendicular to the channels unless approved otherwise by the manufacturer. Gypsum wallboard has a different stiffness across the width than lengthwise. The result will be a potential compromise to the sound or fire rating and/or a sagging ceiling.

CHANGES OVER THE YEARS

The original resilient channels are still produced today exactly as they were 30 years ago. If installed properly, the performance will still be the same. What has changed is the gypsum wallboard and acoustical batts. Gypsum wallboard manufacturers have lowered the weight and improved fire resistance. Some STC and IIC ratings may no longer be valid today because most resilient channel acoustical tests were conducted on gypsum wallboard that was twice the weight of today's product. According to the mass law theory in acoustics, there is a direct line relationship between mass and sound transmission loss performance. While the effect in terms of mass may be small, reductions of 1 to 3 STC are likely.

Likewise, sound-absorbing insulation batts have different densities and thickness for the same R value. For example, glass fiber is typically ½-lb density while mineral wool may be as high as 3-lb density for the same thermal insulation characteristic. Research results indicate that insulation thickness is much more important to the acoustical property than density. Therefore, glass and mineral fiber insulation of the same thickness generally offer the same STC ratings. Slight nuances may occur, and it is best to utilize the same material as used in the test reference and install it in the same fashion (see Figure 7-4).

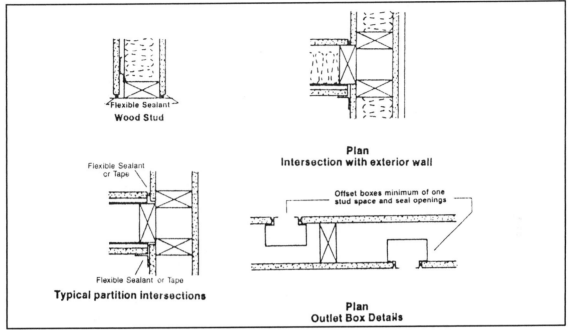

Figure 7-4 Typical details for resilient furring channels.

The most disturbing change to resilient channels is the introduction of copycat products. Original patents have expired, allowing other roll-forming operators to introduce their own version. Unfortunately, those products that have found their way to the marketplace have not been properly tested. A flagrant example is a product actually named RC-1 channel. It appears very similar to the original USG RC-1 channel, but is made from a much heavier gauge material and has no cutouts on the leg. It no longer has the proper resilience. Since USG no longer makes the original channel, there appears to be no one policing the industry. When challenged by a knowledgeable acoustician, some suppliers have provided bogus test reports or no test data to support their claims. Many examples exist where field applications fell far short of expected sound attenuation characteristics. Good acoustical performance with resilient channels can be assured by:

- Installing the system using the configuration and material used in sound tests. (Note: A copy of the test report can be quite revealing.)
- Making sure construction details are the same for fire and sound ratings.
- Using the correct length and location of screws.
- Making sure there are no flanking paths.
- Being liberal when installing acoustical sealant.
- Using the original resilient channel designs

THE BOTTOM LINE – Resilient furring channels are still providing superior performance at the lowest decibel per dollar, even after 30 years. **Still going, still going,** still going.

7.5 QUIET LIGHTWEIGHT FLOOR SYSTEMS

Reprint from Sound & Vibration Magazine, July 1992
By
David A. Harris, CFM, Principal
Building & Acoustic Design Consultants
680 Rainier Lane, Port Ludlow, WA 98365, U.S.A., Phone: (360) 437-0814

ABSTRACT

Lightweight floor systems over wood and metal joists are notoriously noisy. They annoy both the occupant and his neighbor below. Present solutions require top quality carpet and pad and/or expensive and massive floor systems. A floor system has been developed that allows the use of hard surface flooring materials such as vinyl, hardwood or ceramic tile to be applied over a plywood subfloor. The key component is a plywood subfloor material having a viscoelastic core. Utilizing the principles of constrained layer damping, the result is a quiet floor that is only moderately more expensive than conventional plywood and requires no special installation techniques. **This paper discusses the testing of several versions of the damped plywood applied to lightweight wood and steel joist floor systems. Damped plywood is ideal for quieting residential and commercial buildings including: single and multifamily dwellings, stores, offices, schools, clinics and modular buildings.** This paper presents initial evaluations on modular commercial structures for noise levels in the occupied space and recent tests of wood floor/ceiling systems for both airborne and impact noise.

BACKGROUND

In North America, most residential construction projects utilize wood joists with a plywood subfloor. A growing trend is prefabricated wood truss joists. Lightweight commercial buildings are similar construction with some, particularly modular buildings, using lightweight steel joists covered with plywood or a similar lightweight subfloor. A significant user complaint with these floor systems is, they are noisy. Both the source and the receiver below complain about impact noise. Receivers may also be concerned about airborne noise transmission. Both the source and the receiver have the perception of someone walking on a hollow drum head. The stiff taut membrane of a thin plywood subfloor amplifies sounds from footsteps, other impact sounds and airborne noise. In many instances, non-carpeted floors will cause significant negative occupant reaction. A typical comment by both the source and receiver is "This floor sounds like cheap construction". They perceive the floor has "a lack of solidity".

A well known and effective solution is to add a carpet and pad to the occupied source or floor surface. A good quality, thick carpet and a thick resilient pad of foam latex or hair felt are excellent at reducing impact noise. However, the design community, owners and occupants want the aesthetic appeal of ceramic tile, hardwood and vinyl floors. Hard surface materials such as vinyl flooring are common in kitchens, baths, recreation rooms, hospitals, medical offices, schools, stores, etc. These hard surface flooring materials cause noise control problems that traditionally have no practical solution. While inventors have developed systems that work, they typically require considerable mass such as concrete or complicated system solutions. Both add up to significant costs for the quieting effect attained.

NEW MATERIAL USING CONSTRAINED LAYER DAMPING PRINCIPALS

A new material has evolved in recent years that has demonstrated considerable promise. The material is essentially a conventional plywood subfloor panel formed with a noise damped core (Figure 7-5). The core is a viscoelastic material having elastomeric properties specifically formulated to provide a damped panel. Placed at the centroid of a structural subfloor panel, and bonded to the outer layers, the viscoelastic core material creates a constrained layer damped membrane. Resonant frequencies of this membrane are changed, resulting in a sandwich panel with significantly reduced impact noise and airborne sound transmission.

Many materials and systems have been designed to provide a quiet floor. A key to the performance of damped subfloor systems is their ability to be acoustically effective without adversely altering existing structural characteristics. Span and stiffness characteristics of a subfloor will determine joist spacings, floor deflections, shear values and ultimate floor

Figure 7-5 A closeup view of a typical damped plywood product. The viscoelastic damping material, shown in black, is located at the center and adhered to both faces.

loads. Already well established and required by code criteria, any reduction in these characteristics will cause the designer and builder to redesign the system and upgrade if necessary to satisfy well established minimum requirements. To achieve floor systems with improved impact and airborne noise characteristics, designs may be considerably thicker or heavier to compensate for their lack of structural performance. Damped plywood panels, on the other hand, are carefully tuned to be acoustically effective with only minimal changes to structural stiffness. Analysis has shown that the stiffness characteristics of the constrained layer damped materials will be reduced by approximately 10% for equal thickness and grades of plywood. Since most systems are over designed by more than 10% there may be no need for design alterations. Factory lamination and tongue and groove joints may compensate for the lower stiffness.

TECHNOLOGY

Damped floor system technology is an outgrowth of constrained layer damping principals for metals. Conceived originally to quiet vibrating metal panels on industrial equipment, or to reduce impact noise, constrained layer metals have made it possible to design quiet machines. For example, constrained layer damped sheet metal is used to quiet chutes where metal parts are ejected from a forming process. These materials are used on all sorts of equipment from damped resonant panels to resilient machine mounts. The auto industry has capitalized heavily on this technology in recent years. At least three major auto manufacturers have TV commercials that feature the quiet effect gained by using constrained layer damping materials. A notable case in point is the demonstration that compares the sound of a drummer tapping on a damped and undamped cymbal.

MATERIAL PROPERTIES

A casual visual inspection will rarely detect the presence of viscoelastic damping materials in plywood panels. Only on careful examination of a cross section of the material or by tapping on it is it evident there is a difference from normal materials. In plywood, the numerous lamination lines effectively conceal the use of damping materials. Depending on the material used and how the panels are formed, there may be a slight variation in the overall thickness. This is only of concern to those matching floor levels at a juncture between damped and conventional plywood.

Damped subfloors yield a valuable application plus. **The application is identical to regular plywood.** For example, damped subfloor panels, applied over wood or steel joists, are attached with standard nails or power driven fasteners in the same pattern used for a conventional subfloor. As a result, no additional labor is required. Furthermore, the damped panels may be used in conjunction with conventional materials. By locating them where the greatest need for quieting is required, such as hallways, and rooms with hard surfaced floors, increased material costs may be optimized.

TYPICAL APPLICATIONS OF DAMPED PLYWOOD

Typical applications of constrained layer plywood are shown in Figure 7-6. They include residences, apartments, condominiums, modular buildings both commercial and residential, speaker enclosures, buses, trucks and boats. Uses in residences and condominiums include the floor in kitchens, baths and high impact areas where hard surface flooring materials are the norm, stairs, equipment rooms and special purpose rooms. In commercial buildings damped plywood emulates poured concrete floors making them ideal for medical facilities, stores, schools, offices, strip malls, gymnasiums and jazzercise/dance studios. They are particularly desirable in the modular building industry where heavy brittle materials such as concrete are not practical.

Figure 7-6 Damped plywood may be used in place of nondamped construction wherever sound reduction is desired.

Damped floor panels have also been used in buses, trucks and boats to reduce impact noise and transmission of both structure borne and airborne noise.

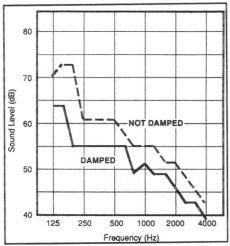

Figure 7-7 Footfall noise levels of source floor with hard surfaces is significantly reduced with damped plywood. Tests conducted by author using modified AMA I-1 procedure and a live male walker.

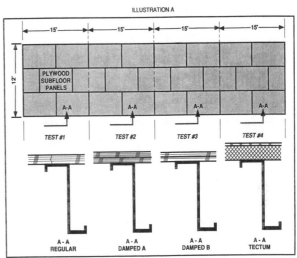

Figure 7-8 Test specimens for four subfloor systems on lightweight steel joist floor assembly typically used in modular buildings. Damped plywood panels are attached with conventional fasteners.

TESTING

The acoustical performance of damped plywood on lightweight floor systems typical in the modular building industry is quite dramatic. See Figure 7-7. These systems perform best as a deterrent to impact noise. However, they also exhibit improved resistance to airborne noise transmission. A damped plywood floor is nearly as effective in controlling impact noise as heavy assemblies 4 times their weight.

Modular floor systems:

The best demonstration of the performance of damped plywood is a series of tests conducted by the author on a typical modular building floor system composed of Z shaped steel joists. A direct comparison of a conventional plywood floor (not damped) next to a damped subfloor is shown in Figure 7-7. A simple subjective walking test confirms the results and is dramatic enough to convince even a severe critic. Damped plywood systems exhibit the sound and feel of a floor having a far greater mass. The perception is one of a solid and expensive floor. By comparison, a conventional plywood floor sounds and feels like walking on a drum head.

Test procedure AMA I-1, developed by Geiger and Hamme Laboratories, provides a convenient technique to evaluate the sound of impact noise on the source room side of the test specimen. Developed originally to evaluate the sounds produced by a live walker on floor ceiling assembly constructed between an isolated source and receiving room chamber, the procedure utilizes a carefully defined male or female walker traversing a set pattern. The author utilized a modified version of AMA I-1 where measurements of sound levels produced by a male walker were recorded in the source room. See Figure 7-7. This technique was used to evaluate a series of floor systems shown in Figure 7-8. Direct subjective comparisons were also made by a number of individuals who walked directly from specimen to specimen. In addition, the specimens were evaluated utilizing vibration analyzing techniques. A comparison of the results of all three techniques; AMA I-1 modified male walker, subjective walking and vibration analyzer correlated very well. In all cases the damped plywood showed a marked improvement over conventional floor systems. **Test results indicate that damped subfloor systems will reduce noise levels by over 8 dB$_A$ when compared with a similar thickness of regular plywood. Subjective and objective analysis shows the damped plywood floor feels and sounds more like a concrete floor and imparts a sense of "solidarity".**

Airborne and impact sound tests:

Airborne and impact sound transmission loss tests also correlate well with the results of footfall sound levels in the occupied space. In early 1993, full scale laboratory tests were conducted on a series of typical floor/ceiling assemblies at Riverbank, an independent NAVLAP approved acoustical testing agency. Airborne, impact and footfall sound tests were conducted per industry accepted procedures as follows:

- **Airborne Sound Transmission Loss:** per American Society of Testing and Materials (ASTM) E-90, Standard Test Method for Laboratory Measurement of Airborne Sound Transmission Loss of Building Partitions. The Sound Transmission Class (STC) was determined in accordance with ASTM E-413, Classification for Rating Sound Insulation.

- **Impact sound:** per ASTM E-492, Standard Test Method for Laboratory Measurement of Impact Sound Transmission Through Floor-Ceiling Assemblies Using the Tapping Machine. The Impact Insulation Class (IIC) was determined per ASTM E-989 Classification for Determination of Impact Insulation Class (IIC).

- **Footfall sound:** AMA I-1, Impact Sound Transmission Test by the Footfall Method using a live female walker. Sound levels were measured in both the receiving room and the source room.

Results show that a floor/ceiling assembly with wood joists, damped plywood, vinyl floor and a ceiling of resilient furring channels, ½-in. gypsum wallboard and a core of mineral fiber insulation has sound transmission class (STC) and impact isolation class (IIC) ratings similar to 8-in. thick concrete and wood joist systems with concrete subfloors. See Figures 7-9 and 7-10.

Airborne and impact tests of a floor/ceiling assembly having prefabricated truss joists in place of 2 by 10 wood joists were also conducted. The results show an STC of 57 and an IIC of 40 with conventional vinyl floor covering and an IIC of 64 with carpet and pad. These results compare favorably with a similar floor system having a plywood and poured gypsum concrete subfloor. Detail comparisons are shown in Figures 7-11 and 7-12.

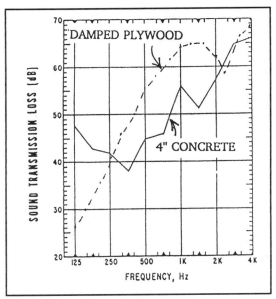

Figure 7-9 Airborne (E90) test of wood joist/damped plywood floor (51 STC) versus 4-in. solid concrete (44 STC).

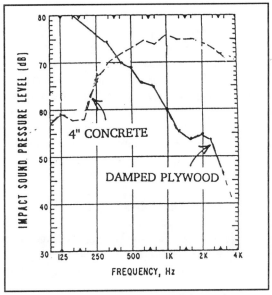

Figure 7-10 Impact (E492) test of damped plywood, wood joist floor/ceiling (34 IIC) versus 4-in. concrete (31 IIC).

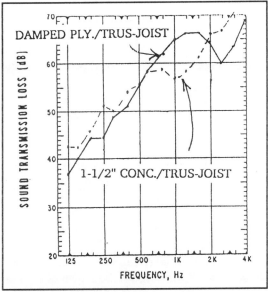

Figure 7-11 Airborne (E90) test; Trus Joist/damped plywood floor (57 STC) versus 1½-in. concrete (58 STC).

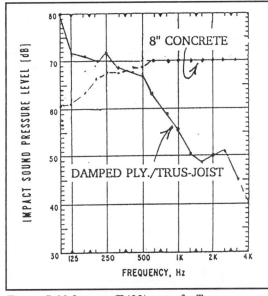

Figure 7-12 Impact (E492) test of a Trus Joist/damped plywood floor (40 IIC) versus 8-in. concrete floor (35 IIC).

257

SUMMARY:

There is substantial evidence that damped plywood systems are very effective in reducing the noise of impact sounds as well as airborne sound transmission. While these findings are acoustically significant, the main advantage is economic. Designers can now specify materials that are quiet without adversely affecting installed costs, and structural performance.

Damped plywood significant attributes are:

■ Breakthrough in Acoustical Technology:

- A substantial reduction of footfall noise levels in lightweight floor systems.
- Improved airborne sound transmission loss characteristics. STC of damped plywood systems is better than 8-in. concrete and wood frame/concrete floors/ceilings.
- Improved impact sound transmission loss characteristics. IIC is equivalent to 8" concrete floors and wood floors with concrete subfloors.

■ No additional labor:

- Application of damped plywood is the same as regular plywood.
- Installation labor is not adversely affected.

■ Structurally equivalent: Damped plywood strength is similar to regular plywood.

■ Economical: A damped plywood floor/ceiling system is only moderately more expensive than a conventional plywood floor/ceiling system. Systems having equivalent acoustical performance are double the cost.

****

ACKNOWLEDGMENTS:

The author is grateful to: Greenwood Forest Products Inc., Lake Oswego, Ore. Which provided test materials, technical support, and funding for this evaluation. Trus Joist MacMillan supplied the trus joists for the tests and test reports for comparisons. Initial evaluations on steel joist floor systems were conducted by Building & Acoustic Design Consultants (BADC), D. A. Harris, Principal. Airborne and impact tests were supervised by BADC and conducted at Riverbank Acoustical Laboratories.

7.6 NOISE ASSESSMENT AND NOISE CONTROL RECOMMENDATION
FOR
IMPERIAL PALMDALE MANUFACTURED HOUSING DEVELOPMENT
45th at R-8 Avenue, Palmdale, Calif., Project DBA 1-90-3
(Paragraph numbers are those of original document)

1.0 INTRODUCTION

The site plan for a 143-lot manufactured housing residential development located in Palmdale, Calif. has been reviewed to determine compliance with the City of Palmdale noise control requirements. Known as Imperial Palmdale Manufactured Housing Development, the following assessment and recommendations are provided as a result of a review and analysis of the noise impacts by the proposal offeror. Figure 7-13 shows the project site vicinity. The site is located on the northwest corner of the intersection of 45th Street East and R-8 Avenue adjacent to the Grecian Isles Mobilehome Park (existing). The terrain of the property is a gradual declining slope (8 ft in 1200 ft) extending eastward from Avenue R-8. The proposed layout of the development appears in Figure 7-14.

Noise-quieting recommendations include construction of a noise barrier wall to block street noise plus upgrades to the sound transmission loss attributes of windows and doors. The resultant acoustical environment inside the dwellings should satisfy Palmdale Noise Ordinance criteria.

2.0 NOISE CONTROL STANDARDS

To protect the health and welfare of the residents, the City of Palmdale requires an exterior community noise equivalent level (CNEL) of 65 A-weighted decibels [dB(A)][32], and 45 dB(A) for the interior CNEL in new single-family dwellings.

3.0 EXISTING AND FUTURE NOISE LEVELS

3.1 Traffic Volumes
From projected traffic volumes for the year 2010 (Horne, 1991), Avenue R-8 has been identified as the major source of noise for the subdivision. This roadway is presently a two lane road with a traffic volume of 1800 vehicles per weekday. In 2010 it is projected that Avenue R-8 will be a major four-lane divided roadway with an estimated traffic volume of 15,000 vehicles per weekday.[33]

3.2 Noise Analysis
Roadway noise levels were analyzed by the Federal Highway Administration Highway Traffic Noise

---

[32] Many use the abbreviation dB(A) to mean dB_A

[33] Telephone conversations with Horne, traffic engineer, Palmdale Traffic Engineering Department, May 23, 1991 and June 4 and 5.

Prediction Model[34] (FHWA, 1978) and traffic data provided by the City of Palmdale. Results are listed in Table 7-1 and show the mitigated CNEL contour distances from the center line of Avenue R-8.

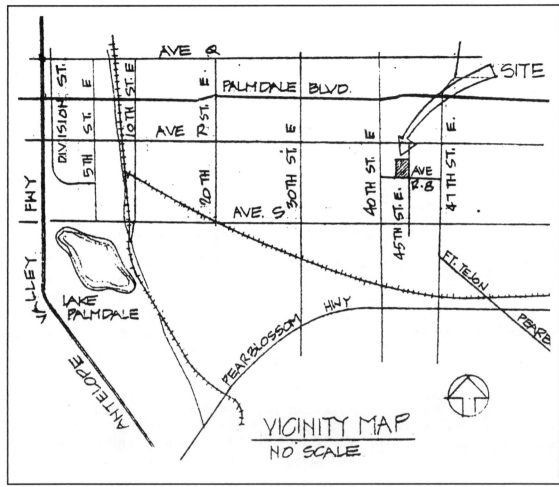

Figure 7-13 Project site vicinity.

[34] T. M. Barry and J.A. Reagan, "Federal Highway Administration Highway Traffic Noise Prediction Model," FHWA 72-RD-108. Washington, D.C.: Federal Highway Administration, December 1978.

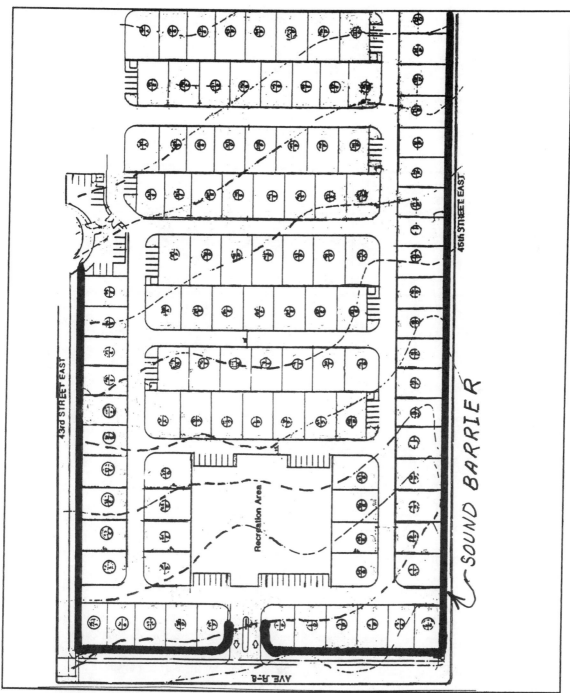

Figure 7-14 Proposed layout of development.

TABLE 7-1 CNEL Location on Property (Feet from roadway centerline)

Roadway	Year	70 CNEL	65 CNEL	60 CNEL
Ave. R-8	2010	60	190	610

For the year 2010, exterior noise levels generated by traffic on Avenue R-8 will exceed 65 CNEL at all the first-row lots adjacent to the roadway. This includes lots 1 to 6 and 55 to 59. These lots will require noise mitigation to achieve the exterior CNEL criteria as required by the City of Palmdale. Other lots farther away from the arterial will be shielded from the traffic noise by the closer rows of housing and will not require noise mitigation.

4.0 NOISE CONTROL

4.1 Exterior
Using the FHWA methodology, a barrier analysis was performed to determine the exterior CNEL at the impacted lots with barrier walls of various heights. Exterior noise levels were calculated with 5 to 9-ft-high barrier walls located on the property line of the subject property. Table 7-2 lists the results of this analysis.

TABLE 7-2 2010 Predicted CNEL levels, dB(A)*
(With barrier at lot line)

Lot No.	No barrier	5 ft. Barrier	6 ft. barrier	9ft. barrier
1– 6 and 55–59	70	65	64**	62**

* Calculated at 5 ft above the pad elevation of the lots.
** Complies with City of Palmdale exterior noise criteria.

To achieve the noise level criteria at the impacted lots nearest to Avenue R-8, the following noise barrier is required: A wall with a height of 6 ft along the property lines of lots 1 to 6 and 55 to 59 as they border Avenue R-8. This barrier should also continue along the property line of lot 6 for the first 80 ft along 45th Street. Likewise, the barrier should continue along the property line of lot 55 as it borders on 43d Street. This same barrier should continue along the project entrance drive on both sides along the entire 80 ft property line of lots 1 and 59.

In order for the barrier wall to achieve the noise reduction that the analysis has indicated, it is essential that the wall be a solid, continuous structure for the full required length and height with no holes, cracks, gaps or other openings that would compromise the noise barrier performance. The barrier should be constructed of materials that are impervious to noise with a minimum design weight of 3.5 lb per square foot of face area. The wall materials and structural design of the barrier should withstand outdoor conditions for a period of time equal to the project's expected existence. Suitable materials for the barrier may be, but are not limited to, the following:

- Masonry or concrete block
- Stucco over wood framing or foam core
- 1-in. thick weather-treated wood with overlapping or tongue-and-groove joints on posts in concrete footings
- ¼-in. thick framed-in safety glass, ⅜-in. thick acrylic plastic or other material with sufficient weight per square foot
- Earthen berm
- Any combination of the above materials (e.g., An 8-ft-high-wall could comprise a 4-ft-high transparent section on top of a 4-ft-high stucco wall, or a 6-ft-high masonry wall on top of a 2-ft-high earthen berm)
- Any of a number of proprietary sound walls having STC 30 or greater

The final performance of the barrier will depend on its location and the finished grades of the building pads, barrier base, and roadway surface. The 6-ft-high block wall shown on the preliminary plans satisfies the above requirements.

4.2 Interior

Doors and windows must be closed in order to meet the City of Palmdale noise ordinance criterion of an interior noise level of 45 dB(A), CNEL. Estimated future noise levels from traffic and the sound transmission loss characteristics of typical residential building shell components indicate that the interior noise criterion of 45 dB(A), CNEL can only be met when the entire building system, including closed windows and doors, has an STC 25. Normal wall and roof constructions should satisfy these criteria. However, walls are only as good as their weakest link. In normal instances doors and windows are the limiting factors. For this reason all doors and windows throughout the development should maintain the STC 25 performance of the exterior envelope. Normally, windows and doors that meet the International Congress of Building Officials (ICBO) Uniform Building Code (UBC) thermal insulation requirements in the closed position will satisfy the acoustical criteria. Doors and windows may be operable but they must be closed to achieve the noise control standards of the City of Palmdale and provide residents with a quiet interior environment.

As a result of the requirement to keep the windows and doors closed, applicable building codes require that mechanical ventilation be provided. For example, UBC Chapter 12, Section 1205(a) requires "In lieu of required exterior openings for natural ventilation, a mechanical ventilating system may be provided (i.e.; such as an air conditioning system). Such system shall be capable of providing two air changes per hour in all guest rooms, dormitories, habitable rooms and in public corridors. One fifth of the air supply shall be taken from the outside."

Fresh air ducting should not penetrate the buildings on any facade with a line of sight to Avenue R-8. If this is not possible, a noise-attenuating type of ducting should be used, such as fiberglass lined duct.

In order to achieve an interior CNEL of 45 dB(A), all second-story construction will require the exterior envelope to provide an STC of 35 or greater. Wood-stud walls with gypsum board interior facings, 1-in. stucco exterior, and insulation is usually sufficient base construction. All penetrations in the acoustical envelope must have equivalent acoustical performance (i.e., STC 35). To satisfy this criterion, windows and penetrations typically require double glazing. Table 7-3 lists manufacturers that provide acceptable windows.

TABLE 7-3 Window Manufacturers*

Guaranteed Products 355 Vineland City of Industry, CA 91749-1265 (213) 968-5573/(714) 529-0169	DeVac, Inc. 4520 Vega Del Rio Fair Oaks, CA 95628 (916) 967-3373
Premier Products, Inc. 750 West 182d Street Gardena, CA 90248-3486 (213) 321-4040/(714) 523-0720	Pella Products 3026 S. Orange Avenue Santa Ana, CA 92707 (714) 957-1455
Rolleze, Inc. 12177 Montague Street Pacoima, CA 91331 (213) 875-0364/(818) 899-9561	International Window Corp. 5625 E. Firestone Blvd. Southgate, CA 90280 (213) 928-6411/(714) 523-8755

* The above list of manufacturers does not constitute an endorsement of their products and is provided for convenience only.

Conventional roof construction techniques are usually sufficient to attenuate traffic noise sufficiently to meet the City of Palmdale criteria. Roof/ceiling systems utilizing a sheet-metal, composite, or shingle roof over standard roof decking, with thermal insulation in the joist cavity and a ceiling of ½-in. gypsum wallboard are generally acceptable.

5.0 SUMMARY

An analysis has been performed to determine compliance of the proposed manufactured home residential development of a 20-acre site located at the corner of Avenue R-8 and 45th Street with the City of Palmdale noise standards. At present there are no intrusive noise sources that require mitigation efforts. However, the City of Palmdale has identified Avenue R-8 for major expansion. According to year 2010 projections, R-8 will become a major artery with at least four lanes and a divider carrying 15,000 vehicles per day. At this level of traffic, noise mitigation efforts are required to satisfy the City of Palmdale noise ordinance requirements.

The analysis shows that impacts on the project will not exceed the 65 CNEL exterior noise standard at the ground level of the impacted lots if noise barriers are placed on the property lines to shield the noise from Avenue R-8 when it becomes a major arterial. If two-story units are to be allowed, the upper story should be outfitted with sound-rated exterior envelopes including doors, windows, and openings that are in a direct line of sight with Avenue R-8. Since doors and windows must be closed to satisfy the indoor noise levels of CNEL 45 dB(A) as required by the City of Palmdale Noise Ordinance, each residence in the development must have mechanical air handling to meet UBC/ICBO building requirements.

7.7 GREEK THEATER AREA SOUND IMPACT ANALYSIS
Submitted to
City of Los Angeles Department of Recreation and Parks
Griffith Administrative Services Division
3990 West Chevy Chase Dr., Los Angeles, California
(Paragraph numbers and table numbers are those in original document)

Description from request for proposal (RFP): "The City of Los Angeles is soliciting proposals for an analysis of the impact on the nearby residential neighborhood of sound generated by and around the Greek Theatre in Griffith Park. The Greek is an outdoor amphitheater with seating for about 6,000, and is scene to about 80 events annually in a season which runs from May through November. Although the theatre is on Park Property, residences circle the Theatre area on nearby hills, and homes line the few narrow streets which carry most of the traffic into the Theatre area. The residents of the homes around and adjacent to the Greek Theatre area of Griffith Park hear sound from the Theatre's amplification system, as well as from crowds enjoying the shows or entering and leaving the Theatre. The Department seeks a thorough study of the types and levels of sound generated by the Greek Theatre and its patrons, along with specific plans to limit, buffer, or otherwise reduce the sound disturbance in the neighborhood."

3.2 TASKS AS DEFINED IN RFP
Offeror will perform the specific tasks outlined in the RFP using in-house staff. We understand these tasks to be:
1. Assemble historic information and current data for the analysis.
2. Define an approach for identifying and ranking the locations and types of sound impact, and diagnosing the root causes of the sound impact problems. Perform noise measurements and analysis.
3. Identify, describe, and illustrate potential remedies or improvements that are economically feasible. Plan and design the acoustical treatments.
4. Provide feasibility and cost/benefit analysis of alternative treatments and recommend a strategy for implementation.

3.3 TASK NARRATIVE
TASK 1 – Assemble Information
Our first task will be to collect and review all available reports and documents that identify the present and future noise impacts from the Greek Theater on the surrounding environs. This will include data and previous reports generated for the City of Los Angeles related to Griffith Park and specifically the Greek Theater area. To implement an efficient communication and interpretation of the documents and project specifics, offeror will schedule an initial meeting with City of Los Angeles Department of Recreation and Parks project leaders.

TASK 2 – Define Approach and Diagnose Problem
The next effort involves the identification and analysis of residences likely to be directly impacted by performances at the Greek Theater:
A. Prepare an inventory of building types and noise attenuation characteristics of the housing located in the direct line of sight to the theater and those that may be impacted by sounds from the theater. An example of the information we will obtain from selected homeowners is shown in Table 3-1

265

Greek Theater Environs Sensitive Housing Inventory.

B. Prepare a digitized analysis of the terrain in the vicinity of the Greek Theater using computer aided design (CAD) and our in-house geographical information system (GIS). Offeror will identify and display preliminary noise impact on nearby sensitive receivers. This analysis will provide a preliminary listing and rank ordering of potential receivers. The hardware and software utilized in our GIS is ideally suited for analyzing the noise propagation from a source to multiple receivers in hilly terrain. By using a grid elevation electronic file of the terrain surrounding the Greek Theater we will efficiently and accurately identify all direct lines of sight to potential receivers. This information will be used to prioritize detailed sound level measurements.

C. Conduct noise measurements to quantify the degree of noise impact at selected receiver locations using the priority established from the CAD investigations. This information will be compared with the predictions to aid in quality control. Offeror will coordinate with the owners or designated representative of the selected homes and the project coordinator to schedule noise measurements at a time to minimize the disruption to occupants. The noise measurements will include equivalent steady-state A-weighted sound levels at sufficient locations to clearly identify parameters and trends. Where appropriate, one-third octave band measurements from a standard sound source will be conducted. The time will be selected to provide ambient levels that are low enough to not be intrusive. This data will be presented in both tabular and graphic form. A computer-generated topographical overlay will be prepared to allow quick and easy identification of specific problem locations and their associated degree of impact.

266

TABLE 3-1 Example Form – Greek Theater Environs Sensitive Receiver Inventory

RESIDENCE # _____

Address:_____

Residence Size: 0 – 1000 sq. ft   _____
                 1000 – 1500 sq. ft   _____
                 1500 – 2000 sq. ft   _____

Over 2000 sq. ft   _____
Residence type: Single story _____, Two story_____
(Single family)   (Specify)_____
Structural type: Wood frame with lap siding _____
                  Wood frame with masonry/stucco _____
                  Masonry _____
                  (Specify type – block, stone, brick)

SITE CHARACTERISTICS:
Orientation to north (direction front of house faces) _____
Orientation to Theater _____
Angle of incidence   _____
Shielding (describe, give size and location)_____
                  _____
                  _____

Background noise level _____
% glazing to wall area: front _____, back_____, side 1_____, side 2____
Aesthetic considerations: _____
                  _____

Historic preservation: _____
Occupant expectations: _____
Other   _____
                  _____

Respondent:_____
    Homeowner ___, family member____, renter ____, other ____
    Age:_____, health _____
Number of noise intrusions: (occurrences during each time period)
          Daytime_____
          Evening_____
          Nighttime_____
Loudness of intrusions: (annoyance level, speech communication, sleep disturbance,
  concentration, relaxation)
Which rooms have the most noise intrusion? _____
What is your greatest noise concern? _____
Special considerations and comments: _____

TASK 3 – Remedies and Economic Feasibility
Potential remedies that include berms, barriers, selective absorption, directivity control, amplification speaker placement, and other solutions conceived during the analysis will be modeled using the offeror computer capability. Potential barriers in a range of sizes will be analyzed and modeled for their noise mitigation characteristics using information obtained from the CAD/GIS program. In a similar manner we will model the use of specially placed sound absorption materials as well as the selective placement of sound amplification speakers. Combining the results for all the potential solutions, we will be able to accurately identify an economic scenario and provide a detailed, prioritized listing of project solutions.

TASK 4 – Feasibility and Cost/Benefit Analysis
The cost of mitigation measures will vary depending on construction details, materials, and size. By assigning a cost factor to the acoustical models developed in Tasks 2 and 3, we will generate a specific cost/benefit for each barrier configuration, sound absorption application, speaker location, or other potential solution that may be conceived during analysis. It is likely that recommended solutions will best be implemented in phases. Those items that are the least disruptive and/or least expensive (i.e., with the greatest decibel-per-dollar benefit) can be implemented in Phase 1. Other items will be assigned to Phase 2 and beyond as deemed prudent. Our recommendations will be provided in a draft report for review by the Los Angeles Department of Recreation and Parks. On approval of the draft report a final Technical Noise Impact Report will be issued. Recommendations outlined in this report can be used to prepare specifications and construction drawings.

TASK 5 – Project Management
Project management and administrative activities to be performed during this study:

- Supervise development of recommendations to ensure consistency with appropriate state, regional, and local standards while employing the latest in noise applications.
- Maintain project files.
- Coordinate study effort with all team members and Department of Recreation and Parks staff.
- Prepare and monitor progress schedule and budget.
- Attend three (3) meetings.
- Prepare monthly progress report and invoice.

7.8 TECHNICAL PROPOSAL
FOR
RESIDENTIAL SOUND-INSULATING PROGRAM
DEPARTMENT OF AVIATION
City of DAYTON, OHIO
(Paragraph numbers and figure numbers are those in original document)

3.1 General

In assessing the needs of this project, the offeror staff visited the project study area, attended the Preproposal Meeting and reviewed the "Addendum to Dayton International Airport, Part 150 Noise Study (1986)." Basically, the project is to assess noise levels near the airport and propose noise quieting measures.

We understand the entire soundproofing project study area may have 156 or more residential homes located in the aircraft noise impacted zone not less than approximately 70 L_{dn} noise contour. We also understand that the Phase I program will involve the acoustical insulation of not more than 10 private residential dwellings located in the vicinity of Dayton International Airport, Dayton, Ohio.

The offeror will perform the specific tasks outlined in the Request for Proposal using in-house staff and subcontractors. We understand these tasks to be:

1. Prioritize areas within the 70 L_{dn} contour for sound insulation implementation.
2. Inventory units in the designated area and recommend candidate units for participation.
3. Develop and implement a public information program.
4. Specify criteria for evaluating applications.
5. Perform noise measurements and analysis.
6. Provide feasibility and cost/benefit analysis of alternative treatments and recommend strategy for implementation.
7. Plan and design the acoustical treatments.
8. Prepare detailed plans and work specifications to be included in bid documents for general contractors (using a minority architectural firm registered in the state of Ohio).
9. Provide construction inspection services during the construction phase (using a professional engineer registered in the state of Ohio).
10. Provide acoustical insulation training to the selected general contractors.

The foregoing general work scope is presented in greater detail in the following sections.

3.2 PHASE I– Identify Project Implementation Priority Areas Priority Phasing, Recommend Participants, and Establish Acoustical Criteria

3.2.1 Project Start-up and Residential Selections – Phase I. Conduct Research, perform Surveys, Review and Analyze Background Information to Develop Implementation Priority Areas and Phasing Priority.

Our first task will be to collect and review all available reports and documents that identify the present and future noise impacts on the Dayton Airport surrounding environs. This will include the Request for Proposal and the "Addendum to Dayton International Airport, Part 150 Noise Study (1986) Technical Report and Technical Report Coordination" prepared by Aviation Planning Associates, Inc., Cincinnati, Ohio, June 1988; Dayton City "Equal Employment Opportunity" and "Affirmative Action Assurance" forms, and other pertinent information gathered during the research efforts. To implement an efficient communication and interpretation of the documents and project specifics, the offeror will schedule an initial meeting with Dayton Airport and project leaders.

The initial client/consultant meeting will be coordinated to allow the offeror team to make a preliminary investigation of potential homes for consideration. (i.e., those within the 70 L_{dn} contour). A walk-through will provide general information about the construction types, availability and accessibility of the homes, and aid in the final selection and categorization of homes. Noise specialists may take initial sound level readings in and near the airport environs to verify the validity of existing and predicted noise levels. If necessary, offeror has the capability to generate aircraft noise contours, predict the number of people impacted by aircraft noise, and present this information in a computer-aided drawing format. We have the latest versions of the Federal Aviation Administration (FAA) Integrated Noise Model (INM), Heliport Noise Model (HNM), Area Equivalent Method (AEM); the U.S. Air Force NOISEMAP model, and NOISEFILE, which is the database for military aircraft. Noise specialists at the offeror who are committed to airport projects are in constant communication with FAA and Air Force personnel as they develop new features, and improve the accuracy of their models. In addition, the offeror has noise measurement instrumentation which is used to measure aircraft overflights and calibrate the computer models. This approach allows for the most accurate assessment of aircraft noise.

An analysis of the foregoing information and comparison with earlier analysis/predictions will be conducted. A narrative report will be prepared and submitted for consideration by the Dayton Airport Noise Abatement Coordinator. A special meeting will be scheduled to explain the findings, respond to questions, and assure a clear understanding of parameters.

A final list of "priority areas" for project implementation, including recommended participating residences with associated cost estimates, will be submitted to the Dayton International Airport Noise Abatement Coordinator and his staff for approval. This information will be in sufficient detail to facilitate utilization in future applications to the Federal Aviation Administration for federal funding of subsequent phases and budget.

3.2.2 Public Relations Program and Solicitation of Candidates

Utilizing the services of a local Dayton communications firm, the offeror, under the direction of Dayton Department of Aviation, will identify appropriate media to promote the program and prepare the promotion text and camera-ready material. Recommended media, timing, frequency, and baseline copy will be based on the most effective cost/benefit ratio. Several options will be presented for consideration and approval by the Noise Abatement Coordinator. The offeror recommendation will contain a written narrative including a description of the overall program, detailed examples of the presentation materials, and the rationale for selecting the recommended plan.

On acceptance of the plan by the Noise Abatement Coordinator, the offeror will prepare applications for property owners desiring to take part in the program. Each interested party will be given a detailed statement of the program goals and how the project will be implemented. A "Program Summary and Guide" will be prepared in a format and words suitable for presentation at public briefings and pertinent public official meetings and as a mass mailer to the general public. This brochure will be informative, concise, and aesthetically tuned for public review and will utilize state-of-the-art techniques. An adjunct 35-mm slide and overhead presentation complete with a script will be prepared. Two sets of the presentation will be submitted to the Noise Abatement Coordinator.

3.2.3 Existing Construction Inventory

The next efforts at the selected residences will be to prepare a detailed inventory of building types and noise attenuation characteristics of the housing located in the "Priority One Eligibility Area." The report will be titled "Airport Environs Housing Inventory" and will contain a description and identification of the following (See example form in Figure 2):

RECOMMENDATIONS:

1. Residence size will be established and categorized into four groups: Under 1000 sq. ft, 1000 to 1500 sq. ft, 1500 to 2000 sq. ft, and over 2000 sq. ft.

2. Structural type of housing will be identified as single- or multistory. Subcategories will be: wood frame with wood or aluminum lap siding, wood frame with masonry siding (i.e., brick or stucco), masonry (i.e., block or poured), other (i.e., describe). Within each of the structural types of housing, an estimate of glazing area as a percentage exterior wall area will be provided. Glazing is a critical component of exterior noise control in residential dwellings.

3. Geographical site characteristics will be noted including orientation to north, orientation to airport, aircraft type that will affect house, elevation of aircraft at maximum effect on house with corresponding angle of incidence (both horizontal and vertical), notation of potential shielding, background or ambient noise levels, and identification of potential non-ircraft noise sources.

4. General types of dwellings will be identified in terms of single-family or multifamily residences. Multifamily will be identified in terms of the number of units by category: 2 to 4, 4 to 8, 8 to 16 and 16+ units per building. Only single-family dwellings will be considered in the 10 initial units for noise mitigation measures.

5. A numerical distribution by percent, of each of the categories listed in items 1 to 4 above will be made.

6. Each general category will be analyzed to determine suitability for participation in project. Analysis will include consideration of potential costs, acceptability of structure to accommodate proposed changes (i.e., structurally, within building code, and applicable local regulations such as homeowners associations), availability of resources (i.e., is there power for air conditioning etc.), and the impact modifications will have on aesthetics, space utilization, and occupant wants, needs, or demands.

An evaluation of the above information will be presented in a report that will also include recommendations on which sites are desirable for implementation of Phase I of the Residential Noise Insulation Program. This report will provide the rationale for selecting the recommended homes. A minimum of two alternative sites for each of the 10 primary recommended sites will be provided. All questions raised by the Coordinator or public officials will be duly responded to in a concise and timely manner.

Figure 2 - Example Form – Airport environs Housing Inventory

RESIDENCE # _____

Address:_____

Residence Size: 0 — 1000 sq. ft. _____

                 1000 — 1500 sq. ft _____

                 1500 — 2000 sq. ft _____

                 Over 2000 sq. ft _____

Residence type: Single story _____, Two story_____

(Single family)   (Specify)_____

   Structural type:

   Wood frame with lap siding _____

   Wood frame with masonry _____

   Masonry _____

   (Specify type — bock, stone, brick)

SITE CHARACTERISTICS:

Orientation to north (direction front of house faces) _____

Orientation to airport _____

Aircraft type (typical) _____

Angle of incidence   _____

   (Approaching, overhead, leaving)

   Shielding (describe, give size and location)_____

   _____

   _____

Background noise level _____

Nonaircraft noise (describe and give level) _____

% glazing to wall area: Front _____,back_____, side 1_____,side 2____

PROJECT SUITABILITY:

Local restrictions: (homeowners association, architectural controls, etc.)

_____

Code violations: (identify) _____

_____

Toxic materials or condition: (existence of radon, poisons, etc.)

_____

Aesthetic considerations: _____

_____

Historic preservation: _____

Occupant expectations: _____

Other_____

When the candidate sites are approved by the Coordinator, the offeror will prepare and distribute a "Preconstruction Occupant Attitudinal Survey" (see Figure 3 for example form) This survey will ascertain the occupants' "perceived" noise exposure. Determination of the occupant perception will be based on an interview with the site owner and/or primary occupant. Survey questions will be established to provide a succinct determination of the actual perceived noise. In addition to the establishment of site specifics from the Airport Environs Housing Inventory Report, the respondents will be asked to identify the following: (1) the number and duration of intrusions they experience during the daytime, evening, and sleeping hours; (2) a description of the noise they find objectionable in terms that can relate to frequency content; (3) the direction from which they perceive the sounds emanate; (4) their description of the loudness in terms that can be related to decibel level; and (5) a determination of their hearing sensitivity and proficiency.

Figure 3 Example Form – Preconstruction Occupant Attitude Survey

PURPOSE: Determine occupant attitude and "perceived" noise exposure

RESIDENCE # _____

Address:_____

Owner: _____

Respondent:_____

    Homeowner ___, Family member____, renter ____, other ____

    Age:_____, Health _____

Number of noise intrusions: (frequency during each time frame)

        Daytime_____

        Evening_____

        Nighttime_____

Loudness of intrusions: (annoyance level)

    (speech communication, sleep disturbance, concentration, relaxation))

Respondent hearing sensitivity and proficiency:

How would you rate the ability of this residence to abate airport noise?

Which rooms have the most noise intrusion?

What is your greatest noise concern?

Special considerations and comments:

Recommendation for program suitability

Note: A similar questionnaire will be developed to measure the occupant response after mitigation measures are completed.

3.2.4 Acoustic Criteria Findings and Recommendations

Using the information of the foregoing reports, the offeror team will recommend and define the noise metrics that are most suitable for the project. A letter report with recommended noise metrics and related definitions will include the offeror rationale for selecting the specific measurement techniques. Metrics to be considered will include, but not be limited to L_{dn} and L_{eq}. Evaluation of all types of aircraft in typical use will be made by measuring arrival and departure overflights, sideline, reverse thrust, back blast, and related noise parameters. A full description of all possible techniques will be provided along with technical and lay person definitions of pertinent terminology. It is anticipated that several methods will be utilized. Speech interference and sleep interference are so divergent in their acoustical makeup that different measurement techniques will be required to provide a concise and understandable evaluation. In addition, an analysis of the sound signature in one-third octave bands will provide insight into specific remedial measures.

3.2.5 Draft and Final Reports with Findings, Conclusions, and Recommendations

All of the information and recommendations will be compiled and summarized into a draft report for discussion at a "milestone" meeting with the Coordinator prior to preparation and issuance of the final report. The offeror team will endeavor to respond to all questions and comments and incorporate them into the final report as deemed prudent.

3.3 PHASE II

3.3.1 Perform Exterior/Interior Noise Level and Sound Attenuation Measurements and Analysis

The offeror will coordinate with the owners or designated representative of the selected homes and the project coordinator to schedule the noise measurements at a time to minimize the disruption to occupants.

Using the metrics recommended and approved, the offeror will conduct existing noise reduction testing and identify weaker elements in the construction which will require acoustic treatments or replacement. The inventory will comprise a list of material types and respective surface areas associated with the exterior walls, windows, doors, and roof. All interior surfaces of each test room will also be inventoried. The inventory of building elements will assist in calculating the sound transmission loss and absorption characteristics of each test room.

Noise reduction testing will be performed by measuring a number of aircraft flyovers at the exterior of the house while simultaneously measuring at locations throughout the house. Three to six rooms with the highest perceived noise intrusion will be tested in each house (i.e., bedrooms, living room, den, kitchen, dining room).

Exterior noise will be measured at a minimum of three points, which has been conceptualized in Figure 1-4: (1) A point near building on the side being impacted by the noise from and approaching aircraft, (2) a similar point on the opposite side of the building where the noise impact is primarily due to sounds from an aircraft leaving the area, and (3) a point above the roof that measures noise impacting from an aircraft directly overhead. Sound levels from these locations, and others as deemed prudent, will be compared with the sound levels within the house with the aircraft at the same relationship. Noise reduction measurements will be calculated, thereby providing a distinct analysis of the sound attenuation characteristics of specific building elements, including exterior wall systems and roof assemblies.

Throughout the noise reduction testing, diagnostic investigations will be conducted to determine the weakest building elements which transmit the greatest amount of noise. These investigations will be performed by inspection, localized listening with diagnostic equipment, and/or one-third octave band sound level measurements for discrete analysis. If determined necessary by insufficient or infrequent flyover noise, the use of an artificial noise source with spectrum analysis equipment may be utilized.

To ensure the accuracy and repeatability of the noise reduction testing, one or more homes will be retested and verified using the same procedure as described above prior to any acoustical modifications. This is intended to demonstrate a high level of confidence in the procedures.

3.3.2 Noise Control Analysis and Design

Conceptual details of alternate designs of noise mitigation will be prepared and reviewed by the team to determine their feasibility and cost effectiveness. Alternative designs will be selected that will satisfy the degree of sound isolation required on the basis of data from the field investigations and research of the most economical suppliers of noise control products and systems. The most cost-effective means of acoustical modifications which will maximize the improvement in noise reduction will be recommended for construction. Three alternatives that meet the sound attenuation criteria will be identified. (Typical examples are shown in the accompanying illustrations.) A written analysis of the present noise attenuation characteristics of each dwelling will be prepared. This analysis will include a summary of average single event noise level reduction, the interior L_{dn}, and acoustical data relevant to sound leak identification. Insofar as is possible by observation and inquiry, the energy loss or thermal insulation characteristics will be determined and reported. A preliminary determination of the existence of potential hazards such as presence of formaldehyde and radon will be recorded.

Examples of the analysis process that the offeror will employ may be found in previous studies. For example, aircraft flyover noise almost always leaks through the fireplace and chimney when present. A typical method of treatment is to provide a costly absorptive lining to the inside of the chimney usually consisting of perforated metal panels over a fireproof, sound-absorbing pack material. With research and testing, it was found that the costly absorptive lining could be eliminated by installing a standard glass door cover over the fireplace opening and a good tight fit around the flue damper. This resulted in a significant cost savings in both construction and energy while acoustically outperforming the absorptive lining method.

Another example of cost-effective design is illustrated with mechanical ventilation. All homes which are impacted by exterior noise must keep the windows closed in order to keep the noise out. However, many homes in Ohio do not have summertime cooling provisions since ventilation through an open window is usually adequate. Therefore mechanical ventilation systems are required which can be a significant expense in the overall cost of a soundproofing program. Many homes have a forced-air heating system which allows an economical alternative. By installing a "summer switch" to the existing forced-air heating systems, the fan unit can be operated independently of the heating system and thereby provides a flow of air throughout the interior rooms while all windows and doors remain shut.

A standardized method of acoustical recommendations will be developed and finalized on the basis of construction types so that the modifications will be well suited for homes which are treated after the initial study. These recommendations will be presented as "Construction Documents" including drawings and specifications in a Construction Specifications Institute (CSI) format. The drawings and specifications will be packaged so that each home may be bid on an individual basis. Alternative solutions will be presented so that they may be bid separately at the contractor's choice. This approach will allow the project coordinator and the homeowner to select those solutions that best fit the project economics, contractor capability, and aesthetic considerations.

Prior to proceeding with the final drawings and specifications, the offeror Team will review all findings and proposed solutions with the homeowner and/or homeowner's representative. Homeowner's comments will be documented on each of the 10 sites and submitted in report form to the project Coordinator. Final drawings and specifications will reflect the desires and concerns of the homeowner as best as possible within acoustical guidelines. With the approval of the project Coordinator, final plans and specifications will be prepared in a CSI format suitable for bid purposes.

The offeror Acousticians will work as a team with project members from the Architectural Firm in the development of the construction plans and specifications. The acousticians will provide conceptual details, and provide review of specific drawings and specifications. The architects will participate in the development of noise control specifications to assure they are in compliance with all building codes, are energy-efficient, utilize optimum construction technology, are feasible to construct, and are aesthetically acceptable. The final report will contain three alternative, feasible treatments for each element of the dwelling envelope. Alternatives will be presented as First Choice, Second Choice, and Third Choice. Each alternative will have a rationale and, if appropriate, cautions for the recommendation plus an estimate of the cost. This report will be submitted to the Project Coordinator prior to finalization for review and comment.

3.3.3 Construction Inspections

Contractor submittals will be reviewed by a Home Inspection Engineer (HIE), a member of the offeror project team, to determine their suitability, compliance with applicable code criteria and understandability in construction of the acoustical recommendations. After the contractor has been selected, the HIE will inspect the work to assure compliance is achieved. A meeting will be scheduled with the construction site foremen, the offeror design team including the Architect, and the project coordinator and/or City of Dayton Building Inspector to assure all parties understand the drawings, specifications, acoustical criteria, and specialized designs required to achieve the acoustical goals. Specialized training of the contractors to ensure proper installation will be provided as deemed prudent.

Near the end of the construction at each participating residence, inspection will be performed by the offeror to insure that all treatments/modifications are installed as intended and according to plans and specifications. The assigned City staff member and representative of the project coordinator will be expected to be present at all inspections. The findings of each inspection will be reported and provided to the Architectural Consultant and Project Administrator. Where defects or problems are observed they will be reported with a summary of corrective actions to be taken. Reinspection will be performed when required by a specific case because of its critical nature and/or complexity.

3.3.4 Noise Reduction Performance Evaluation Testing

To determine the improvements after the noise control recommendations have been implemented, each residence will be retested. The results of the retests will be compared with the initial tests to determine the noise reduction improvements provided by the acoustical modifications. If the City of Dayton or Airport Staff intend to conduct ongoing tests on their own after completion of the program, it would behoove them to appoint an appropriate staff member to be present during all retests for continued training and to assist in the testing.

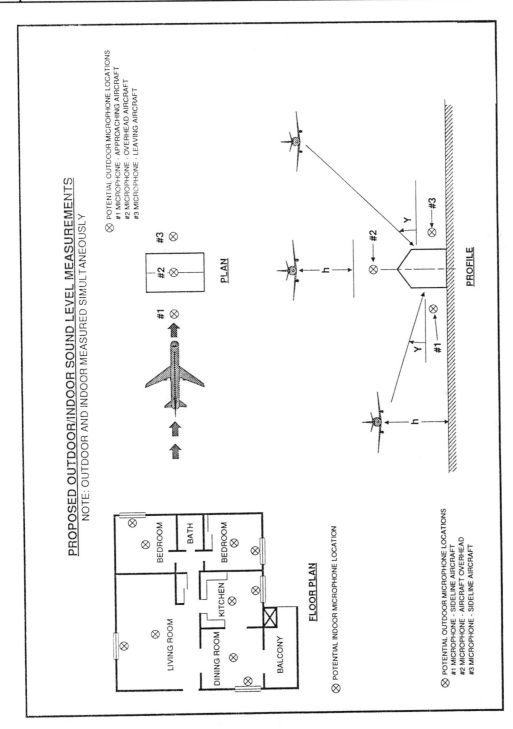

Figure 1-3. Proposed Outside Microphone Locations

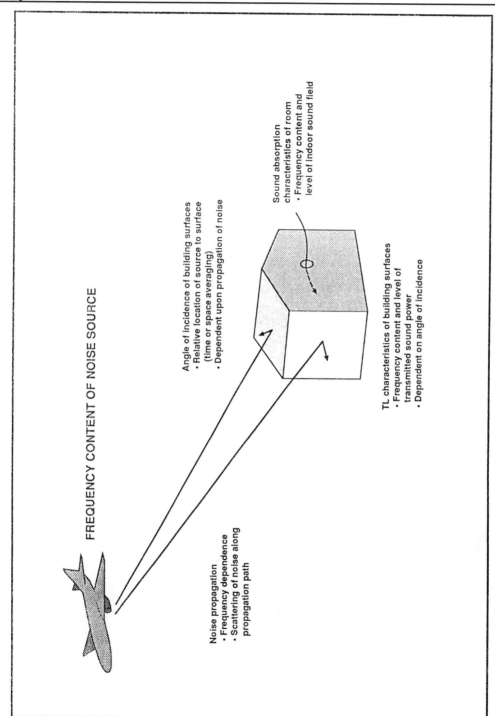

FREQUENCY CONTENT OF NOISE SOURCE

Angle of incidence of building surfaces
· Relative location of source to surface
  (time or space averaging)
· Dependent upon propagation of noise

Sound absorption
characteristics of room
· Frequency content and
  level of indoor sound field

TL characteristics of building surfaces
· Frequency content and level of
  transmitted sound power
· Dependent on angle of incidence

Noise propagation
· Frequency dependence
· Scattering of noise along
  propagation path

Figure 1-4. Factors Affecting the A-weighted Noise Level Reduction of the Building Envelope

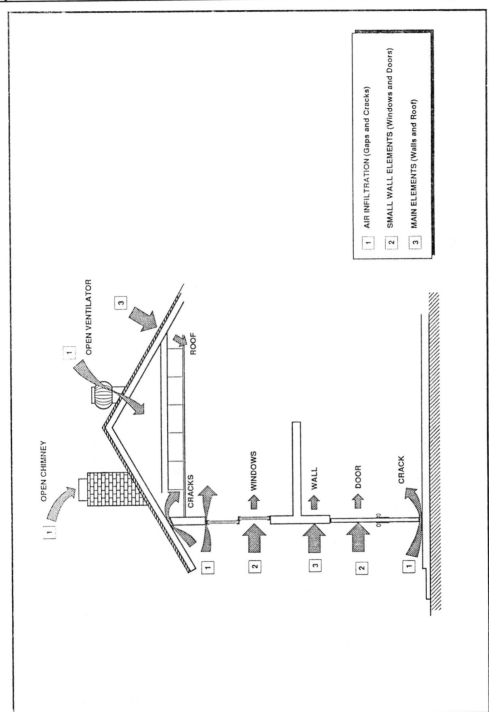

Figure 1-5 Conceptual Illustration of the Three Major Types of Paths by which Noise is Transmitted to Building Interiors.

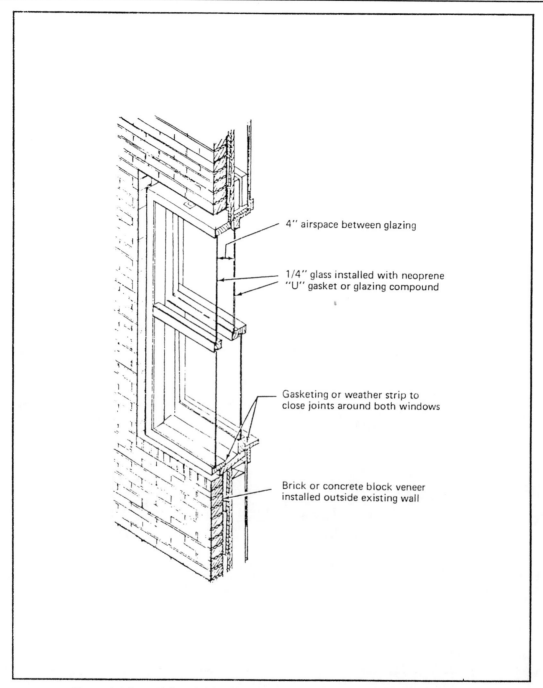

4" airspace between glazing

1/4" glass installed with neoprene "U" gasket or glazing compound

Gasketing or weather strip to close joints around both windows

Brick or concrete block veneer installed outside existing wall

Figure 1-6. Installation of Brick Veneer and Double 1/4-Inch Window Glazing

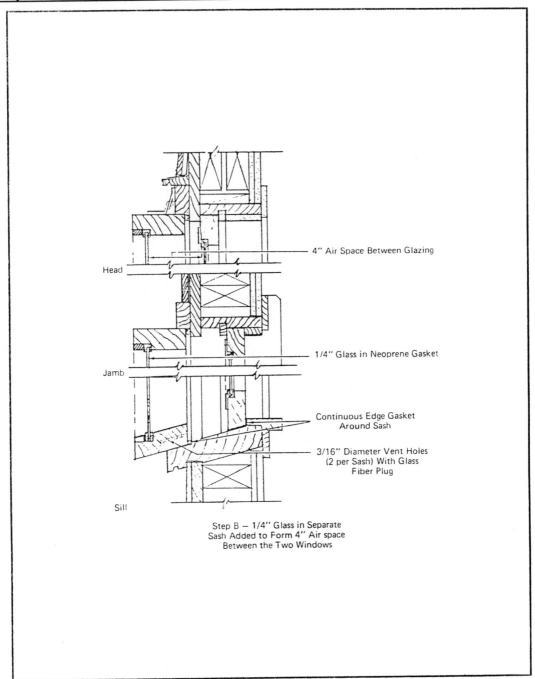

Head

4" Air Space Between Glazing

Jamb

1/4" Glass in Neoprene Gasket

Continuous Edge Gasket
Around Sash

3/16" Diameter Vent Holes
(2 per Sash) With Glass
Fiber Plug

Sill

Step B — 1/4" Glass in Separate
Sash Added to Form 4" Air space
Between the Two Windows

Figure 1-7. Details of Double 1/4-Inch Window Glazing

281

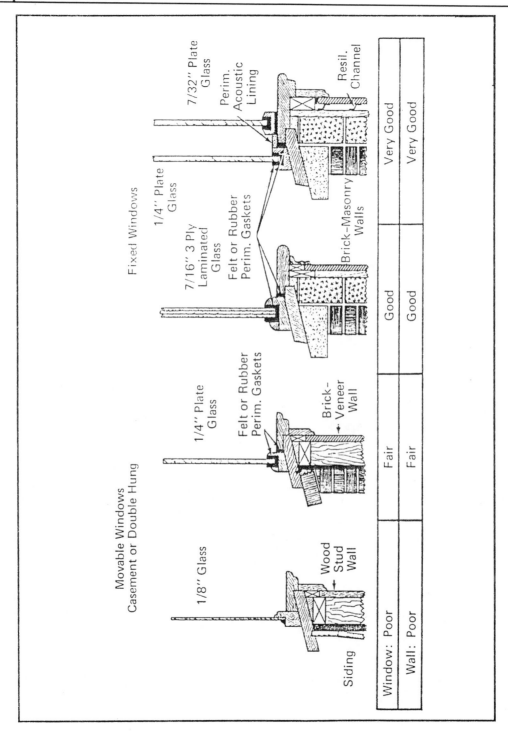

Figure 1-8. Sound Insulation Effectiveness of Various Window-Wall Assemblies

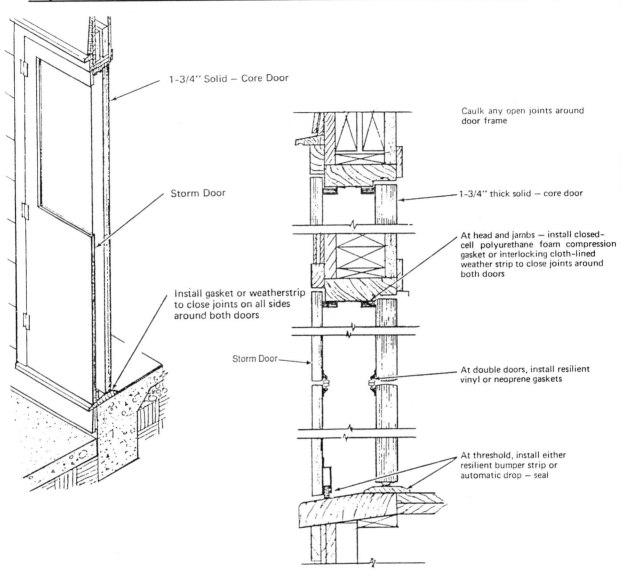

1-3/4" Solid — Core Door

Storm Door

Install gasket or weatherstrip to close joints on all sides around both doors

Caulk any open joints around door frame

1-3/4" thick solid — core door

At head and jambs — install closed-cell polyurethane foam compression gasket or interlocking cloth-lined weather strip to close joints around both doors

Storm Door

At double doors, install resilient vinyl or neoprene gaskets

At threshold, install either resilient bumper strip or automatic drop — seal

Figure 1-9. Door Modifications

Figure 1-10. Details of Door Modifications

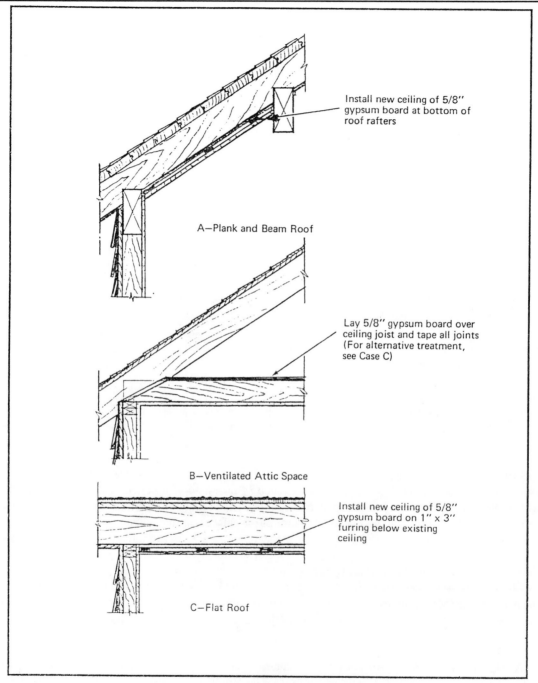

Install new ceiling of 5/8''
gypsum board at bottom of
roof rafters

A—Plank and Beam Roof

Lay 5/8'' gypsum board over
ceiling joist and tape all joints
(For alternative treatment,
see Case C)

B—Ventilated Attic Space

Install new ceiling of 5/8''
gypsum board on 1'' x 3''
furring below existing
ceiling

C—Flat Roof

Figure 1-11. Roof and Ceiling Modifications

284

7.9 **ENVIRONMENTAL IMPACT STUDY (EIS)**
"NOISE ELEMENT"
For
ALBUQUERQUE INTERNATIONAL AIRPORT
(Paragraph numbers, table numbers, and figure numbers relate to section of EIS)

3.4 NOISE

3.4.1 Background

With present air activities near capacity, the Albuquerque International Airport will be undergoing a major construction program to upgrade runways and taxiways. **The objective of this study is to assess the noise impact of these operations and identify mitigation measures.** Noise sources include aircraft run-up, flyover noise, and construction activities. Significant noise impact for on-base and community receptors result from the temporary shutdown of Runway 8-26. During a four month period Runway 17-35 is expected to carry all air traffic. The accompanying noise will impact considerable sensitive receivers.

3.4.2 Noise Terminology

Noise is most often defined as unwanted sound. Sound levels are easily measured, but the variability in subjective and physical response to sound complicates the analysis of its impact on people. People judge the relative magnitude of sound sensation by subjective terms such as "loudness" or "noisiness." Physically, sound pressure magnitude is measured and quantified in terms of a logarithmic scale in units of decibels (dB).

The human hearing system is not equally sensitive to sound at all frequencies. Because of this variability, a frequency-dependent adjustment called A-weighting has been devised so that sound may be measured in a manner similar to the way the human hearing system responds. The use of the A-weighted sound level is abbreviated "dB_A". Figure 3.4.1-1 provides typical A-weighted noise levels measured for various sources and responses of people to these levels.

When sound levels are recorded at distinct intervals over a period of time, they indicate the distribution of the overall sound level in a community during the measurement period. The most common parameter derived from such measurements is the energy-equivalent sound level (L_{eq}); this is a noise descriptor that represents the average sound energy level produced when the actual noise level varies with time.

For airport noise, the Federal Aviation Administration (FAA) and the Air Force have adopted the day-night average sound level (L_{dn}). L_{dn} is the A-weighted L_{eq} over a 24-hour period, with a 10-dB nighttime penalty applied to noise events from 10:00 p.m. to 7:00 a.m. The penalty for nighttime noise events accounts for the increased sensitivity of most people to noise in the quiet nighttime hours. Developed by the U. S. Environmental Protection Agency (EPA), L_{dn} is the "standard metric for determining the cumulative exposure of individuals to noise." Regulations of the U.S. Department of Housing and Urban Development (HUD) include L_{dn} as the standard for measuring outdoor noise environments.

Effective perceived noise level (EPNdB) takes into account both the duration and the tonal components of the noise spectra for varying types of non-sonic-boom aircraft flyover signals. This measure is used by the Federal Aviation Administration in aircraft certification.

3.4.3 Setting

Albuquerque International Airport serves commercial aircraft, and general aviation (GA) aircraft, and also has a substantial presence of military aircraft. The airport is situated close to the city with major residential, commercial, and public properties to the north, east, and west. Major landmarks to the east and south of the airport are Kirtland Air Force Base (AFB), the Cibola National Forest, and Isleta Indian Reservation. Noise-sensitive receptors are located on Kirtland AFB and in the surrounding area. On-base sensitive receptors are listed in Table 3.4-1 and shown on Figure 3.4.1-2.

Present air operations are near capacity. Runways and taxiways need repair and turnouts are unacceptable for higher performance aircraft. Some taxiways cannot support heavier aircraft. The military also has a major helicopter training facility at this location. Training missions include a considerable number of low-level flights that use a helipad approximately 4 miles south by southwest of the airport. Helicopter activity uses different air corridors from fixed wing aircraft.

Figure 3.4.1-3 shows the existing condition noise contours as developed by the Department of Air Force investigations. Comparing the contours with the existing Land Use Map prepared by the Kirtland AFB Planning Department indicates there are numerous sensitive receptors near and within an L_{dn} of 65 dB_A or greater contours. These sensitive receptors include residential areas, schools and hospitals located under or near the flight path (Table 3.4-1).

**Table 3.4-1 Kirtland Air Force Base On-Base Sensitive Receptors
for Existing Air Patterns**

(Source: USAF, 1990 and Kirtland AFB Land-Use Map, 1982)

Sensitive receptor	Estimated L_{dn}, dB_A
1. Veterans Administration Hospital (off-base)	70
2. Family housing — enlisted personnel	70
3. Public school	65
4. Dorms — military	70
5. Correction facility — military	80
6. Hospital	65
7. Chapel	65

Figure 3.4.1-1

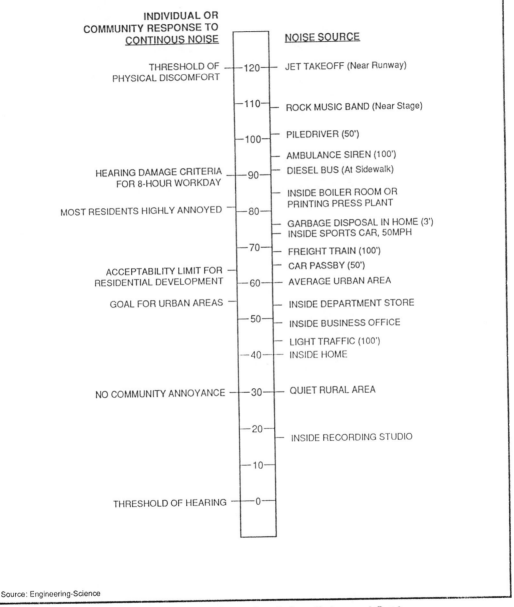

**Figure 3.4.1-1 Typical Sound Levels from Indoor and Outdoor
Noise Sources and their Effect on People**

287

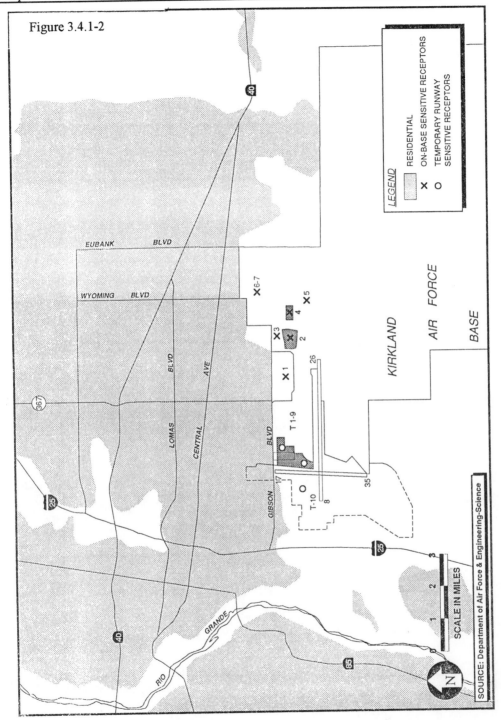

Figure 3.4.1-2

Figure 3.4.1-2 Kirtland AFB Sensitive Receptor Location

Figure 3.4.1-3

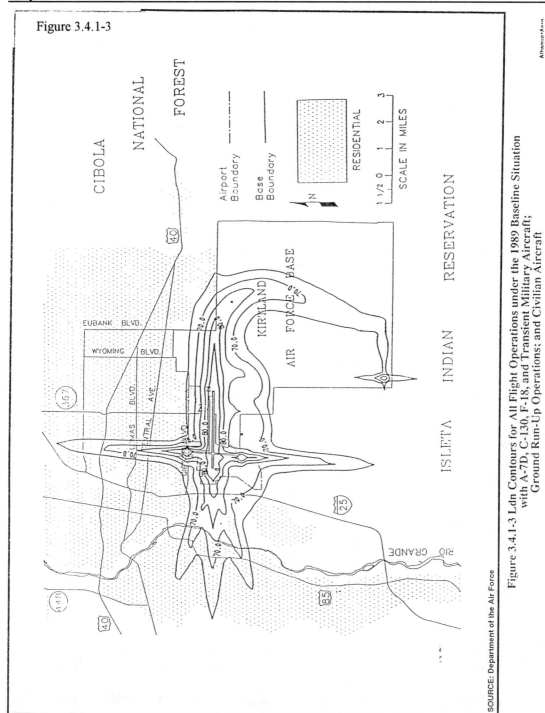

Figure 3.4.1-3 Ldn Contours for All Flight Operations under the 1989 Baseline Situation with A-7D, C-130, F-18, and Transient Military Aircraft; Ground Run-Up Operations; and Civilian Aircraft

SOURCE: Department of the Air Force

3.4.5 Significance Criteria

According to HUD, FAA, and Air Force criteria, residential units and other noise-sensitive land uses are "clearly unacceptable" in areas where the noise exposure exceeds 75 L_{dn}, "normally unacceptable" in regions exposed to levels of 65 to 75 L_{dn}, and "normally acceptable" in areas exposed to noise levels of 65 L_{dn} or less.

The following subsection provides a brief explanation of noise policies used by agencies having jurisdiction over this project.

Federal Regulations: The Federal Aviation Administration (FAA) regulates noise levels at airports. Federal Aviation Regulation (FAR) Part 36 sets noise certification levels for all aircraft designed after 1970. Foreign-manufactured aircraft are subject to International Civil Aviation Organization (ICAO) Annex 16, which is essentially identical to FAR Part 36. It is expected that both ICAO and FAA will further lower noise certification limits for future aircraft designs.

The initial goal of FAR Part 36 is to reduce existing noise levels by 10 dB. An aircraft retrofit and replacement rule has been adopted by the federal government. Since 1974, all newly manufactured U.S. aircraft have been required to meet FAR Part 36 standards.

The FAR Part 36 compliance program requires that at least 50 percent of aircraft over 75,000 pounds in the total aircraft fleet, powered by four engines, with no bypass ratio, or with a bypass ratio less than 2, be replaced or retrofitted by January 1, 1983, and continuing thereafter. Further, 100 percent of all other aircraft 75,000 pounds or more must be in compliance with FAR Part 36, with the exception of two-engine aircraft engaged in small community service which will be replaced or reengined by 1988. Lastly, the FAA requires that 100 percent of the aircraft be in compliance by January 1, 1985, with the exception of two-engine aircraft engaged in small community service.

According to new national legislation, no Stage 2 aircraft can be used after December 31, 2003. The legislation also requires that airlines reduce the number of Stage 2 planes to 15 percent of their fleet by 1999, and that the Department of Transportation issue a national policy by July 1, 1991. Stage 3 aircraft means an aircraft which meets at least one of the following definitions: Certificated aircraft — For aircraft which have been certificated under Federal Aviation Regulation Part 36, an aircraft which, at the time of its manufacture, or, if modified, subsequent to its modification, has been formally and officially certificated by FAA as a "Stage 3 airplane" within the meaning of Section 36.1(f)(6) of Part 36 of the Federal Aviation Regulations (14 C.F.R. (36.1(f)(6) [1987]). Uncertified aircraft — For aircraft not certified by FAA under Part 36 of the Federal Aviation Regulations, and aircraft which is identified under FAA Advisory Circular ("AC") 36-2E as generating noise levels at the Part 36 takeoff measuring point of 89 EPNdB or less.

To aid the airport operator in attaining noise/land compatibility, the FAA promulgated Part 150, "Airport Noise Compatibility Planning," which originally became effective on February 28, 1981, and was updated effective March 16, 1988. Part 150 contains standards for airport operators who voluntarily submit noise exposure maps and airport noise compatibility planning programs to the FAA. This regulation was based on Title I of the "Aviation Safety and Noise Abatement Act" (ASNA Act) of 1979, which adopted modified EPA recommendations for airport noise compatibility planning. Included in the regulation is the establishment of a single system for determining the exposure of individuals to airport noise, and a single system for measuring airport (and background) noise. The regulation also prescribes a standardized airport noise compatibility planning program, which includes: (1) the development and submission of noise exposure maps and noise compatibility programs to the FAA by airport operators; (2) standard noise methodologies and units for use in assessing airport noise; (3) the identification of land uses that are normally compatible (or incompatible) with various levels of airport noise; and (4) the procedure and criteria for FAA evaluation, and approval or disapproval, of noise compatibility programs by the FAA Administrator. The FAR Part 150 Noise Exposure Maps and Noise Compatibility Plan for the Albuquerque International Airport is dated April 1989.

FAR Part 150 contains a table entitled Land Use Compatibility with Yearly Day-Night Average Sound Levels, identifying land uses that are "normally compatible" or "noncompatible" with various levels of noise exposure. The levels of noise exposure, in yearly day-night average sound level (L_{dn}) correspond to the contours developed for each airport. All land uses may be considered as normally compatible with noise levels less than 65 L_{dn}.

U. S. Air Force Regulations: Land use guidelines for the U.S. Air Force are similar the FAA regulations. Thirteen Compatible Use Districts (CUD) are used to classify noise zones from an L_{dn} of 65 to 70 (CUD 13) to an L_{dn} of 85 and above (CUD 1). For example, it is recommended that no residential uses such as homes, multifamily dwellings, hotels and mobile home parks should be located where the noise levels are expected to exceed L_{dn} 65. Some commercial and industrial uses are considered acceptable where noise levels do not exceed L_{dn} 75.

Truck Noise Regulations: The Federal Highway Administration (FHWA) has established noise standards for traffic noise on federal highways. When these standards or "noise abatement criteria" (NAC) are approached or exceeded, noise impact occurs. The NAC for most sensitive receptors (including parks, residences, schools, churches, libraries, and hospitals) is 67 dB_A at receiver location or the boundary (FHWA, 1982).

Construction Noise Regulations: The City of Albuquerque Noise Ordinance has noise limits for construction activities (Ordinance No. 21-1975). According to this ordinance, "It shall be unlawful for any person within a residential zone, or within a radius of 500 feet therefrom, to operate or cause to be operated, any equipment used in construction, repair, alteration or demolition work on buildings, structures, streets, alleys or appurtenances thereto, with sound control devices less effective than those provided on the original equipment, or in violation of an regulations of the United States Environmental Protection Agency; or to operate or cause to be operated any such equipment during the nighttime, except in emergency situations as defined in Section 2 of this ordinance, in such a manner that the sound produced exceeds 50 dB_A, or 10 dB_A above the ambient noise level, whichever is higher, when measured at the residential property line."

The significance of increased noise levels is based on the ability of people to detect changes in their noise environment. If the construction or operation of the proposed project is expected to cause an increase of 5 dB_A or more, resulting in a residential receptor environment of 55 dB_A or more, then the impact is considered significant. Where the ambient is already 50 dB_A or above, an increase of 5 dB_A above the ambient would be significant. Levels 10 dB_A higher than the ambient are illegal in the City of Albuquerque.

Local Land Use Noise Regulations: The City of Albuquerque has adopted Ordinance Number 21-1975 "relating to the control of noise, by establishing noise levels, for the protection of public health and welfare and providing penalties." Known as the "Noise Control Ordinance," this document covers most items normally associated with noise and identifies types of land uses and the associated noise criteria. In general "it is unlawful for any person to make or continue, cause to be made or continued, or allow to be made or continued, any noise in excess of 50 dB(A), or 10 dB(A), above the ambient noise level, whichever is higher, at any residential property line." Aircraft engine noise during takeoff, landing, or ground aircraft movements is exempt.

Albuquerque Airport Noise Abatement: An airport tower "Letter to Airmen" (No. 87-1) discusses informal noise abatement procedures to minimize aircraft noise exposure.

4.4 NOISE

4.4.1 Construction Impacts

Noise impacts from construction activities at the project site are a function of the noise generated by construction equipment, the location and sensitivity of nearby land use, and the timing and duration of the noise-generating activities.

Although construction noise is limited in duration for a given project, adverse impacts due to construction noise are common. Heavy-earth moving and construction equipment is a recognized noise source with potential adverse impacts to sensitive receptors (see Table 4.4.1-1). To assess potential impacts from construction noise, the procedures and guidelines of the Construction Engineering Research Laboratory (CERL, 1978) have been utilized.

Normally, construction activities are carried out in stages, each of which has its own mix of equipment and noise characteristics. The maximum construction noise is expected to be generated during the demolition of old facilities, and the earth moving stages. A typical mix of construction equipment has been identified for use at the various stages of construction. Proposed equipment and the allowable maximum and predicted noise levels are shown on Table 4.4.1-1.

At 50 ft, during normal operating conditions, noise emissions from the equipment to be employed at the site should not exceed the allowed levels indicated on Table 4.4.1-1. These limits have been established for construction vehicles at all federal government structure sites (GSA, 1975) and are appropriate for this project. All contractors doing work at the airport construction site should comply with these regulations.

Table 4.4.1-1 also gives predicted noise levels at 50 ft for each equipment type. These data are based on numerous noise measurements by others and are cited in CERL and EPA documents. In the event the particular equipment chosen for the project does not comply with the allowed limits in Table 4.4.1-1, or is in excess of 75 dB_A at 50 ft, the contractor must provide temporary barriers or other appropriate noise suppression measures having sufficient attenuation characteristics to sufficiently reduce the intruding noise at the airport property line and at all affected sensitive receptors on base.

According to these estimates, demolition and earth-moving construction noise at the nearest sensitive receptor sites would be well below the normal noise levels created by aircraft operations (Figure 3.4.1-3). Therefore, no significant impacts to the surrounding environment are expected.

Table 4.4.1-1 Construction Equipment Noise Level Limits at 50 Ft

EQUIPMENT TYPE [1]	NUMBER USED	HOURS USED	dBA ALLOWED [2]	dBA PREDICTED [3]
Bulldozers (track)	5	630 hrs.	85	90
Front loaders (track)	3	367	75	75
Graders	4	415	75	75
Scrapers	5	1200	80	80
Off-highway trucks	8	300	75	88
Wheeled loaders	4	570	75	75
Rollers	4	370	75	75
Crane	1	200	75	88
Pavers	2	400	80	80

Source: Engineering-Science.

[1] Estimates of the number of pieces of equipment to be used and total hours each will be used. These values represent the worst-case situation with construction at its peak during remodeling Runway 8-26.

[2] These dBA limits cited by the U.S. General Services Administration. have been established as required criteria for this project (Harris, 1979).

[3] Predicted levels based on data cited in CERL and EPA documents and the Handbook of Noise Control, 2nd Ed. (Harris, 1979).

4.4.2 Operation Impacts

During the four-month shutdown of Runway 8-26 for construction of the new runway, departure ways, and taxiways, the major noise sources will include aircraft takeoff noise, construction vehicles, and increased highway noise. The airplane flyover sources are anticipated to be significant and will impact a different area than at present. Many sensitive receptors both on-base and in the surrounding area will have high noise levels that will interrupt sleep, cause speech interference, and tend to increase stress in sensitive individuals. It is expected that a portion of the population of Albuquerque will fall inside the L_{dn} 65 contour including schools, hospitals, churches, and residential communities. Significant portions of the population will have increased awareness of flight operations. Temporary or permanent noise mitigation measures should be developed, as necessary.

Existing on-base sensitive receptors are shown in Table 3.4-1. Day and night loudness levels (L_{dn}), caused by aircraft activities for each location, have been established by overlaying existing and anticipated L_{dn} contours on a map of the airport and local environs (Land Use Map prepared by the Department of the Air Force). The base hospital and chapel are located within the L_{dn} 65 contour. Residents of the base correction facility are within the L_{dn} 80 contour.

Due to increased volume, higher payloads, and faster aircraft, the foregoing noise levels, or slightly higher, will be experienced by these same receptors after the construction is complete and aircraft patterns return to normal.

During the 4-month shutdown of Runway 8-26 there will be significant noise impacts for a large segment of on-base facilities not previously affected. The shift of air traffic patterns to Runway 17-35 during construction will also cause a significant increase in noise levels to a large group of sensitive receptors. Examples of on-base sensitive receptors are shown in Table 3.4-1. Under existing conditions most of these receptors have been exposed to excessive noise levels only during short periods. During the duration of construction these receptors will be subject to unacceptable noise intrusions. Receptors that fall within the 80 or greater L_{dn} contour include the technical library, a new group of family housing, youth center, public school and guest housing. At this level of noise, all activities at these sensitive receptor locations will be interrupted. Normal building construction practices will not be sufficient to attenuate these sounds to an acceptable indoor level. Similar noise levels will impact highly populated areas off-base. Each on-base sensitive receptors that will be impacted by aircraft using Runway 17-35 during construction is listed in Table 4.4.2-1, along with an estimate of the L_{dn} noise level that will impact the site.

Table 4.4.2-1 Kirtland Air Force Base On-Base Sensitive Receptors for Temporary Air Traffic Patterns During Construction

Sensitive Receptor	Estimated L_{dn}
1. Technical library	80-85
2. Family housing — enlisted Personnel	80-85
3. Youth center	80-85
4. Public school	80
5. Guest housing	80
6. Church	75
7. Officers club	75
8. Dorms — military	65
9. Flight training	65
10 Weapons research	85

Source: Engineering Science.

4.4.3 Cumulative Impacts
4.4.3.1 Construction
Construction noise due to the proposed project would occur during the same time frame as will demolition and reconstruction of Apron E in the hanger area. Construction activities associated with Apron E will occur over a 5-year period, so that 20% of that project could be expected to be carried out annually. As a result, additional impacts due to construction and vehicular traffic would be expected. Due to the distances involved, there would be no significant noise impact increase at sensitive receptor locations due to proposed increased construction operations. Likewise an expected increase in vehicular traffic would not increase the L_{eq} more than 5 dB_A; thus, there are no significant traffic noise impacts.

4.4.3.2 Operations after Construction
Noise levels will revert to existing conditions after construction is complete. However, since F-16 aircraft will be replacing A-7D aircraft used by the National Guard and additional flights by both military and commercial airlines are likely to occur, noise may increase both the size of area impacted and the level of noise at sensitive receptors (Table 4.4.3-1).

Table 4.4.3-1 Kirtland Air Force Base Military Aircraft Noise Levels

Military designation	Civilian equivalent*	Noise levels (approach/takeoff)
A-7D	B 707	84/94
C-130	DC10/L-1011	88-97/84-96
F-18	Learjet 25	88-94/80-83
F-16**	Learjet 35	82-83/66-72

* Conversion from military to civilian equivalent per FAA Advisory Circular 36 and NOISEMAP Program Manual.
** New aircraft replacing A-7D aircraft after proposed construction.

6.4 Noise

6.4.1 Mitigation Measures

6.4.1.1 Construction Noise Mitigation

Mitigation measures would be required to minimize construction noise impacts to on-base and nearby off-base sensitive receptors. Especially sensitive receptors, both on-base and in the local community, such as hospitals, schools, convalescent homes, and residences that fall within the L_{dn} 65 contours, would need the implementation of noise mitigation measures to reduce interior noise impacts.

Short-term impacts would be reduced by the following measures:

- Before construction activities begin to affect residential, commercial,and noise-sensitive receptors, they should be given advance notice of the construction scheduled for their area, advised of the likelihood of high-noise levels and informed of the measures taken to reduce noise impacts.
- Require the contractor to use the quietest types of equipment available. As a minimum, manufacturer-recommended silencers, mufflers, and acoustical enclosures and hoods must be properly installed and in good condition. It is recommended that all construction equipment be properly maintained according to the manufacturer's suggested procedures, including the proper fitting and use of noise suppression features and devices.
- Limit the hours of noisiest activities to the daytime during weekdays when within 500 ft of identified sensitive receptors that are actively functioning during evenings, nighttime and/or weekends. It is recommended that the contractor be made aware of the noise standards and construction noise time-of-day limits for the City of Albuquerque and the Air Force. The contractor should be required to adhere to the noise standards or obtain the proper permits and/or variances that allow the holder to exceed noise levels or noise time limits.
- Require the contractor to provide temporary noise barriers to reduce construction noise impacts to sensitive receptors within 100 ft of noisy construction.
- Provide noise monitoring of sensitive receptor areas periodically during construction activities (all shifts) and report any excessive noise-producing activity. Require contractor to investigate and report on measures taken to properly reduce noise impacts.

6.4.1.2 Temporary Operation Noise Mitigation

The new flight patterns from Runway 17-35 to the north used by aircraft during the 4-month Runway 8-26 shutdown would cause significant noise impacts to nearby on-base and local community sensitive receptors. Noise abatement takeoff and landing procedures should be used. Flight tracks should be analyzed to minimize noise impacts to on-base and community receptors. Flight safety concerns should be addressed as well in evaluating noise mitigation measures.

6.4.1.3 Future Operation Noise Mitigation

Long-term noise reduction measures should be considered for future on-base operational noise impacts. These include sealing sound leaks in on-base structures plus providing upgraded doors and windows and sound-attenuating air vents (see illustrations in Section 7.8). These measures would reduce noise impacts as more powerful military aircraft are assigned to the base and commercial activity continues to increase.

References:

Albuquerque, City of, 1975. Environmental Health Department, Consumer Protection Division, Chapter 6, Article 22, Ordinance No. 21-1975, Noise Control Ordinance.

CERL, 1978 — Construction Engineering Research Laboratory, Interim Report N-36, "Construction-Site Noise: Specification and Control, Construction-Site Noise Control Cost-Benefit Estimating Procedures", by F. M. Kessler, P. D. Schomer, R. C. Chanaud, E. Rosendahl, January 1978.

Department of the Air Force, 1990. Environmental Assessment of the Realignment of Units at Kirtland Air Force Base, New Mexico — Department of the Air Force Headquarters, Military Airlift Command, Scott Air Force Base, Illinois, April 1990, Figure 3.3.

FAR Part 150, 1989. Albuquerque International Airport — Noise Exposure Maps and Noise Compatibility Plan, April 1989, by Greiner, Inc.

FHWA, 1977. "Federal Highway Construction Noise: Measurement and Prediction", May 2, 1977, red by J. A. Reagan and C. A. Grant.

GSA, 1975. United States General Services Administration, "Construction Equipment Practices," January 1, 1975.

Harris, C. M., 1979. Handbook of Noise Control, 2nd Ed., McGraw-Hill pp. 31— 4.

Kessler, F. M., R. C. Schomer, R. C. Chanaud, and E. Rosendahl, 1978. Construction Engineering Research Laboratory, (CERL) Interim Report N-36, "Construction-Site Noise: Specification and Control, Construction-Site Noise Control Cost-Benefit Estimating Procedures."

Kirtland Air Force Base, 1982. Colored land use map provided by KAFB Planning Department.

Molzen-Corbin & Associates, 1990. Preliminary Engineering Report, "lbuquerque International Airport, Proposed Improvements to Taxiways A and E."

7.10 **ACOUSTICAL ENVIRONMENT – A HEALTH HAZARD?**
REPORT ON WORLD WORKPLACE-95
ROUND TABLE DISCUSSION
September 19, 1995, Miami Beach, FL

Panel Moderator - D. A. Harris, CFM
Panel Members: David G. Fagen, Acoustical Consultant
John Flood, Eckle Industries
John Kopec, Riverbank
Ken Roy, Armstrong
Alice Suter, Consultant
Richard A. Versen, Schuller

Panel moderator Harris opened the session by indicating that the purpose of the session was to investigate allegations that:

- Acoustical materials are a hazard to your health
- Noisy workplaces cause hearing loss, stress, and loss of life
- Poor workplace acoustics causes significant loss of productivity (the office and factory are at risk)
- ADA acoustical requirements significantly impact building costs

Members of the panel included acoustical experts from ASTM Committee E33 – Environmental Acoustics, Acoustical Consultants; agencies that promulgate acoustical standards/criteria; manufacturers of acoustical materials; concerned facility managers; and naysayers. Their prime concerns were: Is the workplace acoustical environment hazardous to your health? Will acoustical materials made of fiber cause cancer? Are open-cell foams a fire hazard? If so, what can you, as a building owner, facility manager, architect, contractor, or designer do about it and what impact will there be on building design.

PANEL PRESENTATIONS
The following statements were prepared by the panelists and represent the views of the presenter.

IS THE WORKPLACE ACOUSTICAL ENVIRONMENT HAZARDOUS?
David A. Harris, CFM, Consultant – Panel Moderator

The question "Is the workplace acoustical environment hazardous to you health?" has many implications and different responses depending on your point of view. YES, noise can kill and cause permanent injury. However, unlike a chemical spill, noise must be extreme in order to kill outright. Sounds and vibrations from a space ship launch could kill outright if one were close enough. However, this situation is rare.

More typically, hazards due to noise are of a secondary nature. Too much noise causes hearing loss, stress, and possible deafness. Noise also becomes a killer when it masks the sound of a warning from a fellow worker. Noise mitigation efforts deserve the same concern as other health hazards. Unfortunately, this approach is not been the case in our efforts to improve our environment. While we have many laws about noise levels in the workplace, our legal efforts to enforce them have been fraught with inconsistency, lack of funds, and outright misunderstandings.

Most acoustical materials appear benign. However, efficient sound absorbers are typically composed of fibers and open-cell foams. Recent allegations that many fibers can cause cancer have raised serious questions by specifiers of building materials, building owners, and users. In the case of open-cell foams, the issue centers on their flammability and release of toxic fumes. These concerns are real and to ignore them could be costly. By contrast, many have considerable test data to argue the concerns are overblown. What is the answer? Unfortunately the jury is still out. What is at risk, however, is that ignorance of material performance can produce considerable health risk and litigation.

As with any contentious issue, knowledge, the dissemination of test data, and proper application of the test results to actual situations may be the only safeguards for the building owner, designer, architect, contractor, and facilities manager.

Who Regulates Acoustical Issues?

The United States Occupational Safety and Health Administration (OSHA) and the Environmental Protection Agency (EPA) have strict guidelines on the amount of noise that is acceptable in the workplace. For example, OSHA limits the noise that a worker may be exposed to in an 8 hour work day to 90 dB$_A$. Sound levels over 85 dB$_A$ mandate that all workers so exposed shall have audiometric tests conducted annually. Higher noise levels require limited worker exposure (see Figure 7-15). OSHA requires one of three techniques to mitigate a noisy workplace: engineering controls that lower the level below 85 dB$_A$, management controls that limit the time a worker may be in a noisy environment, and temporary measures that can be utilized until acceptable engineering or management controls are installed. Note: ear protectors are a **temporary** solution.

Noise Level, dB$_A$	Allowable Daily Exposure
85	8 hours (audiometric testing required)
90	8 hours
92	6 hours
95	4 hours
100	2 hours
102	1.5 hours
105	1 hour
110	0.5 hour

Figure 7-15 Allowable OSHA noise exposure

The Environmental Protection Agency (EPA) regulates noise levels at property lines. These limits vary depending on the type of noise, the type of receiver, occupancy and the time of day. Generally, noise that is intrusive such as transportation and factory noise is considered unacceptable. State, local, and model building codes have noise ordinances that impact all aspects of our acoustical environment. Criteria may vary from community to community.

Permanent assistive listening system is required in space with:

- Fixed seating and 50 persons; or
- Fixed seating and audio amplification system.

Permanent or portable assistive listening system is required in space with:

- No fixed seating; or
- less than 50 persons and no amplification system.

Figure 7-16   ADA assistive listening requirements

The Americans with Disabilities Act (ADA) has a section dealing with acoustical criteria including the requirements for assistive listening systems (ALA). ALAs are one item in an array of what the ADA calls "auxiliary aids." Various auxiliary aids and other measures may be required for any one facility. ALS requirements apply to assembly areas in which audible communications are integral to the use of the space such as facilities where there is live speech or music (e.g., lecture halls and theaters) or recorded programs (e.g., movie theaters). All public accommodations and commercial facilities designed and occupied after January 26, 1993 require compliance (See Figure 7-16).

The cost of complying with ADA requirements can vary depending on the original plans for the building. If acoustics was not a part of the original planning, ADA criteria could increase costs by $10 per square foot. Addition of sound-absorbing materials, sound barriers, and sound amplification systems are the major cost items. A well-designed space with good acoustics may only require additional wiring for individual amplifiers for a small portion of the space.

Architectural acoustics deals primarily with the sound quality of a space. In the case of auditoriums, concert halls, and public spaces, acoustical designs center on the use of sound-absorbing materials to control the reverberation time in the occupied space to provide the desired listening environment. If speech understandability is a key attribute, a low reverberation time, with no echoes, is desired. Longer reverberation and focused sounds may be desired for some musical performances where no amplification is utilized. Containment of noise from a source signal is needed between occupancies in hotels, residences, offices, and manufacturing facilities. Sound absorption attributes of materials are measured by their noise reduction class (NRC). Barriers between occupied spaces are rated by their ability to block sound and their sound transmission class (STC). Selection of proper acoustical materials and systems of construction is best specified by a qualified acoustician or member of the National Council of Acoustical Consultants. Criteria are established by the user need. Building codes generally provide minimum recommendations.

Open office acoustics is a special field where speech privacy is the key issue. With no barriers of significance, the degree of success in achieving normal or confidential privacy, is a balance and compromise whereby all the elements of the acoustical environment act together as a system. To be effective in acquiring confidential speech privacy the occupied space must have highly sound absorbent ceilings and vertical surfaces (i.e., walls, columns, windows, etc.), sound barriers that block direct sound, and highly sound-absorbent faces to reduce reflected sound, and a background masking system. New rating systems are speech privacy potential (SPP) for an entire working acoustical system and articulation class (AC) for rating individual system components such as a ceiling material, wall treatment, and the like. Criteria are primarily based on user needs and should be established by acoustical experts. The bottom line to achieving a successful open office acoustical environment can also be measured in "worker productivity" and "user satisfaction."

Retrofit acoustics are notoriously expensive and may cost 5 times the expense of an original installation. For this reason it is very cost-effective to require professional acoustical designs and assure they are properly installed in all new or renovated buildings. In addition, legal remedies can range from court-mandated corrections to fines for building code and/or noise ordinance violations up to $100,000 in certain cases.. In the case of rental space such as multifamily dwellings, hotels, and offices, poor sound barriers between living units will cause considerable loss in rent and/or sales. Poor office acoustics likewise lowers property values and worker productivity. Auditoriums with bad acoustics are shunned by potential performers, causing a loss of revenue.

An extensive family of acoustical standards exists. Most have been prepared by the American Society of Testing and Materials (ASTM) Committee E33 – Environmental Acoustics. Expert volunteers from manufacturers, general interest and users have prepared industry-wide standards for testing and standardizing the acoustical properties of materials and systems of construction, architectural and environmental acoustics, industrial noise control, and office acoustics. ASTM Designation E 1433 – Standard Guide for Selection of Standards on Environmental Acoustics, Annual Book of Standards Volume 4.06, contains a detailed list of all the applicable acoustical standards and a description of the method, use, result, and summary for each standard. ASTM Committee E33 sponsors a 2-day Seminar on Architectural Acoustics each year. [Call (360) 437-0814 or (215) 299-2610 for details.]

PRESENTERS VIEWPOINTS:

CONCERNS ENCOUNTERED WITH ACOUSTICAL MATERIALS
David G. Fagen, Acoustical Consultant

A present "hazard" in the acoustical environment is a lack of understanding of product performance and limitations. Often this is the result of improper use of terminology. Other times it is the result of misleading marketing literature and naivety of the sales reps. Some examples discussed are misapplication of sound insulation versus sound isolation, and published sound absorption data, impact isolation data, and sound transmission loss data. [Editors note: Dave presented several humorous stories and examples to make his point. Next time you see him, ask him to tell his "stressed out alligators" story.]

THE USE OF FIBERGLASS FROM A MANUFACTURERS PERSPECTIVE
John Flood, V.P. Acoustic Division, Eckel Industries, Inc.

For over 40 years Eckel Industries has been manufacturing products for industrial and architectural noise control. Eckel also designs, manufactures and installs large acoustical test facilities such as anechoic chambers. The basic absorption material we use in these products is fiberglass, or mineral wool. As a manufacturer of products containing fiberglass we have to be concerned with the potential health hazards posed from the use of this material.

Since fiberglass has been listed since June 1987 by the International Agency for Research on Cancer (IARC) as a potential carcinogenic material, we have had many customers question the use and application of this material. Because of these concerns we have recently had an independent laboratory perform tests on some of our products to determine the actual migration of fiberglass fibers from the product into the air. Even if the test results proved negative customers, will still shy away from the use of this material.

300

We have altered the design of our products to meet the customer demands. In some cases the fiberglass is contained in hermetically sealed polyethylene bags, and in some other instances the fiberglass wedge units are encased in fiberglass cloth or some other cloth-like material.

We provide Material Safety Data Sheets (MSDS) on the fiberglass material with all shipments of material that contain loose or exposed fiberglass. We not only have concerns for our customers but also for our workforce, who are exposed daily to the fiberglass and mineral wool as it is processed for application in our products. We carefully follow the handling recommendations in the safety data sheets.

Whether or not fiberglass is truly a carcinogenic material, we have had to take measures to protect our customers concerns and to also cover ourselves for any potential liability. This is done through redesign of our products and education of our customers. In our experience, fiberglass is one of the most economical and highest-performance materials available for acoustical absorption. We hope to continue using this material in a safe and responsible manner.

HAZARDOUS PRODUCTS OLD AND NEW:
A LABORATORY MANAGERS PROSPECTIVE
John Kopec, Manager, Riverbank Acoustical Labs (RAL)

Many old products used for noise control applications are not acceptable today because they supposedly contained materials that either are possibly flammable, toxic, or a health hazard. Since 1918 the RAL staff has handled many specimens that later were said to contain such materials. To avoid any reoccurrences the staff has developed some quick tests that have proven most successful.

Although some new products meet present environmental and health specifications and achieve excellent acoustical results, proper knowledge of their respective makeup, application, and installation is essential. What new product applications are good and what are not? What materials are manufacturers avoiding? How has competitive advertising created a dilemma for users? How does laboratory accreditation fit into this dilemma? The answers to these questions could prove most interesting and useful to all concerned.

[Editor's note: John provided personal insights on each question that may be obtained by reviewing the audio tape of the session, available from International Facility Management Association (IFMA) staff.]

SOUND/NOISE/HUMAN RESPONSE
Kenneth P. Roy, Ph.D., Armstrong Innovation

Since the early 1950s a great deal of investigation has been pursued through many disciplines relative to the effects of sound and the acoustic environment on man. Excessive noise is generally acknowledged to have an adverse effect on building occupants, ranging from fatigue to reduced productivity to impaired speech communications to hearing loss. The acoustic environment is a primary concern almost everywhere – in the office, in schools, hospitals, public buildings, and in our factories.

Three specific acoustic environments are presented which span the breadth of workplace situations that may be of interest to us relative to sound and health. It has been shown that high noise levels in the industrial workplace will have a detrimental effect on our sense of hearing, either through a stress-induced or acute-onset hearing loss, both of which are addressed by OSHA standards in the workplace. Even in the much less noisy office environment, where hearing loss is generally not an issue, noise induced annoyance and lack of privacy can be shown to cause a loss in productivity. And finally, what about the "flip side" of noise as a health hazard – in the healthcare environment sound and music are being used as a tool in health and healing.

INDUSTRIAL WORKPLACE:

Federal legislation pursuant to the public health or welfare relative to occupational noise began with the Walsh-Healy Public Contracts Act (1960), was expanded to the Occupational Safety and Health Act (1970), and the Noise Control Act (1972). These put into place criteria for health hazards and established limits for noise exposure of industrial workers. The OSHA Noise Standard was later amended (1982) to require audiometric testing in cases of exposure to noise levels of 85 dB_A or higher for an 8-hour day. The original standard of 90 dB_A, and the time/level exposure limits remain in effect for the purposes of engineering and/or administrative controls. It is recognized that an excessively noisy environment is a health hazard, and has a direct correlation with hearing loss.

OFFICE WORKPLACE:

A series of five Steelcase (Lou Harris and Associates) research studies were published from 1978 to 1991 which related the effects of worker attitudes and office environments on such things as job performance and productivity. These found that facilities managers rank productivity first on a list of office design issues, with 78% rating it very important. When asked to rate the seriousness of office comfort problems, office workers identified privacy issues (distractions, noise, lack of privacy) just after temperature, and 62% claimed it is very important to have privacy to work effectively. Two volumes of Using Office Design to Increase Productivity were published by Michael Brill, Buffalo Organization for Scientific and Technological Innovation (BOSTI) and Westinghouse Furniture Systems in 1984–1985. These studies discuss the relationship between the degree of enclosure and its effects on accessibility, control over distractions and interruptions, and privacy. Major findings of the BOSTI research relative to job performance or satisfactions were, in order of importance, degree of enclosure, internal layout, type of furniture, noise flexibility, floor area, etc.

It is relatively difficult to measure productivity in terms of the type of work which we generally do in offices as opposed to in the factory. However, from papers given at the June 1994 meeting of the American Society of Heating, Refrigeration and Air Conditioning Engineers (ASHRAE), H. G. Lorsch and O. A. Abdou presented information from a 1988 Building Owners and Managers Association (BOMA) study based on a survey of 400 executives involved with space planning and use, which indicated an expected productivity gain of 26% if the worst problem were related to acoustics and noise were eliminated. The authors cautioned that this number is necessarily crude and may be exaggerated by the respondents due to frustration with the issue; however, it should stand as an indication of relative importance. A paper presented by W. M. Kroner and J. A. Stark-Martin found that improved indoor architectural and environmental design contributed to an overall productivity increase of 16% for an insurance company's move to a new facility. Additionally, this study indicated that the use of environmentally responsive workstations (ERW) which included task masking was responsible for a 2% increase in productivity.

HEALTHCARE:

If we take the premise that excessive noise is stressful, and thus injurious to our health, then the reverse of this must be that music is soothing, and thus advantageous to our health. While browsing the association bookstore at the Symposium on Healthcare Design 1994, I found many books and tapes on just this subject.

In the book <u>Music Physician for Times to Come</u>, an anthology by Don Campbell, you will find many interesting aspects of sound and music for health and healing, From <u>The Music Therapist</u> by Joseph J. Moreno, "in considering the historical development leading to the current applications of music therapy as a professional discipline, it is crucial to remember that music therapy is not a recently developed modality. Rather, music therapy as a profession is only a recently and specialized line of development of the continuing 30,000 year old shamanic traditions of music and healing still being practiced throughout the world." We can stand to learn from this, so that we can start to transform a workplace filled with noise and stress into a workplace characterized by music and harmony and productivity.

ADVERSE EFFECTS OF WORKPLACE NOISE
Alice H. Suter, Hearing Consultant

Noise has long been recognized as one of the most prevalent occupational hazards. Approximately 9 million U.S. workers are exposed to potentially hazardous levels of noise, 5½ million of them in manufacturing and utilities. Hazardous levels are considered to be time-weighted average noise levels above about 85 dB_A, although some of the more sensitive workers may be affected at lower levels.

Noise-induced hearing loss is a critical health problem. Hearing impairment from noise may lead to a handicap that seriously degrades the quality of life, especially in one's retirement years. It can cause withdrawal from social situations, dependency upon one's spouse for all communication needs, and ultimately, loneliness, isolation, and depression.

Although hearing impairment is the most well-known adverse effect of noise, noise also causes other adverse effects, such as interference with speech communication and the perception of warning signals, which can lead to job inefficiency and safety hazards. Certain noise conditions can also lead to productivity decrements, especially when noise levels are greater than 95 dB_A, the noise conditions are unpredictable, or the job being performed is complex.

Noise exposure does not kill people, at least directly, but there is recent evidence that high levels of noise and the resulting hearing losses contribute to industrial accidents. There is also reason for concern that hearing protection devices, which are worn to prevent hearing loss, may actually impair work safety under certain conditions. In addition, there is growing evidence that noise adversely affects general health, although there is some controversy about this issue in the scientific community.

FIBERGLASS HEALTH AND SAFETY:
MOVING BEYOND THE SCIENCE
Richard A. Versen, MPH, CIH, Schuller International

For years, acoustical engineers and architects have relied upon fiberglass materials to provide the sound reduction benefits that they need to properly attenuate noisy environments. Today, these professionals are responsible for creating an acceptable indoor environment – one that not only provides acceptable acoustical controls but good indoor air quality as well. While the acoustical benefits continue to be realized, concerns about possible health risks associated with the use of fiberglass insulation as an acoustical treatment have arisen. Fiberglass is one of the most widely tested and researched materials available in the construction industry. Studies dating back to the early 1940s and continuing today provide tremendous information from which benefit/risk analyses can be undertaken. While the weight of this scientific information is reassuring, confusion and controversy continue to confound the issues. Scientific disagreement about which types of studies are most relevant to people unfortunately cloud the issues. For acoustical applications there really is no confusion. Once installed, there is essentially no exposure to building occupants. Because of this, numerous agencies and review groups have concluded that there is insignificant risk associated with the use of fiberglass materials in an indoor environment. Our industry is committed to continuing to add additional research on fiberglass safety to assure that the many benefits of fiberglass are realized safely. By moving beyond the science, and the ever-present dueling scientists, acoustical specialists can make rational and commonsense decisions about their continued use of fiberglass insulation materials.

[Note: A more detailed version of Mr. Versen's presentation was published in the Proceedings of WorkPlace 95. A copy can be obtained by calling Mr. Verson at (800) 654-3103 or D. A. Harris at (360) 437-0814.]

References and additional information sources:
- *Noise Control Manual*, D. A. Harris, Van Nostrand Reinhold (VNR), 1991.
- *Planning and Designing the Office Environment*, D. A. Harris et al VNR, 1991.
- American Society for Testing and Materials (ASTM) *Annual Book of Standards* Vol.4.06.
- *Acoustics for the Workplace of Tomorrow, IFMA Proceedings*, October 1993 and November 1994. D. A. Harris.
- *ASTM Seminars on Architectural Acoustics*, Sept. 21–22, 1995, Tampa, D. A. Harris.

Panelists:

David A. Harris – Moderator: A Certified Facility Manager (CFM) from Port Ludlow, Wash., Dave is the author of *Planning and Designing the Office Environment,* 1st and 2d eds., *Noise Control Manual*, and many technical papers on building systems, materials of construction, and acoustics. He is an inventor with 11 patents, has over 30 years of experience in R&D and market development, is past president and Executive Director of the Noise Control Association (Quiet Man Award), member of NCAC, ASA (Silver Certificate), member and officer of ASTM E33 (Certificate of Appreciation). Harris has a B.S. in Light Building Industries (e.g., construction technology) from the University of Wisconsin, with graduate work in building codes and business administration. Phone: (360) 437-0814.

Alice H. Suter, Consultant, Ashland, Ore.: Alice Suter has worked in the area of noise effects and hearing conservation for more than 20 years. She has an undergraduate degree in liberal arts, an M.S. in education of the deaf, and a Ph.D. in audiology. She has been influential in noise criteria development, regulation, and public policy, first at the U.S. EPA's Office of Noise Abatement and later at OSHA in the U. S. Department of Labor. At the EPA she participated in the development of criteria for noise effects, including the psychological, extra-auditory physiological, performance, and communication effects, as well as the effects on hearing. As Senior Scientist and Manager of the Noise Standard at OSHA, she was principal author of the hearing conservation amendment to the standard for occupational exposure to noise. She has also held positions of Visiting Scientist and Research Audiologist at the National Institute for Occupational Safety and Health. She is now a consultant in industrial audiology and community noise in Ashland, Ore. Among her clients have been the World Health Organization, private corporations, individuals, citizen groups, and government agencies on the federal, state, and local level. Dr. Suter is a Fellow of the Acoustical Society of America and the American Speech-Language-Hearing Association. Phone: (503) 488-8077

Richard A. (Rick) Versen, MPH, CIH, Director, Product Stewardship and Occupational Health, Schuller International Inc., Mountain Technical Center, Littleton, Colo: Versen has over 14 years of experience in occupational safety and health with Schuller International Inc., a subsidiary of Manville Corporation. His current responsibilities include domestic and international product stewardship program development for all corporate divisions. He has given many presentations to employees, distributors, contractors, and end users about the health and safety aspects of Schuller products. Additionally, he has responsibility for the Toxicology, Epidemiology, and Product Safety programs for the company. Versen is also directing the indoor environmental quality research activities. He previously managed Schuller's product safety responsibilities and its corporate industrial hygiene program, directing its AIHA-accredited laboratory. He has conducted numerous industrial hygiene surveys in a variety of manufacturing processing and end user operations domestically and internationally. Versen has also provided technical expertise in exposure assessments for laboratory animal testing and has assisted with epidemiologic studies. He has a bachelors degree in chemistry, Carthage College, and a masters degree in Public Health–Industrial Hygiene, University of Illinois School of Public Health. Phone: (800) 654-3103.

John Kopec, Director, Riverbank Acoustical Laboratories, Geneva, Ill.: John directs the execution of acoustic and aerodynamic testing, including test design, instrumentation, procedures, documentation, and evaluation of acoustical materials and systems of construction. Riverbank, a subsidiary of the Illinois Institute of Technology, is a world-renowned independent acoustical testing laboratory. Mr. Kopec has participated in numerous community noise impact studies, served as project engineer on noise control programs for HVAC equipment, author and/or coauthor of over a dozen technical papers on acoustics, is an officer of ASTM E33 Committee on Environmental Acoustics, has been a major presenter at ASTM E33 Seminars on Architectural Acoustics, and was recently named a Fellow of the Acoustical Society of America. Phone: (708) 232-0104

John Flood, Vice President of Eckel Industries, Cambridge Mass. and President of the Noise Control Association: John has developed and marketed many innovative materials and systems for the building materials and noise control industry. Eckel Industries manufactures and markets anechoic wedges, acoustical test chambers, and specialty architectural acoustical systems. Phone: (617) 491-3221

David G. Fagen, Principal, Fagen Assoc. Inc., Acoustical Consultants, St. Petersburg, Fla. An acoustical consultant, and an architect, Dave has been an officer of ASTM Committee E33 on Environmental Acoustics, and a long time member of National Council of Acoustical Consultants (NCAC) and has experienced first hand the value of good acoustical planning and the problems to be expected from poor or no acoustical planning. Phone: (813) 823-3564

Kenneth Roy, Ph.D., Senior Research Associate, Armstrong World Industries Inc., Lancaster, Pa. Ken has a Ph.D. in Acoustics from Pennsylvania State University and over 20 years' experience testing and developing acoustical materials/systems. He has presented numerous papers and technical articles on architectural and open-office acoustics and has been an invited speaker at IFMA, AIA and other industry organizations. He is an officer of ASTM E33 and a long-time member of ASA. Phone: (717) 396-5700.

APPENDIX 1 - GLOSSARY OF ACOUSTICAL TERMS

(Note: Most of these definitions are quoted directly from ASTM C634, Standard Definitions of Terms Relating to Environmental Acoustics. Those noted by an * are provided by the author.)

<u>Acoustical material</u> – any material considered in terms of its acoustical properties. Commonly and especially, a material designed to absorb sound. (Note; Early definitions required a minimum of .50 Noise Reduction Coefficient (NRC). This criteria has been dropped in recent years.)

<u>Airborne sound</u> – sound that arrives at the point of interest, such as one side of a partition, by propagation through air.

<u>Ambient noise</u> – the composite of airborne sound from many sources near and far associated with a given environment. No particular sound is singled out for interest.

<u>Articulation Class</u> (AC)* – the sum of the weighted sound attenuations in a series of 15 test bands between 200 & 5000 Hz. per ASTM E1110 - Standard Classification for Determination of Articulation Class.

<u>Average sound pressure level</u> – of several related sound pressure levels measured at different positions or different times, or both, in a specified frequency band, ten times the common logarithm of the arithmetic mean of the squared pressure ratios from which the individual levels were derived. Note: an average sound pressure level obtained by integrating and averaging during certain time periods is often called equivalent sound pressure level and, when A-weighted, equivalent sound level.

<u>A-Weighted sound level</u> (dB_A)* – The most common single number rating system for measuring the loudness of a noise. It may be read direct on most sound level meters by selecting the designated scale. It is obtained by applying the A-weighted frequency response curve to the measured sound. The response curve is indicative of the way humans respond to different frequencies. (e.g. Two sounds having equal energy or intensity but of different frequencies do not sound equal to the human ear. We are bothered by high frequency noise more than low frequency sound. The A scale was developed to compensate for this anomaly.)

<u>Background noise</u> – noise from all sources unrelated to a particular sound that is the object of interest. Background noise may include airborne, structure borne, and instrument noise.

<u>Composite loss factor</u>* (CFL) – A measure of the damping characteristics of a material or system. Acceptable damping has a composite loss factor of 0.05.

<u>Composite Noise Rating</u> (CNR)* – used to measure aircraft noise it is a modification of the perceived noise level (L_{pn}) with adjustments for the number of events and the time of day. CNR is replaced by NEF.

<u>Community Noise Equivalent Level</u> (CNEL)* – A modification of the day-night average sound level (L_{dn}) with an additional factor of 3 applied to the evening noise levels between 7 pm. and 10 pm. CNEL has been adopted by the State of California to establish environmental noise regulations. A similar regulation is called

the Environmental Designation for Noise Abatement (EDNA) by the State of Washington.

<u>Damp</u> – to cause a loss or dissipation of the oscillatory or vibrational energy of an electrical or mechanical system.

<u>Day-Night Average Sound Level</u> (L_{dn})* – A modification of the Equivalent Sound Level (L_{eq}) whereby the time period is defined as 24 hours and the weighting factor of 10 is applied to the measured night time noise levels between 10 pm. and 7 am. It was developed by the U. S. Environmental Protection Agency (EPA) and is the standard metric for measuring aircraft noise by the Federal Aviation Agency (FAA).

<u>Decay rate</u> – for airborne sound, the rate of decrease of sound pressure level after the source of sound has stopped; for vibration, the rate of decrease of vibratory acceleration, velocity, or displacement level after the excitation has stopped.

<u>Decibel</u> (dB) – the term used to identify ten times the common logarithm of the ratio of two like quantities proportional to power or energy. (See level, sound transmission loss.) Thus, one decibel corresponds to a power ratio of (10 to the 0.1 power) to the n power. Note: since the decibel expresses the ratio of two like quantities, it has no dimensions. It is, however, common practice to treat "decibel" as a unit as, for example, in the sentence, "The average sound pressure level in the room is 45 decibels."

<u>Diffraction</u> – a change in the direction of propagation of sound energy in the neighborhood of a boundary discontinuity, such as the edge of a reflective or absorptive surface.

<u>Diffuse sound field</u> – the sound in a region where the intensity is the same in all directions and at every point.

<u>Direct sound field</u> – the sound that arrives directly from a source without reflection.

<u>Equivalent Sound Level</u> (L_{eq})* – also known as the equivalent-continuous sound level or average sound level this descriptor is based upon the measured time varying A-weighted sound level for a given period of time.

<u>Field sound transmission class</u> (FSTC) – a single-number rating derived from measured values of field transmission loss in accordance with Classification E 413, Determination of Sound Transmission Class. It provides an estimate of the performance of the partition in certain common sound insulation problems.

<u>Field transmission loss</u> (FTL) – of a partition installed in a building, in a specified frequency band, the ratio, expressed on the decibel scale, of the airborne sound power incident on the partition to the sound power transmitted by the partition and radiated on the other side.

<u>Flanking transmission</u> – transmission of sound from the source to a receiving location by a path other than that under consideration.

<u>Frequency</u>* – the number of cycles per second measured in units of Hertz (Hz). A frequency of 1000 Hz means 1000 cycles per second.

<u>Impact insulation class</u> (IIC) – a single-number rating derived from measured values of normalized impact sound pressure levels in accordance with Annex A1 of ASTM Method E 492, Laboratory Measurement of Impact Sound Transmission Through Floor-Ceiling Assemblies Using the Tapping Machine. It provides an estimate of the impact sound insulating performance of a floor-ceiling assembly.

<u>Impact noise rating</u>* (INR) – a single-number rating of a floor/ceiling assembly derived from measured values of impact noise when tested in accordance with procedure ISO R140 (Tapping Machine) This rating is similar to and replaced by Impact Isolation Class (IIC).

<u>Insertion loss</u> (IL) – of a silencer or other sound-reducing element, in a specified frequency band, the decrease in sound power level, measured at the location of the receiver, when a sound insulator or a sound attenuator is inserted in the transmission path between the source and the receiver.

<u>Level</u> (L) – ten times the common logarithm of the ratio of a quantity proportional to power or energy to a reference quantity of the same kind. (See sound power level, sound pressure level.) The quantity so obtained is expressed in decibels.

<u>Level reduction</u> (LR) – in a specified frequency band, the decrease in sound pressure level, measured at the location of the receiver, when a barrier or other sound-reducing element is placed between the source and the receiver. Note: level reduction is a useful measure in circumstances when transmission loss, insertion loss, or noise reduction are not measurable.

<u>Metric sabin</u> [L^2] – the unit of measure of sound absorption in the meter-kilogram-second system of units.

<u>Noise</u> – unwanted sound.

<u>Noise Exposure Forecast</u> (NEF)* – is used exclusively for aircraft noise and allows for adjustments based on the number of events and time of day.

<u>Noise isolation class</u> (NIC) – a single-number rating derived from measured values of noise reduction, as though they were values of transmission loss, in accordance with ASTM Classification E 413, Determination of Sound Transmission Class. It provides an estimate of the sound isolation between two enclosed spaces that are acoustically connected.

<u>Noise reduction</u> (NR) – in a specified frequency band, the difference between the space-time average sound pressure levels produced in two enclosed spaces or rooms by one or more sound sources in one of them. Note: it is implied that in each room there is a meaningful average level; that is, that in each room the individual observations are randomly distributed about the average value, with no systematic variation with the position within the permissible measurement region. Noise reduction becomes meaningless and should not be used in situations where this condition is not met.

<u>Noise reduction coefficient</u> (NRC) – a single-number rating derived from measured values of sound absorption coefficients in accordance with ASTM Test Method C 423, for Sound Absorption and Sound

Absorption Coefficients by the Reverberation Room Method. It provides an estimate of the sound absorptive property of an acoustical material. NRC values range from near 0 for hard reflective materials such as glass and gypsum board to 1.2 for several inches of highly efficient fiberglass boards.

Normalized noise reduction (NNR) – between two rooms, for a specified frequency band, the value that the noise reduction in a given field test would have if the reverberation time in the receiving room were 0.5 seconds.

Normal mode – of a room, one of the possible ways in which the air in a room, considered as an elastic body, will vibrate naturally when subjected to an acoustical disturbance. With each normal mode is associated a resonance frequency and, in general, a group of wave propagation directions comprising a closed path.

Outdoor-indoor transmission loss (OITL) – of a building facade, in a specified frequency band, ten times the common logarithm of the ratio of the airborne sound power incident on the exterior of the facade to the sound power transmitted by the facade and radiated to the interior. The quantity is expressed in decibels.

Perceived Noise Level (L_{pn})* – a measurement of the "noisiness" of a sound. It is calculated from octave or 1/3 octave band sound level data and is based on standardized properties of the human hearing sensitivity as determined by psycho acoustic methods. (e.g. $L_{pn} = L_a + 13$) The Effective Perceived Noise Level (L_{epn}) allows for adjustments due to the presence of a prominent pure tone and event duration.

Pink noise – noise with a continuous frequency spectrum and with equal power per constant percentage bandwidth. For example, equal power in any one-third octave band.

Receiving room – in architectural acoustical measurements, the room in which the sound transmitted from the source room is measured.

Reverberant sound field – the sound in an enclosed or partially enclosed space that has been reflected repeatedly or continuously from the boundaries.

Reverberation – the persistence of sound in an enclosed or partially enclosed space after the source of sound has stopped; by extension, in some contexts, the sound that so persists.

Reverberation room – a room so designed that the reverberant sound field closely approximates a diffuse sound field, both in the steady state when the sound source is on and during decay after the source of sound has stopped.

Sabin [L^2] – the unit of measure of sound absorption in the inch-pound system. (i.e. 1 sabin = 1 dB/sq. ft.)

Sound absorption – (1) the process of dissipating sound energy. (2) the property possessed by materials, objects and structures such as rooms of absorbing sound energy. (3) A; [L_2]; metric sabin - in a specified frequency band, the measure of the absorptive property of a material, an object, or a structure such as a room. Note: sound energy passing through a wall or opening may sometimes be regarded as being absorbed.

<u>Sound absorption coefficient</u> (α) (dimension less); metric sabin/m^2 - of a surface, in a specified frequency band, the measure of the absorptive property of a material as approximated by the method of ASTM Test Method C 423, for Sound Absorption and Sound Absorption Coefficients by the Reverberation Room Method. Ideally, the fraction of the randomly incident sound power absorbed or otherwise not reflected.

<u>Sound attenuation</u> – the reduction of the intensity of sound as it travels from the source to a receiving location. Sound absorption is often involved as, for instance, in a lined duct. Spherical spreading and scattering or other attenuation mechanisms.

<u>Sound insulation</u> – the capacity of a structure to prevent sound from reaching a receiving location. Sound energy is not necessarily absorbed; impedance mismatch, or reflection back toward the source, is often the principal mechanism. Note: sound insulation is a matter of degree. No barrier is a perfect insulator of sound.

<u>Sound intensity</u> (I): [MT^3]; W/m^2 – the quotient obtained when the average rate of energy flow in a specified direction and sense is divided by the area, perpendicular to that direction, through or toward which it flows. The intensity at a point is the limit of that quotient as the area that includes the point approaches zero.

<u>Sound isolation</u> – the degree of lack of acoustical connection. There are, in general, two ways to achieve a degree of sound isolation: by insulation, preventing the sound from reaching a receiving location, and by attenuation, reducing the intensity of sound as it travels toward a receiving location.

<u>Sound level</u> – of airborne sound, a sound pressure level obtained using a signal to which a standard frequency weighting has been applied. Note 1: three standard frequency-weightings designated A, B and C are defined in ANSI S1.4, Specification for Sound Level Meters. Note 2: the frequency-weighting and method of averaging must be specified unless clear from the context.

<u>Sound power</u> (W); [ML^{2}T^3]; (W) – in a specified frequency band, the rate at which acoustic energy is radiated from a source. In general, the rate of flow of sound energy, whether from a source, through an area, or into an absorber.

<u>Sound power level</u> (L_w) – of airborne sound, ten times the common logarithm of the ratio of the sound power under consideration to the standard reference power of 1 pW. The quantity is expressed in decibels.

<u>Sound pressure</u> (p); [ML^{-1}T^{-2}]; Pa – a fluctuating pressure superimposed on the static pressure by the presence of sound. In analogy with alternating voltage, its magnitude can be expressed in several ways, such as instantaneous sound pressure or peak sound pressure, but the unqualified term means root-mean-square sound pressure. In air, the static pressure is barometric pressure.

<u>Sound pressure level</u> (L_p) – of airborne sound, ten times the common logarithm of the ratio of the square of the sound pressure under consideration to the square of the standard reference pressure of 20 μPa. The quantity so obtained is expressed in decibels. Note: the pressures are squared because pressure squared, rather than pressure, is proportional to power or energy.

<u>Sound transmission class</u> (STC) – a single-number rating derived from measured values of transmission loss in accordance with ASTM Classification E 413, Determination of Sound Transmission Class. It provides an estimate of the performance of a partition in certain common sound insulation problems.

<u>Sound transmission coefficient</u> (τ) [dimension less] – of a partition, in a specified frequency band, the fraction of the airborne sound power incident on the partition that is transmitted by the partition and radiated on the other side. Note -- Unless qualified, the term denotes the value obtained when the specimen is exposed to a diffuse sound field as approximated, for example, in reverberation rooms meeting the requirements of Test Method (ASTM) E90.

<u>Sound transmission loss</u> (TL) – of a partition, in a specified frequency band, ten times the common logarithm of the ratio of the airborne sound power incident on the partition to the sound power transmitted by the partition and radiated on the other side. The quantity so obtained is expressed in decibels. Note: unless qualified, the term denotes the sound transmission loss obtained when the specimen is exposed to a diffuse sound field as approximated, for example, in reverberation rooms meeting the requirements of ASTM Test Method E 90 for Laboratory Measurement of Airborne-sound Transmission Loss of Building Partitions.

<u>Source room</u> – in architectural acoustical measurements, the room that contains the noise source or sources.

<u>Speech privacy noise isolation class</u>* (NIC′) – a single number rating system derived from data on a screen, ceiling, wall covering, background masking system or a complete interior environment (i.e. open plan office) using PBS C.2 test procedures. This is an objective measurement that is the forerunner of several ASTM standards for the open plan office.

<u>Speech privacy potential</u>* (SPP) – a single number rating system derived from the data generated by evaluating the sound attenuation characteristics of an office screen, ceiling, wall treatment, background masking system or a complete acoustical system using test method PBS C.1. This is a "subjective measurement" usually used in the field to evaluate the degree of speech privacy between two adjacent work stations.

<u>Structure borne sound</u> – sound that arrives at the point of interest, such as the edge of a partition, by propagation through a solid structure.

<u>Vibration isolation</u> – a reduction, attained by the use of a resilient coupling, in the capacity of a system to vibrate in response to mechanical excitation.

<u>White noise</u> – noise with a continuous frequency spectrum and with equal power per unit bandwidth. For example, equal power in any band of 100 Hz width.

Appendix 2 - DESIGN GUIDE WORKSHEETS

Calculating acoustical predictors is both an art and a science. It has been the desire of every acoustician and person confronted with a noise problem to come up with a magical formula that will provide a good indicator of the resultant noise level after treatment. Unfortunately, there is no universal formula, though many have been postulated. A main concern of practical acousticians is that there are so many conditions that can change the results dramatically that it is nearly impossible to cover all circumstances. The following examples are given as an aid to selecting a design that may be practical for a particular set of circumstances. It is strongly urged that laboratory testing and field analysis be conducted if a specific criterion is to be satisfied. *While the following examples have been utilized successfully, their accuracy will be affected by circumstances unique to the situation. If compliance with a specification, code or government criteria is required, it is recommended that a qualified acoustician[1] be engaged.*

CALCULATING A-WEIGHTED SOUND LEVELS (dB$_A$)

1. Measure the octave band sound pressure levels at 125, 250, 500, 1000, 2000 and 4000 Hz with a sound level meter set on the flat or linear frequency weighting scale.

2. Apply the following correction numbers to the octave band levels to obtain equivalent levels for A-weighted octave band analysis:

Octave band center frequency, Hz	125	250	500	1000	2000	4000
correction factor	-16	-9	-3	0	+1	+1

3. Successively combine each octave band level with the next, using the following difference table:

If the difference in dB between two levels is:	0	1	2-3	5-7	8-9	10 or more
Add to the higher level:	3	2.5	2	1.5	0.5	0

4. Round the final answer to obtain the total dB$_A$ level (Figure A2-1)

---

[1] Qualified acousticians include members of: the Noise Control Association, 680 Rainier Lane, Port Ludlow, WA 98365 (360) 437-0814 & National Council of Acoustical Consultants, 66 Morris Ave., Suite 1A, Springfield, NJ 07081-1409, (201) 564-5859.

How to calculate A weighted sound levels (dBA)

1. Measure the octave band sound pressure levels at 125, 250, 500, 1000, 2000 and 4000 Hz with a sound level meter set on the flat or linear frequency weighting scale.

2. Apply the following correction numbers to the octave band levels to obtain equivalent levels for A weighted octave band analysis.

Octave band center frequency, Hz	125	250	500	1000	2000	4000
Correction factor	-16	-9	-3	0	+1	+1

3. Successively combine each octave band level with the next, using the following difference table.

If the difference in dB between two levels is—	0	1	2-3	4	5-7	8-9	10 or more
Add to the higher level—	3	2.5	2	1.5	1	0.5	0

4. Round the final answer to obtain the total dBA level.

Example:

Frequency (Hz)	125	250	500	1000	2000	4000
Octave band level (dB)	108	103	99	104	101	85
Correction	-16	-9	-3	0	+1	+1
	92	94	96	104	102	86

+ 2 dB 96
+ 3 dB 99
+ 1 dB 105
+ 2 dB 107
+ 0 dB . . . . The result is 107 dBA.

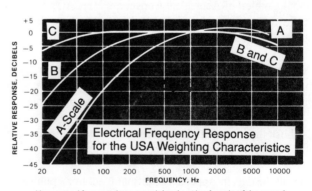

. How sound frequencies are weighted on the A scale of the sound level meter

Figure A2-1 - Example calculation of dB$_A$.

314

REVERBERANT SOUND CONTROL GUIDELINES

Acoustical material manufacturers provide a wide range of products and materials that can be effectively employed to reduce excessive reverberant noise (i.e. reflected sounds that cause echoes). Two approaches may be used.

1.   Where the overall steady state noise level must be reduced, control of the reverberant sound field itself is usually the best solution.

2.   Where reverberant noise produces echoes in such spaces as residences, swimming pools, arenas, gymnasiums, classrooms, factories and auditoriums, or where speech intelligibility from a public address system must be improved, control of the reverberation time is usually the best approach. Reverberation time is the interval required for a sound to decay 60 dB after it has been stopped. Generally this interval should be between 2.0 and 2.5 seconds in order to avoid echoes that interfere with speech intelligibility.

Most reverberant sound problems will involve a combination of these two approaches. For a thorough investigation of the acoustical environment that is to be treated, sound absorption coefficients should be considered at octave band center frequencies of 125, 250, 500, 1000, 2000 and 4000 Hz. Or, if a sound occurring at a particular frequency is known to be the major offender, calculations at that frequency alone may suffice. Also, a rough approximation of the solution to a reverberation noise or reverberation time problem may be accomplished using sound absorption characteristics of the environment at 500 Hz. When complex problems are encountered, it is recommended that the services of an acoustical consultant be engaged who is a member of the National Council of Acoustical Consultants (NCAC). (See Footnote 1 &/or Chapter 4.7)

The following procedure may be used for estimating the amount, type and array of acoustical control products necessary to control the reverberant noise problem:

1. Determine the existing sound absorption coefficients for the walls, floor and ceiling of the room or area to be treated. (See data tables in Appendix 3 for estimates if necessary.) This will establish baseline information for solving the reverberant noise problems.

2. Establish the acoustical design requirements for the room or area, considering such criteria as OSHA noise exposure limits or reverberation time desired.

3. Consult the data in Appendix 3, or manufacturer's published sound absorption coefficients and values for the products under consideration.

A worksheet for determining the reverberation time and amount of material required to be added to a room to achieve the desired reverberation time follows. A second worksheet with a specific example also follows.

Example:

Determine the change in the reverberant sound level and in the reverberation time at 500 Hz in a 200 by 100 by 30 foot room.
Ceiling: wood deck . . . . Sound absorption coefficient 0.14
Floor: concrete . . . . Sound absorption coefficient 0.01
Walls: gypsum
    wallboard . . . . Sound absorption coefficient 0.03

Acoustical treatment will consist of the following:
Suspended Fiberglas
    ceiling . . . . Sound absorption coefficient 0.90
¼" pile carpet
    on floor . . . . Sound absorption coefficient 0.15
Fiberglas nubby glass cloth board (on half the
    wall area) . . . . Sound absorption coefficient 0.73

The sabins of absorption in the room are calculated according to the following procedure:

	WALL	CEILING	FLOOR
UNTREATED ROOM			
1. List areas of room surfaces.	18,000	20,000	20,000
2. List sound absorption coefficient for each room surface.	.03	.14	.01
3. Multiply Line 2 by Line 1 to compute sabins.	540	2800	200
4. Add results on Line 3 for total sabins, all room surfaces.		3540	
5. List sabins for people in room.		—	
6. List sabins for space absorbers.		—	
7. Add Lines 4,5,6 to find total sabins for room.		3540	
ACOUSTICALLY TREATED ROOM —			
1. List areas of room surfaces.*	16,800*	20,000	20,000
2. List sound absorption coefficient for each room surface.	.03,.73	.90	.15
3. Multiply Line 2 by Line 1 to compute sabins, each half of walls	252+6132	18,000	3,000
4. Add results on Line 3 for total sabins, all room surfaces.		27,384	
5. List sabins for people in room.		—	
6. List sabins for space absorbers.		—	
7. Add Lines 4,5,6 to find total sabins for room.		27,384	

A To determine reduction in reverberant noise levels produced by adding sound absorbing material to a room use the following procedure:

1. Determine total sabins for untreated room. — 3,540
2. Determine total sabins for room with added acoustical treatment. — 27,384
3. Divide Line 2 by Line 1. — 7.74
4. Take the logarithm of Line 3. — .89
5. Multiply Line 4 by 10 to obtain reduction in reverberant noise level. (Approximately 9 dB) — 8.9 dB

To determine the change in reverberation time in the room described in this example after it has been acoustically treated as described, use the following procedure:

B UNTREATED ROOM

1. Calculate the volume of the room in cubic feet. — 600,000
2. Multiply Line 1 by 0.05. — 30,000
3. Determine total sabins for room. — 3,540
4. Divide Line 2 by Line 3 for reverberation time in seconds. — 8.47

B ACOUSTICALLY TREATED ROOM

1. Calculate the volume of the room in cubic feet. — 560,000*
2. Multiply Line 1 by 0.05. — 28,000
3. Determine total sabins for room. — 27,384
4. Divide Line 2 by Line 3 for reverberation time in seconds. — 1.02

The reduction of the reverberation time from 8.47 to 1.02 seconds will solve the reverberant sound problem very well.

Worksheet for solving reverberant sound problems

The sabins of absorption in a room are calculated according to the following procedure:

	WALL	CEILING	FLOOR
1. List areas of room surfaces.	___	___	___
2. List sound absorption coefficient for each surface.	___	___	___
3. Multiply Line 2 by Line 1 to compute sabins.	___	___	___

4. Add results on Line 3 for total sabins, all room surfaces ___
5. List sabins for people in room. ___
6. List sabins for space absorbers. ___
7. Add Lines 4,5,6 to find total sabins for room. ___

A. To determine reduction in reverberant noise levels produced by adding sound absorbing material to a room, use the following procedure:

1. Determine total sabins for untreated room. ___
2. Determine total sabins for room with added acoustical treatment. ___
3. Divide Line 2 by Line 1. ___
4. Take the logarithm of Line 3. ___
5. Multiply Line 4 by 10 to obtain reduction in reverberant noise level. ___

The noise reduction in Line 5 can be improved successively by adding more sound absorbing material to the room and repeating Steps 2 through 5. The practical upper limit for reduction of reverberant noise levels is 10 to 12 dB. If estimates are in excess of this amount they should be carefully checked.

B. To determine the reverberation time in a room, use the following procedure:

1. Calculate the volume of the room in cubic feet. ___
2. Multiply Line 1 by 0.05. ___
3. Determine total sabins for room. ___
4. Divide Line 2 by Line 3 to obtain reverberation time, in seconds. ___

C. To determine the amount of sound absorbing material to be added to a room in order to achieve a desired reverberation time, use the following procedure:

1. Calculate the volume of the room in cubic feet. ___
2. Multiply Line 1 by 0.05. ___
3. List desired reverberation time in seconds. ___
4. Divide Line 2 by Line 3 for total sabins required in room. ___
5. Determine sabins for untreated room. ___
6. Subtract Line 5 from Line 4 for sabins of absorption to be added. ___

Additional sabins given on Line 6 will provide desired reverberation time. Select acoustical materials to provide this added absorption from the acoustical data

EXAMPLES OF SPR NOISE CONTROL

Example 1 - Controlling noise at its SOURCE

A large electric motor produces the noise levels shown in line 1 of the table at a nearby receiver station (Figure A2-3).

Step 1(a):
A removable enclosure for the motor is to be built. To determine the approximate degree of noise reduction that can be expected, refer to the data tables in Appendix 3 (or manufacturer's data). Since 1/2" plywood is an economical and practical material with which to build an enclosure, we fill in the insertion loss values for a 1/2" plywood enclosure. These are shown on Line 2 in the table. (Data is from table SIL-2, Appendix 3.)

Subtracting these insertion loss values from the noise levels measured before acoustical treatment of the noise source, we find that the noise levels at the worker's station can be reduced by the enclosure to the levels shown in Line 3 of the table.

The 94 dB$_A$ sound level is determined by applying correction factors for the A weighted levels, and combining octave band levels, as was described in the previous section of this appendix. This level is still above OSHA allowances for exposure during an 8-hour day; in fact, a worker may only be subjected to this level for 4 hours. (See Chapter 7.10, Figure 1, page 320) The sound enclosure must be made more acoustically efficient.

Step 1(b):
Improving the sound attenuation may be accomplished by adding one inch thickness of medium density fiberglass insulation board to the **interior** surface of the 1/2" plywood enclosure. Similar results can be achieved with open cell foam. The increased insertion loss values (from Appendix 3, Table SIL-2, line 2) are shown in Line 5 of the table. Subtracting these insertion loss values from Line 4, we find that the noise levels at the receiver can be reduced to the levels shown in Line 6.

While an 81 dB$_A$ sound level is well within OSHA allowances for 8-hour-day exposure additional measures may be required for noise comfort.[2] This example clearly shows the effectiveness of adding insulation board for source noise control. (Note: These results apply to an enclosure with no holes, seams or other sound leaks. If leaks exist, these insertion loss values will not be achieved.) If required, the inside surface of the enclosure could be covered with a plastic film to protect the insulation from oil or water vapor. However, such a film should not be more than one mill thick or it will have adverse acoustical effects.

Other solutions that address themselves to treating the source of noise might be to:
- Relocate the motor further from the receiver.
- Replace the noisy motor with a quieter one.
- Check, and replace, worn gears or other moving parts which might be the underlying cause of the excessive noise.
- Consider active noise remedies.

---

[2] Chapter 7 - Case Study number one addresses a similar situation for a pool filter/heater unit.

Example 1 Step 1(a)	OCTAVE BAND CENTER FREQUENCIES,Hz						
	125	250	500	1000	2000	4000	dBA*
1. Noise level before treatment	108	103	99	104	101	85	107
2. Insertion loss, ½″ plywood	-13	-11	-12	-12	-13	-15	
3. Noise level after treatment	95	92	87	92	88	70	94
Step 1(b)							
4. Noise level before treatment	108	103	99	104	101	85	107
5. Insertion loss, plywood + insulation	-18	-17	-23	-30	-38	-40	
6. Noise level after treatment	90	86	76	74	63	45	81

*See   dBA calculation method.

Figure A2-2 - Example 1 - SOURCE CONTROL

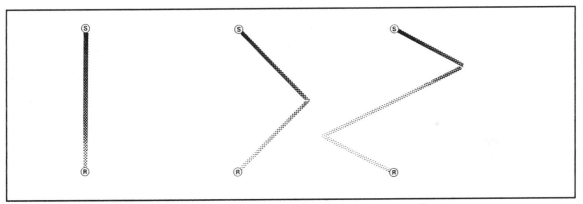

Figure A2-3 - Direct, single and multiple reflection sound paths from source to listener.

Controlling noise along its PATH

Sound, in traveling from a source to a listener, can take two paths. *One*: it may take a direct path, not striking any surface before arriving at the listener's location. *Two:* it may take an indirect path, being reflected from one or more surfaces. In most instances, both direct and indirect sound reaches the listeners position (Figue 2A-2).

The most effective means of reducing indirect sound is to place sound absorptive materials on the surfaces the sound strikes. Thus, when the sound strikes these surfaces, most is absorbed and very little is reflected off the surface. Open cell foam and fiberglass are among the most efficient sound absorptive materials available. They can absorb up to 99% of the sound that strikes their surfaces. Acoustical material manufacturers supply a wide range of products, from building insulation to acoustical ceiling panels, that can be used to absorb

319

reflected sound. Where possible, the installation of an acoustical ceiling in a room is one of the most effective means of reducing sound reflections. Various types and sizes of ceiling panels are available.

If the installation of a ceiling is not feasible because of the presence of pipes, lights, electrical wires, ducts or other systems, then unit sound absorbers (i.e. baffles) may be used in the ceiling area, or acoustical treatments may be applied to side walls or to the underside of the roof deck. As in the case of insulation for enclosures, a wide range of materials can be used for acoustical treatments depending on temperature, humidity, durability, density and surface finish requirements. If desired, these insulations can be covered by porous facings such as pegboard, expanded metal or cloth fabrics with little loss of sound absorption values. (See tables of sound absorption values in Appendix 3 or manufacturer's literature)

The worksheet provided earlier in this appendix can be used to estimate the reduction in reverberant noise level (or in reverberation time) that may be expected when sound absorbing materials are added to the space. Calculations should be done at each octave band to estimate the overall effect of treatment. How to use this work sheet is made clear in the accompanying example.

Direct sound cannot be reduced by the addition of sound-absorptive materials to surfaces, since by definition direct sound does not strike any surface before reaching the listener. **The only effective means of reducing direct sound along its path is to install an acoustically effective barrier** (i.e. a structure that is less than the full height and width of the noise path area) between the noise source and the receiver.

A sound barrier, to be most effective, must have two acoustical properties. *One:* the sound transmission loss or noise reduction capacity of the barrier must be high enough so that sound is attenuated in passing through the barrier. *Two:* it must be sound absorptive so that sound striking the barrier is absorbed and not reflected back into the area of the source. Since by definition a barrier is free-standing (i.e. it does not extend from the floor to the ceiling or roof), sound will be diffracted around the barrier in a similar manner to that in which light is diffracted around the corner of a building.

Depending on the size of the barrier, the location of the noise source and receiver relative to the barrier, and the frequency content of the noise source, the noise reduction across a barrier due to sound diffraction may approach 24 dB – the practical limit that can be expected of such measures. Therefore, it is imperative that the sound transmission loss of the barrier be at least 24 dB so sound doesn't pass through the barrier instead of being diffracted around it.

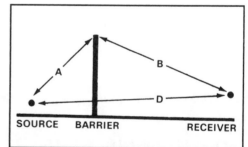

For most barriers, a septum with a weight of more than 1.5 lbs/sq. ft. (such as 1/2" plywood, 1/2" gypsum board, or 20 gauge sheet metal), plus at least two inches of fiberglass insulation on the source side of the barrier, should be sufficient. In some cases, such as a highway noise barrier, it may be necessary to construct a heavier barrier to reduce low frequency noise.

Figure A2-4 Distances involved in determining barrier sound attenuation using nomogram in Figure A2-5.

The nomogram in Figure A2-5, can be used to calculate the amount of sound attenuation in dB provided by a barrier blocking direct transmission of sound along a path from source to receiver. The barrier width should be twice its height to be effective.

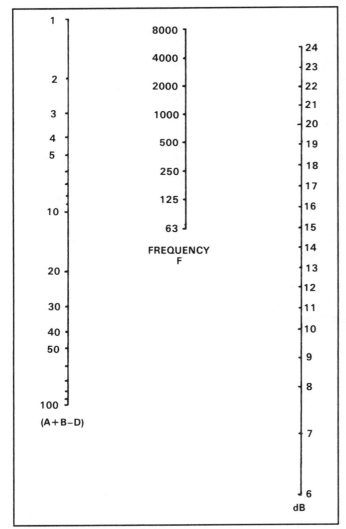

Figure A2-5 - Nomogram for determining barrier sound attenuation in dB.

In the nomogram, values for line (A + B - D) are determined by referring to Figure A2-

■ Line A represents the distance from the noise source to the top of the barrier, in feet.

■ Line B represents the distance from the receiver to the top of the barrier, in feet.

■ Line D represents the straight line distance from the source to the receiver position, in ft.

■ Line F of the nomogram (Figure A2-5) represents the octave band center frequency of the offending noise.

■ The line at the right in Figure A2-5 labeled dB, provides the attenuation in dB that is provided by the barrier.

Use of the nomogram is made clear in the following example.

Example 2 - Controlling noise along its PATH

The same motor, from example 1, is producing the same noise levels at the receiver station (see line 1 in Figure A2-6), but, for reasons of service accessibility, an enclosure is considered impractical. It is decided to treat the path of the noise by building a barrier between the motor and the worker's station adjacent to the motor location.

Noise paths have been studied and it has been determined there is a good likelihood that reflected sound is not presenting a problem. (i.e. walls and ceilings have absorption coefficients of 0.85 or better). Distance D from the motor to the worker's ear is 10.4 feet; distance A from the motor to the top of the barrier is 6.4 feet; distance B from the worker's ear to the top of the barrier is 5.1 feet. The value (A + B - D) is determined to be 1.1 (see position 1 in Figure A2-7), and this value is located on the left hand line of the nomogram in Figure A2-5. Lines are drawn from this point through the octave band center frequency values on the middle line of the nomogram and extended to cross the right hand line, where barrier attenuation values in dB may be read for each frequency and entered in the calculations

321

(Line 2 in Example 2, Figure A2-7). Subtracting these values from the octave band noise levels given in Example 1 and correcting for A scale weighting (per Figure A2-1), the overall result is 91 dB$_A$. This level exceeds OSHA allowances for 8-hour exposure, so the barrier nomogram is now used to determine the effect of placing the barrier closer to the noise source. When located one foot from the source, distance A becomes 4.1 feet and distance B is 9.1 feet. (distance D remains 10.4 feet.) The value (A + B - D) is now 2.8 (see position 2 in Figure A2-6). Using the nomogram, barrier attenuation values are read on Line 5 in Example 2 (Figure A2-7). When motor noise levels are attenuated by these amounts to the levels on Line 6 in Example 2 and corrected for A scale weightings, the level becomes 88 dB$_A$, a level which is within OSHA exposure limits for an 8-hour day.

The nomogram can be used in this manner to optimize barrier height given a fixed location, or to optimize its location given a fixed height.

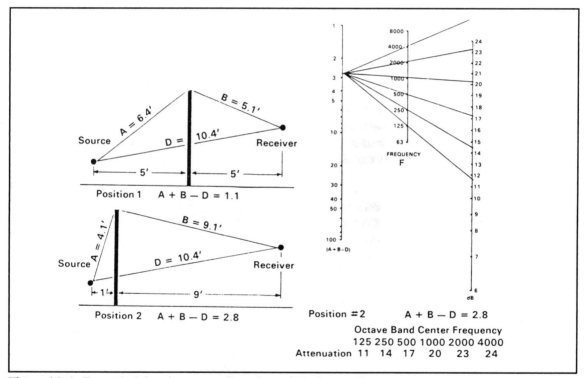

Figure A2-6 - Example 2, barrier attenuation schematic and nomogram.

322

Example 2 Step 1(a)	OCTAVE BAND CENTER FREQUENCIES, Hz						
	125	250	500	1000	2000	4000	dBA*
1. Noise level before treatment	108	103	99	104	101	85	107
2. Barrier attenuation, location 1		-9	-11	-13	-16	-19	-22	
3. Noise level after treatment	99	92	86	88	82	63	91
Step 1(b)							
4. Noise level before treatment	108	103	99	104	101	85	107
5. Barrier attenuation, location 2	-11	-14	-17	-20	-23	-24	
6. Noise level after treatment	97	89	82	84	78	61	88

*See   dBA calculation method.

Figure A2-7 - Example 2, barrier attenuation calculation.

The above result applies to a condition where no reflected sound reaches the receiver. If there is any reflected sound, these transmission loss values will not be achieved unless adjacent reflecting surfaces -- walls, ceiling, or other surfaces -- are highly sound absorbent.

Controlling Noise at the RECEIVER

The most commonly used measures for receiver noise control are ear plugs or ear protectors. These, however, are classified by OSHA as **"temporary measures."** OSHA requires **"permanent solutions"** known as Engineering or Management Controls. And, the only permanent methods of reducing noise at the receiver position is to build a partial or complete enclosure around the receiver/listener or utilize active noise control.[3] An enclosure for a listener is very similar to an enclosure for a noise source.

The basic difference between the two is that a receiver enclosure must provide an environment in which the occupant can function efficiently and comfortably. This usually means that lights, windows, a door, and a ventilation system must be provided. These may degrade the overall acoustical performance of the enclosure due to sound leaks and lower noise reduction values of doors and windows.[4] Therefore, greater effort is required to design and build an occupant enclosure than fabricating a noise source enclosure.

The use of resilient furring channels on wood studs or 25 gage drywall steel studs, with gypsum wallboard with a core of soft fibrous insulation, plus liberal use of caulking to seal sound leaks, is an excellent start in the design of an occupant enclosure. In many instances, controls, dials and gauges may be installed in the enclosure to further reduce the time necessary for the user to spend outside it in the noisy environment. Doors and windows should, if possible, be located on the side away from the noise source, and provision for ventilation should be located and constructed so the system will not conduct noise into the enclosure.

Example 3 - Controlling noise at the RECEIVER

Example 3 Step 1(a)	OCTAVE BAND CENTER FREQUENCIES, Hz						
	125	250	500	1000	2000	4000	dBA*
1. Noise level without enclosure	108	103	99	104	101	85	107
2. Room adjustment factors	-0	-1	-3	-4	-4	-4	
3. Transmission loss, plain wall	-16	-23	-37	-45	-46	-37	
4. Noise level within enclosure	92	79	59	55	51	44	77
Step 1(b)							
5. Noise level without enclosure	108	103	99	104	101	85	107
6. Room adjustment factors	-0	-1	-3	-4	-4	-4	
7. Transmission loss, insulated wall	-24	-35	-47	-54	-57	-42	
8. Noise level within enclosure	84	67	49	46	40	39	68
*See dBA calculation method.							

Figure A2-8 - Example 3, Controlling Noise at RECEIVER

---

[3] Electronic noise canceling is an emerging technology that is not yet practical for typical residential applications. For more on active noise control see the Noise Control Manual by D. A. Harris.

[4] See end of Appendix 2 for a technique to estimate the STC of a composite structure.

The same motor is producing the same noise levels at the receiver station (line 1 in Figure A2-8). A motor enclosure is impractical; the path of sound reaching the receiver station is such that a barrier or partition will not block sufficient sound. Therefore, the receiver station needs to be enclosed.

The effective noise reduction that will be achieved at the receiver station due to the introduction of a total enclosure is the sum of two factors: the sound transmission loss of the enclosure boundaries, plus a room adjustment factor which is equal to ten times the logarithm of the ratio of the room sound absorption to the total surface area of the noise-exposed room boundaries.

Step 1(a):

For the purposes of this example we will assume that the room adjustment factor is given in Lines 2 and 6 of the table. Referring to Appendix 4 of this manual, we find the Sound Transmission Loss values for a wall constructed of 3-5/8" metal studs with 1/2" gypsum board on both sides. These are listed on Line 3 of the table. While the resulting 77 dB_A sound level inside the enclosure (line 4) is within OSHA noise exposure limits for an 8-hour day[5], it is still considered uncomfortably noisy. Therefore, it is decided to add acoustical treatment to the enclosure.

Step 1(b):

Adding 3-1/2" of fiberglass building insulation to the stud cavity of the enclosure will provide additional sound transmission loss. The total transmission loss (metal stud and gypsum board wall plus building insulation), also taken from Appendix 4 of this manual, is listed on Line 7 in the table. Subtracting the total effective noise reduction values from the noise levels before treatment, we find the noise level within the worker's enclosure can be reduced to 68 dB_A (see values on Line 8 of Figure A2-8).

The calculated dB_A reading is well within OSHA exposure limits. It can also be expected to result in a noise level within the enclosure approximating that of a normal, moderately noisy shop. The results apply to an enclosure without sound leaks, without a plate glass window facing the noise source, with an acoustically rated door also facing away from the noise source, and with all ventilation and other openings properly treated to avoid sound leaks. If leaks exist, these sound transmission loss values will not be achieved.

---

[5] See Chapter 4, Figure 4-3 for OSHA Requirements.

CONTROLLING ENVIRONMENTAL NOISE

Keeping outside noises out:

Measuring sound transmission loss and sound absorption coefficients of building materials provides a means of predicting the noise level within a space. The noise in a given space may come from a source in that space or from an adjacent space. This section will deal with noise sources outside a building.

Determining the influence of outside sources:

Use the following equation to predict the noise level in a room exposed to an outside noise source:

$$L_p(int) = L_p(ext) - TL + 10 \log S/A + ADJ$$

$L_p(int) =$ Predicted average sound pressure level in the building interior at a given frequency in dB or dB_A.

$L_p(ext) =$ measured or predicted average sound pressure level at the building exterior for a given frequency band, in dB or dB_A.

$TL =$ sound transmission loss of the exterior wall/roof at a given freq. band, in dB or STC

$S =$ total exposed exterior surface area of room, in sq. ft.

$A =$ total sabins of absorption in the room at a given frequency band or NRC

$ADJ =$ adjustment factor which takes into consideration certain characteristics of the sound source. Example: for aircraft traverses or for continuous traffic, the sound field incident on the building facade is a reasonable approximation to the reverberant field condition in which the TL was measured - in this case ADJ = 3 dB.

The term ADJ is equal to 3 dB only when the sound field incident on the facade approximates a reverberant field condition. When this is not the case, the term ADJ takes a more general form. Thus, ADJ = 3 dB + G where G, stated in dB is an adjustment for the geometrical arrangement of the noise source relative to the building facade. The sound transmission loss of a building component is dependent on the angle of incidence of the sound wave striking the component. Since TL is determined with random incidence sound, adjustments must be made for situations where the sound is incident from fixed angles such as from a stationary source. Figure A2-9 shows values for G to be used for different angles of incidence relative to the building facade.

Angle of incidence, degrees	Adjustment G, dB
0 – 30	-3
30 – 60	-1
Random	0
60 – 80	+2
Greater than 80	+5

Figure A2-9 - Adjustment G to allow for primary angles of incidence

Example:

A small house has an outside measured sound level of 80 dB_A. due mostly to airplane flyover noise. The angle of incidence is estimated at 25 degrees, room of interest is 10' x 20' with an 8' ceiling height and is bare except for a carpeted floor having an NRC of 40. All other surfaces have an NRC of .05. The exterior wall is wood studs with clapboard siding, sheathing, insulation and gypsum board with an estimated STC of 40. There are no windows, doors or other flanking paths. What is the expected interior sound level - L_p(int)?

L_p(int) = 80 dB$_A$ - 40 TL + 10 log 160/114 - 3 (G @ 25∘)
   (room size (S) is 8'x20' = 160,
   (sound absorption (A) = 160 is sum of all room surfaces in sq. feet times NRC of material.
    e.g. floor is 10x20 = 200 x .40 for carpet. All other surfaces are wallboard at .05)
   or
L_p(int) = 80 - 40 + 2 - 3 = 39

Answer: Sound level inside room is estimated to be 39 dB$_A$.

DETERMINING THE SOUND TRANSMISSION LOSS OF A COMPOSITE PARTITION

A partition may have different components, such as doors or windows. The designer needs to know the transmission loss of this composite partition before the overall noise reduction can be calculated.

Design considerations

The equation for predicting the noise level in a building interior can be used in two ways. *One:* if a preliminary exterior wall design exists, the interior noise level can be predicted for that design and compared to code compliance criteria. *Two:* if a preliminary design does not exist, the minimum required sound transmission loss required to meet code requirements can be determined without first determining construction details.

To screen proposed facade constructions, or any other partition system, a simplified equation for predicting interior noise levels uses the STC value of a construction in place of transmission loss (TL) values. In addition, other noise level descriptors such as A-weighted dB levels (dB_A), day-night average level (L_{dn}), or community noise equivalent level (CNEL) may be used for the exterior noise levels $L_p(ext)$ in the equation. This equation is thus written as follows:

$$L(int) = L(ext) - STC + 10 \log S/A + ADJ$$

L(int) = approximate interior noise level in the same unit as used for L(ext)

L(ext) = approximate exterior noise level in dB, dB_A, L_{dn} or CNEL

STC = Sound Transmission Class of the exterior facade construction

ADJ = 3 + G + F, where G = values in Figure A2-9 & F = adjustment for frequency spectrum characteristics of the noise source per Figure A2-10.

10 log S/A = approximated using Figure A2-11 to determine the value A and the chart shown in Figure A2-12 to determine log S/A.

Noise source	Adjustment F, dB
Jet aircraft within 500 feet	0
Train wheel/rail noise	2
Road traffic, few trucks	4
Jet aircraft at 3000 feet	5
Road traffic, over 10% trucks	6
Diesel-electric locomotive	6

The above method is only an approximation and should be used only as a screening tool. Engineering calculations for code compliance may necessitate the services of an acoustical consultant.

The first method is most useful when the designer has a particular construction in mind. It will determine easily whether a particular design will work acoustically by first determining the composite STC of the exterior wall based on the individual STC values and surface area of each element of the composite wall.

Figure A2-10 - Adjustment F to allow for spectrum shape of common outdoor noise sources

328

	Types of furnishings			
	HARD: sound re-flective walls, floor and ceil-ing, no drapes	STANDARD: reflective walls, acousti-cal ceiling, hard floor	SOFT: acoustical ceiling, carpet or drapes	VERY SOFT: acoustical ceil-ing, carpet or drapes and wall furniture
"A" factor	0.3	0.8	0.9	1.0

Figure A2-11 - "A" factors for estimating room absorption from floor area

The composite STC can be estimated with the use of Figure A2-12 as follows (assuming two elements of the composite):

Calculating Composite STC

First: calculate the difference between the STC of the two elements (see Appendix 3, Tables STL 1-7 for exterior wall, door and window values).

Second: calculate the area percentage of the lower STC element. (e.g. door or window)

Third: Determine the adjustment to be subtracted from the higher STC value to give the composite STC of the two elements.

Fourth: Repeat the procedure for additional elements in the composite construction.

Substitution of the values into the facade equation given above will give a prediction of the interior noise level. If lower than design criteria values, the design is acceptable.

Comparison of individual noise contributions indicates which element is the weak link. This component can either be changed in size or in type to reduce its contribution. Optimized design calls for each wall component to be designed so as to contribute equally to solving the problem. However, this usually impractical for reasons of material or architectural design.

Example: Assume a stucco wall with insulation resilient channel construction. Appendix 3, Table STL-6 lists an STC of 57 for such a design. The wall has a solid wood door in it. Appendix 3, table STL-4 lists an STC of 27 for the door. The door area is 13% of the total wall area.

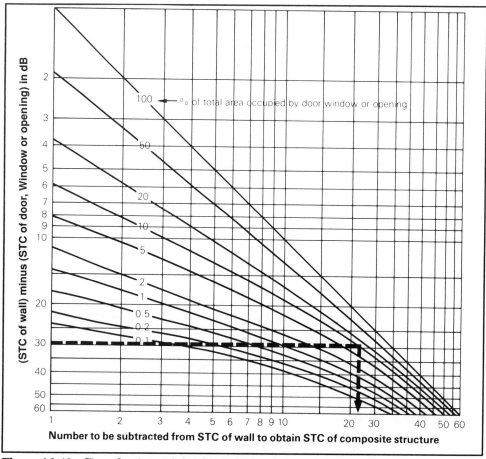

Figure A2-12 - Chart for determining STC of a composite structure

To find the composite STC: 57 - 27 = 30.

Find this STC difference (30) on the vertical axis of Figure A2-12 and move horizontally to 13%, the door area.

Drop to the horizontal axis to read the number to be subtracted from the wall STC.

Thus 57 - 23 = 34. **The composite STC of the partition is 34.**

NOISE REDUCTION WITHIN A ROOM

The amount of noise reduction obtained in a confined area, such as a living room, recreation room, conference room, auditorium, classroom, lunch room, natatorium, meeting rooms, or sanctuary, by adding sound absorbing materials, depends on several factors. The size and geometry of the room, the sound absorbing properties of existing materials, the location of the noise sources, the amount of sound absorbing material added in the space, and the placement of such material are factors to consider. It is therefore impossible to determine precisely the degree of noise reduction that can be expected from acoustical treatments in such a space without taking into account the above factors. An experienced acoustician or member of the National Council of Acoustical Consultants (NCAC), trained in solving noise control problems, should be able to consider these factors and arrive at fairly accurate calculations of the effectiveness of noise control measures.

To calculate the noise reduction when sound absorbing material is added to a room one may use one of two techniques. The first is fully described early in this Appendix in the "Worksheet for solving reverberant sound problems" section A - reduction in reverberant noise levels. A second method, that provides a rough idea of the maximum amount of noise reduction that can be expected, is as follows:

- If the acoustical material in a room has a sound absorption coefficient of 0.85 to 1.00, the maximum amount of noise reduction that can be expected is 10 to 12 dB.

- If the acoustical material has a sound absorption coefficient of 0.65 to 0.85 then the maximum noise reduction will be 7 to 10 dB.

- If the acoustical material has a sound absorption coefficient of less than 0.65 than the maximum noise reduction will be less than 6 dB.

COMMUNITY NOISE AND OTHER NOISE

Noise from freeways, airports, and other community supported sources or events are the focus of considerable attention in our society. These issues are complex both from a technical and social standpoint. The legal implications of these noise problems are beyond the scope of this manual. Where specific noise sources are identifiable and a solution that involves the Source, Path, and Receiver [6] are to be implemented, those techniques are applicable. Local communities have implemented noise regulations of many types. Some are the typical "barking dog" variety, while others have implemented specific limits of sound level from a source at a specific distance. An example is Redondo Beach, CA, where the so called "boom boxes" (loud vehicle music amplifiers) are limited to a specific sound level meter reading at a specified distance from the vehicle. While measures discussed in this manual may be utilized, these regulations are primarily designed to encourage the operator to *control the source using management controls.*

---

[6] Chapter 4 contains regulations, codes and typical examples for a major metropolitan area and a small quiet village. Chapter 6 provides considerable environmental and community noise information. Chapter 7 contains a case study for a new community in a mid size city.

A simple technique that may be utilized to obtain a general idea of the amount of noise reduction is contained within the formula:

$$NR = L_{(source)} - L_{(receiver)}$$

where NR is the Noise Reduction in dB, usually dB A weighted, and
L is the measured loudness at the source and receiver locations.

This measurement is described in detail in ASTM E1014 Standard Guide for Measurement of Outdoor A-Weighted Sound Levels.

Another simple rule of thumb is that a noise produced in the outdoors where there are no sound reflectors will attenuate approximately 6 dB for every doubling of distance.

If there is a barrier between the source and receiver, use the technique for barriers described in Example 2 earlier in this appendix.

Design of highway noise barriers is akin to, but more complex than, design of part-high barriers discussed earlier. Where the noise situation does not clearly fall in one of the sections discussed earlier, namely the Source, Path and Receiver controls, it is recommended that a qualified acoustician be retained or call 360 437-0814 for assistance. A whole technology has evolved for the prediction of highway noise including several computer programs based on traffic volume, type of vehicles by percent, # of traffic lanes and the incline or grade.

Similarly, a whole technology exists for airport noise. Again, this is a specialized field complete with computer simulations and a whole new set of terms specific to the circumstances. Contact the author or the Federal Aviation Authority (FAA) for additional information. Factors affecting the A-weighted noise level reduction of a typical building envelope are shown in Chapter 6.

CONTROLLING NOISE WITHIN PROPERTY BOUNDARIES:

With the passage of federal, state and local noise ordinances, it is becoming more important that objectionable noise be prevented from being transmitted beyond adjacent property boundaries. This is especially true where residential areas are adjacent to a specific noise source.

There are several ways to contain noise within the boundaries of a noisy industrial operation. First, one should attempt to quiet the source using the techniques discussed under source control. Be sure to include purchase of quiet equipment, and active and/or passive noise suppressors (mufflers or noise canceling) in the design effort. If these measures have been pursued extensively and other measures are required, consider locating the noise producing equipment within the central zone of the plant or building. Another way to contain sound is to use exterior building shells with high sound transmission loss values. Be sure to include consideration of sound leaks and flanking paths as discussed in the sections dealing with reducing noise along the path. Where vents and ducts are transmitting the source, consider lining the inside of the ducts with sound absorbing materials or installing duct attenuators. If a fence is to be constructed, utilize the principles outlined in an earlier section for barriers. Where specific sound sources can be defined as the intrusive noise, consider active noise treatment measures.

The most effective means to contain noise within property boundaries is to quiet the noise source. A "Buy

Quiet" action plan will provide the most dB/$ benefits. Quieting a noisy machine or other noise source has spawned an entire industry of noise control products, materials and systems. Known as industrial noise control, other texts written by the author provide extensive measures to reduce noise in an industrial environment[7].

CONTROLLING NOISE IN ADJACENT OFFICE AREAS

Disconcertingly high noise levels are often encountered in office areas adjacent to noisy industrial operations. Even though such noise levels may not approach the exposure limits established by OSHA for an 8-hour day, they may be of sufficient intensity to distract and annoy office occupants, interfering with efficiency and making speech communication difficult. In fact, noise levels of 55 to 60 dB_A are generally considered excessive for the office environment.

Other texts written by the author provide an extensive discussion of the measures that may be taken to provide speech privacy between work stations.[8] Background masking sound, provided it is not so high that it is in and of itself intrusive, has been effective in controlling some exterior noise. When the intrusive noise is greater than 55 dB_A, other measures should be employed. Identification of the specific source and implementing appropriate controls is the most effective. If quieting the source is not practical, then the barrier must be sufficient to block the intrusive noise. Design the partitions to have an effective STC and eliminate all sound leaks and flanking paths.

---

[7] For a comprehensive discussion of industrial noise control see *Noise Control Manual* by D. A. Harris, 1991, published by Van Nostrand Reinhold.

[8] For a comprehensive discussion of office acoustics, refer to Chapter 2 of the book Planning and Designing the Office Environment, Second Edition, Harris, D. A. et. al., Van Nostrand Reinhold, New York, NY, 1991 and Noise Control Manual, D. A. Harris, 1991, Van Nostrand Reinhold. (Now owned by Chapman and Hall.)

Appendix 3 - ACOUSTICAL DATA

Section 1 - SOUND ABSORBING MATERIALS

Sound Absorption Coefficients (SAC) in these tables are listed by type of sound absorbing material/product as follows:

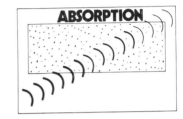

Table SAC-1 - building insulation
Table SAC-2 - insulation batts
Table SAC-3 - fiberglass boards
Table SAC-4 - fiberglass board with facings
Table SAC-5 - metal building insulation
Table SAC-6 - roof form board
Table SAC-7 - mineral ceiling board
Table SAC-8 - fiberglass ceiling board
Table SAC-9 - other ceiling board
Table SAC-10 - wall panels with various facings
Table SAC-11 - fiberglass with perforated covers
Table SAC-12 - floor coverings
Table SAC-13 - miscellaneous building materials and surfaces

Unless otherwise noted, tests were conducted in a NVLAP approved facility in accordance with ASTM C 423, Standard Test Method for Sound Absorption and Sound Absorption Coefficients by the Reverberation Room Method. Sound absorption coefficients for each sample were measured over one-third-octave bands and are reported at the preferred octave band center frequencies. In some cases, the measured sound absorption coefficients are greater than 1.00. As recommended by the test method, these values are reported as measured and not adjusted. The corresponding NRC for a material may also be greater than 1.00 according to the ASTM test method. The sound absorption coefficients of these materials are not significantly affected by coverings such as expanded sheet metal, metal lath, hardware cloth, screening or glass cloth. When other coverings having less open surfaces are used, the description so notes the material. **For additional information, consult acoustical manufacturers technical literature.**

Notes to sound absorption tables:

SAC - Sound Absorption Coefficient, determined per ASTM C423. (^{n}) denotes source listed at end of section.

NRC - Noise Reduction Coefficient; an average of SAC @ 250, 500, 1000 & 2000 Hz.

(1)   Mounting method for specimen per ASTM E795:
        (Note: Identical specimens installed with a different mount will exhibit dramatically different test results.)
        Type A - Test specimen laid directly against test surface.

        Type B - Test specimen cemented to gypsum board and laid directly against the test surface.

        Type C - Test specimen comprising sound-absorptive material behind a perforated, expanded or other open facing.

        Type D - Test specimen on wood or other furring strips. Suffix # indicates distance from test surface in millimeters.

        Type E - Test specimen mounted with an air space behind it. Suffix # indicates distance from test surface in millimeters. (e.g.; E-405 = specimen over 16 inch air space, a typical suspended ceiling system utilized for commercial construction in North America.)

        Type G - Test specimen is a drapery, window shade, or blind hung parallel to the test surface. Suffix # indicates distance from the test surface in millimeters.

        Type H - Test specimen is a drapery suspended away from any vertical surface.

        Type I - Test specimen is a spray- or trowel-applied material on an acoustically hard substrate.

        Type K - Test specimen is an office screen

        Mod. 7 - Material placed against 24 gauge sheet metal over a 16 inch air space. This mounting configuration is typical of a sheet metal enclosure with insulation on one side. Data include facings exposed to sound source, if specified.

        No. 6 - Material placed over 24 gauge galvanized sheet metal with 1 inch air space.

(2)   Facings:

        ■ FRK (Foil Reinforced Kraft): Foil faced laminate with glass fiber reinforced kraft backing.
        ■ ASJ (All Service Jacket): An embossed laminate of white kraft facing with glass fiber reinforcing and a foil backing.

336

Table SAC-1
BUILDING INSULATION[1]

Fiberglass (F/G) is 1/2 lb., Mineral is 1.5 lb. density.
insulation exposed to sound
unless noted otherwise.
Use: sound energy absorber in walls & floors.
Protected with open facings,
they are excellent sound absorbers.

Product Type	Thickness/Mounting		Octave Band Frequencies (Hz.)						
			125	250	500	1000	2000	4000	NRC
F/G, R-11	3.5"	A	.34	.85	1.09	.97	.97	1.12	**.95**
F/G, R-19	6.25"	A	.64	1.14	1.09	.99	1.00	1.21	**1.05**
F/G, R-11	3.5"	E-405	.80	.98	1.01	1.04	.98	1.15	**1.00**
F/G, R-19	6.25"	E-405	.86	1.03	1.13	1.02	1.04	1.13	**1.05**
F/G, R-11 (FRK facing exposed to sound)	3.5"	A	.56	1.11	1.16	.61	.40	.21	**.80**
F/G, R-19 (FRK facing exposed to sound)	6.25"	A	.94	1.33	1.02	.71	.56	.39	**.90**
Mineral	2"+1" airspace	A	.35	.70	.90	.90	.95	.95	**.85**
Mineral (covered with 20% open area perforated sheet metal)	2"+1" airspace	C	.40	.80	.90	.85	.75	.40	**.70**

Table SAC-2
INSULATION BATTS - no facings[1]

F/G is 0.8 lb. density, Mineral is 2 lb. density
Use: energy absorber in walls & floors,
Protected with open facings,
they are excellent absorbers.

Product Type	Thickness/Mounting		Octave Band Frequencies (Hz.)						
			125	250	500	1000	2000	4000	NRC
R-8	2.5"	A	.21	.62	.93	.92	.91	1.03	**.85**
R-11	3.5"	A	.38	.88	1.13	1.03	.97	1.12	**1.00**
R-8	2.5"	E-405	.59	.84	.79	.94	.96	1.12	**.90**
R-11	3.5"	E-405	.73	.98	.98	1.05	1.08	1.15	**1.00**

Table SAC-3
FIBERGLASS BOARDS - plain (no facing)[1]
1, 3, & 5 lb. density,
1" - 4" thick
Use: sound energy absorbers
on surfaces or as liners.

Type/Density	Thickness/Mounting		Octave Band Frequencies (Hz.)						
			125	250	500	1000	2000	4000	NRC
plain, 1#	1"	A	.17	.33	.64	.83	.90	.92	**.70**
plain, 1#	2"	A	.22	.67	.98	1.02	.98	1.00	**.90**
plain, 1#	3"	A	.43	1.17	1.26	1.09	1.03	1.04	**1.15**
plain, 1#	4"	A	.73	1.29	1.22	1.06	1.00	.97	**1.15**
plain, 1#	1"	Mod. 7	.38	.34	.68	.82	.87	.96	**.70**
plain, 1#	2"	Mod. 7	.44	.66	1.07	1.06	.99	1.06	**.95**
plain, 1#	3"	Mod. 7	.53	.96	1.19	1.07	1.05	1.03	**1.05**
plain, 1#	4"	Mod. 7	.61	1.10	1.20	1.11	1.08	1.09	**1.10**
plain, 1#	1"	E-405	.32	.41	.70	.83	.93	1.02	**.70**
plain, 1#	2"	E-405	.44	.68	1.00	1.09	1.06	1.10	**.95**
plain, 1#	3"	E-405	.77	1.08	1.16	1.09	1.05	1.18	**1.10**
plain, 1#	4"	E-405	.87	1.14	1.24	1.17	1.18	1.28	**1.20**
plain, 3#	1"	A	.11	.28	.68	.90	.93	.96	**.70**
plain, 3#	2"	A	.17	.86	1.14	1.07	1.02	.98	**1.00**
plain, 3#	3"	A	.53	1.19	1.21	1.08	1.01	1.04	**1.10**
plain, 3#	4"	A	.84	1.24	1.24	1.08	1.00	.97	**1.15**
plain, 3#	1"	Mod. 7	.33	.28	.62	.88	.96	1.04	**.70**
plain, 3#	2"	Mod. 7	.38	.63	1.10	1.07	1.05	1.05	**.95**
plain, 3#	3"	Mod. 7	.45	.98	1.17	1.06	1.00	1.02	**1.05**
plain, 3#	4"	Mod. 7	.62	1.10	1.15	1.05	.99	1.01	**1.05**

Table SAC-3 (Continued)
FIBERGLASS BOARDS - plain[1]

Product Type	Thickness/Mounting		Octave Band Frequencies (Hz.)						
			125	250	500	1000	2000	4000	NRC
plain, 3#	1"	E-405	.32	.32	.73	.93	1.01	1.10	**.75**
plain, 3#	2"	E-405	.40	.73	1.14	1.13	1.06	1.10	**1.00**
plain, 3#	3"	E-405	.66	.93	1.13	1.10	1.11	1.14	**1.05**
plain, 3#	4"	E-405	.65	1.01	1.20	1.14	1.10	1.16	**1.10**
plain, 5#	1"	A	.02	.27	.63	.85	.93	.95	**.65**
plain, 5#	2"	A	.16	.71	1.02	1.01	.99	.99	**.95**
plain, 5#	3"	A	.54	1.12	1.23	1.07	1.01	1.05	**1.10**
plain, 5#	4"	A	.75	1.19	1.17	1.05	.97	.98	**1.10**
plain, 5#	1"	Mod. 7	.32	.30	.66	.90	.95	1.01	**.70**
plain, 5#	2"	Mod. 7	.39	.59	1.06	1.08	1.05	1.13	**.95**
plain, 5#	3"	Mod. 7	.49	.93	1.15	1.06	.99	1.00	**1.05**
plain, 5#	4"	Mod. 7	.57	1.06	1.13	1.02	.94	1.00	**1.05**
plain, 5#	1"	E-405	.30	.34	.68	.87	.97	1.06	**.70**
plain, 5#	2"	E-405	.39	.63	1.06	1.13	1.09	1.10	**1.00**
plain, 5#	3"	E-405	.66	.92	1.11	1.12	1.10	1.19	**1.05**
plain, 5#	4"	E-405	.59	.91	1.15	1.11	1.11	1.19	**1.10**

Table SAC-4
FIBERGLASS BOARDS with facings[1]
1, 3, & 5 lb. density,
Fiberglass Reinforced Kraft (FRK) or
All Service Jacket with foil (ASJ) facing,
facing exposed to sound, 1" - 4" thick
<u>Use:</u> semi-rigid board for sound absorbing
surfaces that require protective face or vapor barrier.

Type/Density	Thickness/Mounting		Octave Band Frequencies (Hz.)						NRC
			125	250	500	1000	2000	4000	
FRK face, 3#	1"	A	.18	.75	.58	.72	.62	.35	**.65**
FRK face, 3#	2"	A	.63	.56	.95	.74	.60	.35	**.75**
FRK face, 3#	1"	Mod. 7	.31	.45	.62	.65	.51	.28	**.55**
FRK face, 3#	2"	Mod. 7	.38	.51	.83	.73	.53	.37	**.65**
FRK face, 3#	1"	E-405	.33	.49	.62	.78	.66	.45	**.65**
FRK face, 3#	2"	E-405	.45	.47	.97	.93	.65	.75	**.75**
FRK face, 5#	1"	A	.27	.66	.33	.66	.51	.41	**.55**
FRK face, 5#	2"	A	.60	.50	.63	.82	.45	.34	**.60**
FRK face, 5#	1"	Mod. 7	.25	.48	.28	.57	.39	.30	**.45**
FRK face, 5#	2"	Mod. 7	.38	.36	.39	.37	.56	.38	**.40**
FRK face, 5#	1"	E-405	.29	.52	.33	.72	.58	.53	**.55**
FRK face, 5#	2"	E-405	.50	.36	.70	.90	.52	.47	**.60**
ASJ face, 3#	1"	A	.17	.71	.59	.68	.54	.30	**.65**
ASJ face, 3#	2"	A	.47	.62	1.01	.81	.51	.32	**.75**
ASJ face, 3#	1"	E-405	.27	.54	.57	.66	.58	.36	**.60**
ASJ face, 3#	2"	E-405	.53	.44	.93	.77	.55	.35	**.65**
ASJ face, 5#	1"	A	.20	.64	.33	.56	.54	.33	**.50**
ASJ face, 5#	2"	A	.58	.49	.73	.76	.55	.35	**.65**
ASJ face, 5#	1"	E-405	.24	.58	.29	.75	.57	.41	**.55**
ASJ face, 5#	2"	E-405	.42	.35	.69	.80	.55	.42	**.60**

Table SAC-5
METAL BUILDING INSULATION (MBI)
0.5 lb. density fiberglass[1],
3-6 ft. wide insulation blanket
with vinyl facing,
facing exposed to sound
Use: commercial & warehouse buildings
that require exposed insulation & vapor barrier.

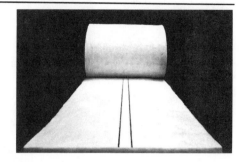

Product Type	Thickness/Mounting		Octave Band Frequencies (Hz.)						
			125	250	500	1000	2000	4000	NRC
Vinyl face	2"	A	.21	.63	1.10	.74	.33	.17	.70
Vinyl face	3"	A	.38	.98	1.20	.62	.42	.24	.80
Vinyl face	4"	A	.56	1.22	1.08	.64	.48	.23	.85
1 mil Tedlar" face	5"	A	.83	1.35	1.06	.85	.76	.54	1.00

Table SAC-6
ROOF FORM BOARD
6 lb. density fiberglass (F/G)[1],
cementious binder wood strand board (Tectum)[2],
gypsum board.
ceiling side exposed to sound
Use: a permanent form for light weight
aggregate concrete and gypsum concrete
pored-in-place roof decks.

Product Type	Thickness/Mounting		Octave Band Frequencies (Hz.)						
			125	250	500	1000	2000	4000	NRC
F/G, economy	1"	A	.04	.24	.70	.98	.99	.95	.75
F/G, deluxe	1"	A	.08	.37	.84	.99	1.01	1.02	.75
Tectum	1.5"	A	.07	.22	.48	.82	.64	.96	.55
Tectum	3"	A	.21	.41	1.00	.75	1.00	.97	.80
Tectum & F/G	4.24"	A	.63	1.13	1.13	.89	1.09	.96	1.05
Tectum & F/G	5.25"	A	1.05	1.17	1.00	.95	.91	.98	1.00
gypsum form board	1/2"	A	.05	.05	.05	.03	.03	.03	.05

Table SAC-7
MINERAL CEILING BOARD
fire resistive mineral (rock wool) base board,
facings generally textured & painted,
ASTM Classification per E1264 is Type III,
2' x 4' boards for lay in suspended ceilings,
or 12" x 12" T&G tile, wgt. .8 to 1.5 lb./sq. ft.
facings: painted (paint), perforated (perf), fissured (fissure).
Use: standard suspended acoustical ceiling systems,
and time design fire rated floor/ceiling assemblies

Product Type	Thickness/Mounting		Octave Band Frequencies (Hz.)						NRC
			125	250	500	1000	2000	4000	
paint/perf/fissure[1]	5/8"	E-405	.25	.34	.41	.56	.65	.63	**.50**
paint/perf/fissure[3]	5/8"	E-405	.42	.34	.44	.72	.67	.64	**.55**
paint/perf/fissure/fire[3]	5/8"	E-405	.34	.30	.53	.78	.73	.69	**.60**
paint/perf/fissure/tile[3]	5/8"	E-405	.42	.29	.52	.68	.64	.69	**.55**
paint/perf/fissure[3]	3/4"	E-405	.34	.36	.71	.85	.68	.64	**.60**
cast/rough texture[3]	3/4"	E-405	.36	.35	.66	.86	.71	.74	**.65**
paint/perf/fissure[4]	3/4"	E-405	.41	.37	.55	.83	.94	.96	**.70**
cast/rough texture[4]	3/4"	E-405	.60	.61	.67	.89	.98	1.02	**.80**

Note: Data selected as representative of mineral ceiling board manufacturers. Variations in patterns, surface treatment, backing and mount will affect SAC values. See individual manufacturers literature for details.

Table SAC-8
FIBERGLASS CEILING BOARD

glass fiber base board with vinyl or glass cloth face,
5/8" vinyl is 0.15 #/sq. ft., 1-1/2" cloth is 0.62 #/sq. ft.
ASTM Classification per E1264 is Type XII,
2'x4'- 5'x5' boards or 12" x 12" T&G tile.
<u>Use:</u> standard suspended acoustical ceiling systems,

Product Type	Thickness/Mounting		Octave Band Frequencies (Hz.)						
			125	250	500	1000	2000	4000	NRC
vinyl not perf.[3]	5/8"	E-405	.59	.45	.63	.80	.78	.63	**.65**
vinyl perf.[3]	5/8"	E-405	.72	.61	.67	.81	.83	.76	**.75**
vinyl not perf.[3]	1"	E-405	.63	.56	.70	.74	.76	.56	**.70**
vinyl not perf.[3]	3"	E-405	.82	.86	.92	.87	.68	.52	**.85**
glass cloth face[3]	3/4"	E-405	.71	.92	.67	.91	.99	.94	**.85**
glass cloth face[3]	1"	E-405	.84	.94	.73	.97	1.02	1.01	**.90**
glass cloth face[3]	1-1/2"	E-405	.80	1.02	.90	1.07	1.09	1.05	**1.00**

Note: Data is representative of faced fiberglass boards. See manufacturers literature for specific performance.

Table SAC-9
OTHER CEILING BOARD

wood fiber base board (wood fiber)[5], Type I & II
cement bonded wood strand board (Tectum)[2],Type XX
open cell melamine foam sheet or wedge design (foam)[6], Type XX
ASTM Classification per E1264 is Type I, II, XX.
2'x2 or 4' boards or 12" x 12" T&G tile.
<u>Use:</u> acoustical ceiling systems for non-fire rated construction,
suspended (E-405) or direct applied (A mount).

Product Type	Thickness/Mounting		Octave Band Frequencies (Hz.)						
			125	250	500	1000	2000	4000	NRC
			na = product no longer available but was popular in 1950's						
wood fiber, painted	1/2"	E-405	.na	.na	na	.na	.na	.na	**.35**
wood fiber, drilled	1/2"	E-405	.na	.na	.na	.na	.na	.na	**.45**
wd. fiber, paint/fissured	1/2"	E-405	.na	.na	.na	.na	.na	.na	**.50**
Tectum™	1"	E-405	.40	.42	.35	.48	.60	.93	**.45**
Tectum	3"	E-405	.70	.50	.52	.76	.92	1.16	**.70**
Tectum/6" F/G batt	1"/6"	E-405	1.01	.89	1.06	.97	.93	1.13	**.95**
foam ceiling board	2"	E-405	.57	.61	.61	.83	.90	.97	**.75**
foam wedge	1"	A	.16	.34	.81	1.00	.99	.97	**.80**
foam wedge/Hypalon™	1"	A	.18	.65	1.04	.73	.68	.56	**.80**
foam sheet	2"	A	.15	.23	.60	.94	.98	.94	**.75**
foam sheet	3"	A	.16	.45	.98	1.00	1.00	1.00	**.85**

Table SAC-10
WALL PANELS

glass fiber base board, 7#, with fabric covering (F/G-cloth)[1],
mineral fiber base board with fabric covering (mineral-cloth)[3]
woven fabric is open weave, nonwoven is needle punched fabric,
cement bonded wood strand board, no covering (Tectum)[2],
open cell melamine foam sheet or wedge design, no covering (foam)[6],
2'-4'x 2-10' boards applied direct to existing wall (A mount).
Use: acoustical wall treatment systems for application to
existing wall surfaces in commercial and residential construction.

Product Type	Thickness/Mounting		Octave Band Frequencies (Hz.)						NRC
			125	250	500	1000	2000	4000	
F/G - woven	1"	A	.04	.29	.76	1.03	1.05	1.08	**.80**
F/G - nonwoven	1"	A	.05	.30	.80	1.00	1.02	.95	**.80**
mineral - woven	3/4"	A	.04	.21	.74	.92	.77	.68	**.65**
mineral - nonwoven	3/4"	A	.04	.21	.82	.93	.85	.84	**.70**
mineral - woven	1"	A	.18	.40	.89	1.06	.99	1.06	**.85**
mineral - vinyl	5/8"	A	.12	.31	.51	.59	.59	.49	**.50**
Tectum™	1"	A	.06	.13	.24	.45	.82	.64	**.40**
Tectum	2"	A	.15	.26	.62	.94	.64	.92	**.60**
foam wedge	1"	A	.16	.34	.81	1.00	.99	.97	**.80**
foam wedge/Hypalon™	1"	A	.18	.65	1.04	.73	.68	.56	**.80**
foam sheet	2"	A	.15	.23	.60	.94	.98	.94	**.75**
foam sheet	3"	A	.16	.45	.98	1.00	1.00	1.00	**.85**

Table SAC-11
FIBERGLASS WITH PERFORATED COVERS

glass fiber base board, 3# density (F/G), with various coverings[1],
coverings are; 1/4" pegboard with 1/4" holes 1" o.c. (pegboard),
24 ga. sheet metal with 3/32" holes, 13% open area (metal),
unfaced values also apply to expanded metal, wire mesh, etc.
<u>Use:</u> acoustical wall treatment systems for application to
existing wall surfaces in commercial and residential construction.

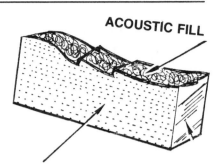

ACOUSTIC FILL

PERFORATED STEEL SKIN

Product Type	Thickness/Mounting		Octave Band Frequencies (Hz.)						
			125	250	500	1000	2000	4000	NRC
F/G - unfaced	1"	A	.06	.20	.68	.91	.96	.95	.70
F/G - pegboard	1"	A	.08	.35	1.17	.58	.24	.10	**.60**
F/G - perf. metal	2"	A	.18	.73	1.14	1.06	.97	.93	**1.00**
F/G - pegboard	2"	A	.80	1.19	1.00	.71	.38	.13	**.80**
F/G - unfaced	2"	A	.22	.82	1.21	1.10	1.02	1.05	**1.05**
F/G - unfaced	4"	A	.84	1.24	1.24	1.08	1.00	.97	**1.15**
F/G - pegboard	4"	A	.80	1.19	1.00	.71	.38	.13	**.80**
F/G - unfaced	6"	A	1.09	1.15	1.13	1.05	1.04	1.04	**1.10**
F/G - pegboard	6"	A	.95	1.04	.98	.69	.36	.18	**.75**

Table SAC-12
FLOOR COVERINGS

carpet with various backings applied over concrete[7],
woven wool loop in pile height noted, (carpet, 1/4")
backing is hair pad of 40 or 86 oz./sq. yd. (pad,40).
vinyl, asphalt and linoleum ratings will be similar (vinyl).
wood floors are regular or parquet (wood, reg.)
concrete, terrazzo and other hard floors are similar (conc.)
wood joist floor may yield higher values at 125 & 250 Hz.
<u>Use:</u> typical floor coverings in commercial and residential construction.

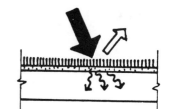

Product Type	Thickness/Mounting (pile height)		Octave Band Frequencies (Hz.)						
			125	250	500	1000	2000	4000	NRC
carpet - no pad	1/8"	A	.05	.05	.10	.20	.30	.40	**.15**
carpet - no pad	1/4"	A	.05	.10	.15	.30	.50	.55	**.25**
carpet - pad, 40	1/4"	A	.15	.17	.12	.32	.52	.30	**.30**
carpet, loop - pad, 86	3/8"	A	.20	.50	.68	.72	.65	.90	**.65**
carpet/foam	3/16"	A	.05	.10	.10	.30	.40	.50	**.25**
carpet/foam	5/16"	A	.05	.15	.30	.40	.50	.60	**.35**
concrete/terrazzo	na	A	.01	.01	.01	.02	.02	.02	**.00**
vinyl, asphalt, rubber	na	A	.02	.03	.05	.03	.03	.02	**.05**
wood - varnished	na	A	.15	.11	.10	.07	.06	.07	**.10**
wood - parquet	na	A	.04	.04	.07	.06	.06	.07	**.05**
marble or glazed tile	na	A	.01	.01	.01	.01	.02	.02	**.00**

Table SAC-13
MISCELLANEOUS BUILDING MATERIALS AND SURFACES[1]

Product Type	Thickness/Mounting		Octave Band Frequencies (Hz.)						NRC
			125	250	500	1000	2000	4000	
brick, unglazed	na	A	.03	.03	.03	.04	.05	.07	**.05**
brick, unglazed painted	na	A	.01	.01	.02	.02	.02	.03	**.00**
concrete block, painted	na	A	.10	.05	.06	.07	.09	.08	**.05**
concrete block, slotted	8"	A	.97	.44	.38	.39	.50	.60	**.45**
concrete blk, slot/filled	8"	A	1.07	.57	.61	.37	.56	.55	**.55**
concrete blk, special	12"	A	.57	.76	1.09	.94	.54	.59	**.85**

(slotted & special concrete block are proprietary[7] with sequential cavities and special core design)

Product Type	Thickness/Mounting		125	250	500	1000	2000	4000	NRC
concrete/terrazzo	na	A	.01	.01	.01	.02	.02	.02	**.00**
vinyl, asphalt, rubber	na	A	.02	.03	.05	.03	.03	.02	**.05**
wood - varnished	na	A	.15	.11	.10	.07	.06	.07	**.10**
marble or glazed tile	na	A	.01	.01	.01	.01	.02	.02	**.00**
gypsum board, 2x4 studs, painted			.10	.08	.05	.03	.03	.03	**.05**
plaster, rough or smooth finish on lath			.02	.03	.04	.05	.04	.03	**.05**
hardwood plywd. panel, ¼", wood frame			.58	.22	.07	.04	.03	.07	**.10**
wood roof decking, T&G cedar			.24	.19	.14	.08	.13	.15	**.15**
glass, 1/4" sealed, large panes			.05	.03	.02	.02	.03	.02	**.05**
glass, 24 oz. operable window, closed			.10	.05	.04	.03	.03	.03	**.05**
water surface, as in swimming pool			.01	.01	.01	.01	.02	.03	**.00**
open sky, (theoretical)			1.00	1.00	1.00	1.00	1.00	1.00	**1.00**

References:

1 - Noise Control Manual, D. A. Harris, VNR, 1991
2 - Tectum Corporation, various technical publications
3 - Armstrong World Industries, various technical publications
4 - United States Gypsum Co., various technical publications
5 - Celotex, various technical publications
6 - Illbruck, Sonex Acoustical Division, various technical publications
7 - Proudfoot technical publications

348

Section II - SOUND TRANSMISSION LOSS OF MATERIALS AND SYSTEMS:

Notes to Section II data:

(1) **STC Ratings:** All system tests such as wall, floor/ceiling, roof/ceiling, doors and window tests were conducted in the laboratory in compliance with ASTM E-90, Standard Method for Laboratory Measurement of Airborne Sound Transmission Loss of Building Partitions. The sound transmission loss for each sample was measured over one-third-octave bands. The Sound Transmission Class (STC) rating was calculated per ASTM E413 Classification for Rating Sound Insulation. Transmission loss data are reported at 1/3rd octave band center frequencies where available.

(2) **NIC Ratings:** Sound insertion loss data in these tables are the difference between sound pressure levels measured at an opening in the wall of a reverberation chamber excited by sound before and after a material is inserted in the opening. The test procedure is generally described in ASTM E90. Data is reported at 1 octave center band frequencies. The Noise Isolation Class (NIC) is calculated per ASTM E413, Classification for Rating Sound Insulation and is used herein to indicate the specimen size is nominally 2' x 2' or otherwise does not fully meet the criteria for determining STC.

Sound insertion loss (SIL) in these tables is listed by type of material as follows:

> Table SIL-1 - typical building materials
> Table SIL-2 - plywood enclosures
> Table SIL-3 - special constructions

Sound transmission loss (STL) in these tables is listed by type of system as follows:
(Note: Impact Noise Ratings are also given for floor/ceiling assemblies)

> Table STL-1 - wood stud partitions, interior
> Table STL-2 - steel stud partitions, interior   **Good Barrier**
> Table STL-3 - floor/ceiling assemblies (STC & IIC)
> Table STL-4 - doors and door assemblies
> Table STL-5 - windows and window assemblies
> Table STL-6 - exterior walls
> Table STL-7 - roof/ceiling assemblies

Data for these tables has been extracted from many sources. Unless noted otherwise, the source are tests conducted by the author. Superscript numbers ([1]) refer to sources at the end of the section.

Table SIL-1
TYPICAL BUILDING MATERIALS[1]

Sheets of building materials tested as a single ply,
surface weight of materials in pounds per square foot (psf),
Materials weighing the same and with similar stiffness
can be expected to provide similar results.
Use: simple sound barrier

5/8" gypsum board.

4" thick flat concrete panel, 54 psf.

Material/Type	Thickness/Weight, psf		Octave Band Frequencies (Hz.)						NIC
			125	250	500	1000	2000	4000	
plywood	1/2"	1.33	17	20	23	23	23	24	21
plywood	5/8"	2.00	19	23	27	25	22	30	24
sheet metal	16 ga.	3.38	18	22	28	31	35	41	31
sheet metal	20 ga.	1.50	16	19	25	27	32	39	27
sheet metal	24 ga.	1.02	13	16	23	24	29	36	25
gypsum board	3/8"	1.3	12	18	23	28	33	23	26
gypsum board	1/2"	1.8	18	22	26	29	27	26	26
gypsum board	5/8"	2.2	19	22	25	28	22	31	26
glass, single strength	3/32"	1.08	15	18	25	26	28	29	26
glass, double strength	1/8"	1.40	16	19	25	29	30	20	24
glass, plate	1/4"	2.78	20	25	26	30	23	30	27
acrylic sheet	1/8"	0.75	14	17	22	24	27	34	24
acrylic sheet	1/4"	1.45	16	19	26	27	30	29	27
acrylic sheet	1/2"	2.75	20	25	26	30	23	30	29
loaded vinyl	-	0.5	11	12	15	20	26	32	21
loaded vinyl	-	1.0	15	19	21	28	33	37	27
loaded vinyl	-	1.25	17	19	28	30	34	39	29
(a limp barrier material with lead or other heavy additives similar to sheet lead)									
FRP	1/8"	1.13	15	18	25	26	29	36	27
FRP	1/4"	2.08	19	22	28	31	32	25	29
FRP	1/2"	4.20	21	27	29	34	27	36	29
(FRP - Fiberglass Reinforced Plastic, typical for building panels)									
concrete panel	4"	54	48	42	45	55	57	67	44
concrete panel	6"	75	38	43	52	55	57	72	55
concrete panel	8"	95	44	48	55	58	53	67	58
gypsum/plywood, lam.	½/⅜"	3.0	16	20	25	29	32	31	28
gypsum/wd. fiber, lam.	½/½"	2.6	17	22	28	32	32	30	30
gypsum/gypsum, lam.	½/½"	3.6	19	25	30	32	29	37	31
gypsum/gypsum, lam.	1/1"	7.2	25	30	32	33	35	43	34
(lam. - laminated with contact cement or gypsum joint compound)									

Table SIL-2 PLYWOOD ENCLOSURES[1]

1/2" thick plywood enclosure with various sound absorbing linings, joints and edges sealed airtight, absorbing material toward source.
Use: enclosure around a noise source

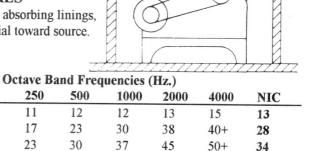

Material/Type	Octave Band Frequencies (Hz.)						
	125	250	500	1000	2000	4000	NIC
plywood unlined	13	11	12	12	13	15	**13**
plywood & 3#, 1" fiberglass board	18	17	23	30	38	40+	**28**
plywood & 3#, 2" fiberglass board	18	23	30	37	45	50+	**34**
plywood & 3#, 4" fiberglass board	19	29	38	47	58	60+	**39**
plywood & 0.5#, 3.6" fiberglass insul.	17	25	29	36	41	45+	**34**

Table SIL-3 SPECIAL CONSTRUCTIONS[1]

vinyl and sheet metal sandwich with various sound absorbing cores, joints and edges sealed airtight.
(Note; adding sound absorbing material on source side may increase values as indicated in Table SIL-2.)
Use: enclosure around a noise source

Material/Type	Octave Band Frequencies (Hz.)						
	125	250	500	1000	2000	4000	NIC
2 ply 1¼# lead vinyl & 2.5" air space	12	34	31	37	43	48	**34**
above + 2½" fiberglass batt in core	25	43	38	43	47	58	**42**
2 ply 16 ga. sheet metal & 2.5" air space	23	33	34	37	38	48	**37**
above + 2½" fiberglass batt in core	26	33	36	38	41	51	**38**
1 ply 16 ga. sheet metal, 1 ply 24 ga.	20	36	37	41	44	52	**40**
(sheet metal, separated by 2½" space filled with R-8 building insulation.)							
2 ply 20 ga. sheet metal/1" 5# F/G core	18	17	30	38	47	55	**32**
above with 2 layers of 1" 5# F/G core	15	18	35	42	51	56	**32**
above except with 3 layers of 1" core	16	23	40	46	52	61	**31**
1 ply 16 ga. sheet metal, 4½" core of	31	45	48	58	64	64	**50**
4 # fiberglass board, with other face of 0.040" aluminum, 4" ribbed.							
above with core of 0.5# fiberglass	25	43	48	56	63	61	**48**
2 - 1" layers of 3# fiberglass board	7	6	5	10	16	20	**11**
above + fiberglass reinforced kraft faces	9	10	7	15	23	31	**14**
26 ga. metal building siding	12	14	15	21	21	25	**20**
above + 2" vinyl faced 1# fiberglass	11	15	16	29	31	37	**24**
above except 3" fiberglass	12	16	18	31	32	39	**25**

TABLE STL-1
WOOD STUD PARTITIONS, interior

load bearing partitions are 2x4 wood studs 16" o.c.,
gypsum wallboard affixed with fasteners 8" o.c.,
all joints are taped & spackled and perimeter sealed
unless otherwise noted. Test data obtained in ⅓ octave
band frequencies. Octave band center data is presented.

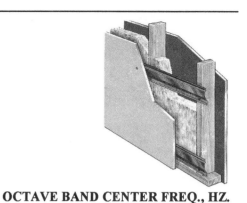

SKETCH	DESCRIPTION	OCTAVE BAND CENTER FREQ., HZ.						
		125	250	500	1000	2000	4000	STC
	2x4 wood studs, 16" o.c., ⅜" gypsum wallboard both sides[3]	13	22	34	43	52	40	**35**
	2x4 wood studs, 16" o.c., ½" gypsum wallboard both sides.	15	27	36	42	47	40	**35**
	2x4 wood studs, 16" o.c., ⅜" gypsum wallboard, 3" fiberglass insulation core	18	25	38	46	56	45	**41**
	2x4 wood studs 16" o.c., ½" gypsum wallboard both sides, 3½" fiberglass (R-11) core.	15	31	40	46	50	42	**39**
	2x4 wood studs, 16" o.c., ⅝" Type X (fire rated) gypsum wallboard.	16	22	35	41	36	48	**34**
	2x4 wood studs, 16" o.c., ⅝" Type X gypsum wallboard, (1 hr. fire rating) 3" fiberglass insulation core.	17	35	37	43	37	47	**36**

2x4 wood studs 16" o.c., ½" gypsum wallboard, 2 ply one side, 1 ply other side.	17	32	40	45	50	45	**38**
2x3 wood studs 24" o.c., ½" gypsum wallboard - 2 ply one side one ply on other.	26	32	36	45	52	42	**41**
2x4 wood studs 24" o.c., ½" gypsum wallboard, 2 ply one side 1 ply on other side, 3" rock wool insulation core.	33	35	40	47	55	44	**44**
2x4 wood studs 16" o.c., ½" gypsum wallboard, 2 ply both sides.	15	35	43	48	53	50	**39**
2x4 wood studs 24" o.c., ½" gypsum wallboard, 2 ply both sides[2]. (⅜"/ ½" is similar)	29	32	37	47	55	47	**43**
2x4 wood studs, 16" o.c., 2 ply ⅝" Type X (2 hr. fire rated) gypsum wallboard each side, nailed 8" o.c.[3]	25	34	34	46	45	51	**40**
2x4 wood studs 16" o.c., 1/2" gypsum wallboard, 2 ply both sides, 3½" (R-11) fiberglass insulation in core. (⅜"/½" is similar)	21	37	45	50	55	51	**45**
wood studs, 16"o.c., ½" wood fiber board base layer, ½" gypsum wallboard faces nailed	18	32	40	46	53	45	**42**

353

2x4 wood studs, 16" o.c., 2 ply ⅝" Type X gypsum wallboard nailed 8" o.c.[3] (2 hr. fire rating).	25	34	34	46	45	51	**40**
2x4 wood studs 16" o.c., ½" wood fiber board base layer, ½" gypsum wallboard faces, strip laminated @ center & edges[3].	23	37	45	55	55	55	**46**
2x4 wood studs, 16" o.c., ¼" gypsum panels base layer ½" gypsum wallboard faces applied vertically & strip laminated at center and edges of face board.	21	35	41	51	55	55	**45**
2x4 wood studs, 16" o.c., ¼" gypsum panels base layer, ½" gypsum wallboard faces, strip laminated, 3" mineral fiber core[2].	30	47	48	57	59	62	**53**
2x4 wood studs 16" o.c., **resilient channels 1 side** 24" o.c., ½" gypsum wallboard both sides.	15	32	40	49	52	45	**39**
2x4 wood studs 16" o.c., **resilient channels 1 side**, 24" o.c., ½" gypsum wallboard both sides, 3½" (R-11) fiberglass insulation core.	22	40	53	57	58	50	**46**

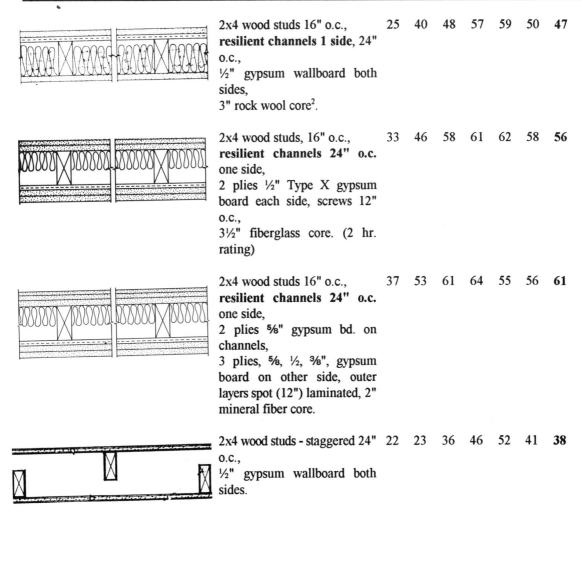

2x4 wood studs 16" o.c., **resilient channels 1 side**, 24" o.c., ½" gypsum wallboard both sides, 3" rock wool core[2].	25	40	48	57	59	50	**47**
2x4 wood studs, 16" o.c., **resilient channels 24" o.c.** one side, 2 plies ½" Type X gypsum board each side, screws 12" o.c., 3½" fiberglass core. (2 hr. rating)	33	46	58	61	62	58	**56**
2x4 wood studs 16" o.c., **resilient channels 24" o.c.** one side, 2 plies ⅝" gypsum bd. on channels, 3 plies, ⅝, ½, ⅜", gypsum board on other side, outer layers spot (12") laminated, 2" mineral fiber core.	37	53	61	64	55	56	**61**
2x4 wood studs - staggered 24" o.c., ½" gypsum wallboard both sides.	22	23	36	46	52	41	**38**
2x4 wood studs - staggered 24" o.c., ½" gypsum wallboard both sides, 3½" (R-11) fiberglass insulation core.	31	37	47	52	56	50	**49**

355

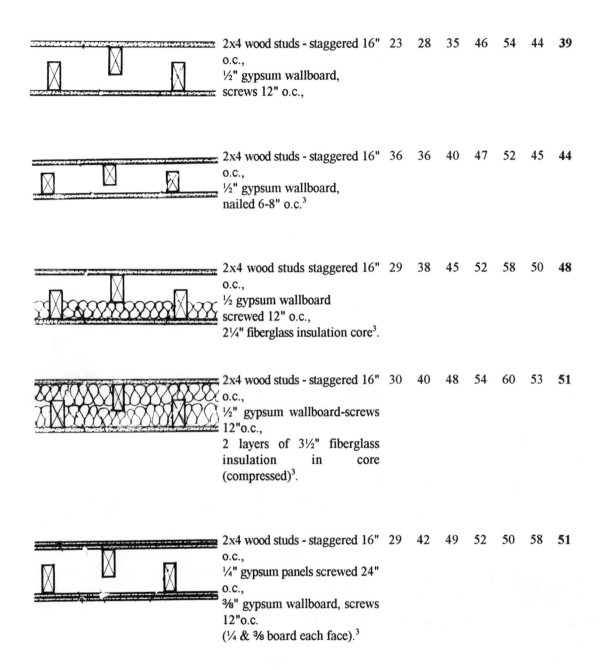

2x4 wood studs - staggered 16" o.c., ½" gypsum wallboard, screws 12" o.c.,

23	28	35	46	54	44	**39**

2x4 wood studs - staggered 16" o.c., ½" gypsum wallboard, nailed 6-8" o.c.[3]

36	36	40	47	52	45	**44**

2x4 wood studs staggered 16" o.c., ½ gypsum wallboard screwed 12" o.c., 2¼" fiberglass insulation core[3].

29	38	45	52	58	50	**48**

2x4 wood studs - staggered 16" o.c., ½" gypsum wallboard-screws 12"o.c., 2 layers of 3½" fiberglass insulation in core (compressed)[3].

30	40	48	54	60	53	**51**

2x4 wood studs - staggered 16" o.c., ¼" gypsum panels screwed 24" o.c., ⅜" gypsum wallboard, screws 12"o.c. (¼ & ⅜ board each face).[3]

29	42	49	52	50	58	**51**

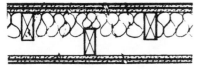

2x4 wood studs - staggered 16" 34 48 57 55 53 62 **56**
o.c.,
¼" gypsum panels screwed 12"
o.c.,
⅜" gypsum wallboard
screwed 12" o.c.,
3½" fiberglass insulation in
core[3].

2x4 wood studs - staggered 16" 31 35 45 49 52 55 **46**
o.c.,
⅜" plywood (shear wall)
nailed 6" on edges, 12" in
field,
½" gypsum wallboard
nailed 12" o.c.[3]

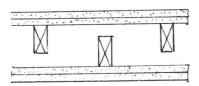

2x4 wood studs - staggered 16" 31 42 45 54 53 59 **51**
o.c.,
2 ply ⅝" Type X gypsum
wallboard each face, nailed 6-
8" o.c.,
(2 hr. fire rating)[3]

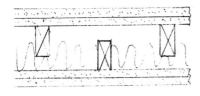

2x4 wood studs - staggered 16" 35 45 52 55 54 57 **53**
o.c.,
2 ply ⅝" gypsum wallboard
nailed 6-8" o.c. each face,
3½" fiberglass insulation core[3].

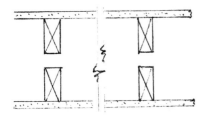

double row (spaced 1") of 2x4 30 41 45 50 55 49 **47**
wood studs 16" o.c.,
½" Type X gypsum wallboard
both sides screwed 12" o.c.[3]

double row (spaced 1") 2x4 wood studs 16" o.c., ½" Type X gypsum wallboard both sides screwed 12" o.c., 3½" fiberglass insulation core[3].	32	48	57	53	54	51	**56**
double row (spaced 1") 2x4 wood studs 16" o.c., ½" Type X gypsum wallboard both sides screwed 12" o.c., 3½" fiberglass insulation in both cavities.[3]	39	51	62	65	57	63	**62**
double 2x4 wood studs **24" o.c.,** 2 layers ½" gypsum board each side, 3½" (R-11) fiberglass in each core.	44	55	63	67	71	71	**64**

358

TABLE STL-2
STEEL STUD PARTITIONS, interior

25 ga. drywall steel studs (non-load bearing) 24" o.c.,
gypsum wallboard applied vertically, affixed with screws
spaced 8" at edges & 12" o.c. in field of board,
all joints are taped & spackled and perimeter sealed
unless otherwise noted. Test data obtained in ⅓ octave
band frequencies. Octave band center data is presented.

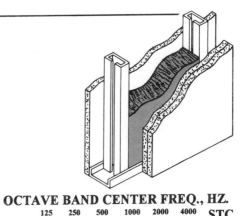

SKETCH	DESCRIPTION	OCTAVE BAND CENTER FREQ., HZ.						STC
		125	250	500	1000	2000	4000	
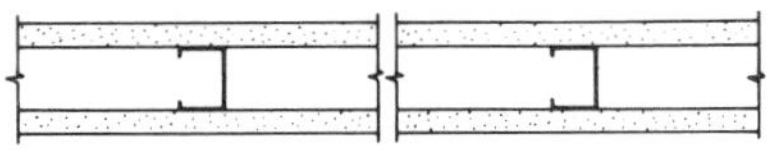	1⅝" steel studs, 24" o.c., ⅝" Type X gypsum wallboard both sides[3] (1 hr. fire rated)	21	41	48	41	48	41	35
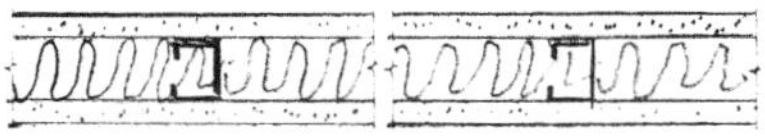	1⅝" steel studs, 24" o.c., ⅝" Type X gypsum wallboard both sides[3] (1 hr. fire rated), 1½" fiberglass or mineral wool core.	22	29	45	49	51	51	45
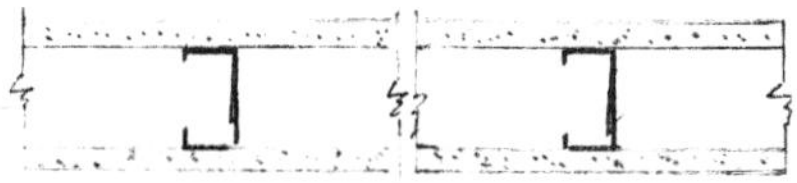	2½" steel studs, 24" o.c., ½" Type X gypsum wallboard both sides[3] (1 hr. fire rated).	18	25	42	47	51	41	37
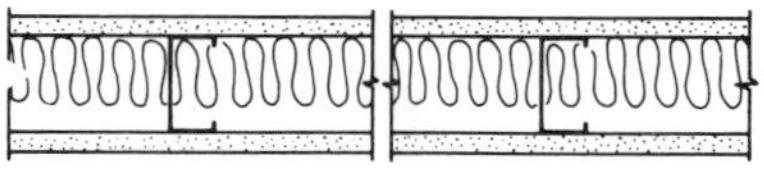	2½" steel studs, 24" o.c., ½" Type X gypsum wallboard both sides[3] (1 hr. fire rated), 2½" fiberglass (R-8) core.	21	35	48	55	56	43	45
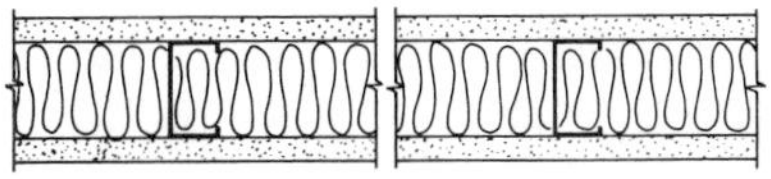	2½" steel studs, 24" o.c., ⅝" Type X gypsum wallboard both sides[3] (1 hr. fire rated), 2½" fiberglass (R-8) core.	26	41	52	54	45	51	45
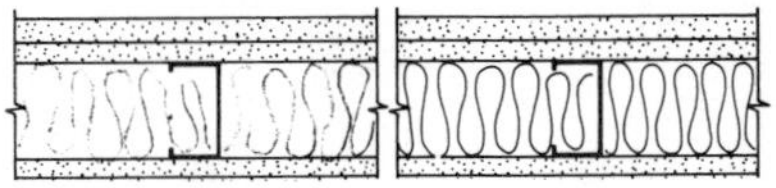	2½" steel studs, 24" o.c., 2 layers ½" gypsum wallboard one side, 1 layer on other side, 2½" fiberglass insulation core.	30	42	51	59	62	51	50

2½" steel studs, 24" o.c., 2 plies ½" gypsum board each side[3]. — 25 35 46 53 55 51 **46**

2½" steel studs, 24" o.c., 2 plies ½" Type X gypsum wallboard each side (2 hr. fire rated), 2½" fiberglass insulation core[3]. — 36 49 60 62 64 58 **58**

2½" steel studs 24" o.c., ¼" gypsum board base both sides, ½" gypsum wallboard, faces, — 31 42 51 55 54 50 **50**

2½" steel studs 24" o.c., ¼" gypsum board base both sides, ½" gypsum wallboard, faces, 3" mineral fiber insulation in core. — 35 48 55 58 55 54 **53**

3⅝" steel studs, 24" o.c., ½" gypsum wallboard both sides. — 16 23 37 45 46 37 **36**

3⅝" steel studs, 24" o.c., ½" gypsum board both sides, 3½" fiberglass or mineral wool core. — 24 35 47 54 57 43 **45**

3⅝" steel studs, 24" o.c., 2 ply ½" gypsum wallboard one side & 1 ply on other side, — 21 29 43 50 51 42 **41**

3⅝" steel studs, 24" o.c., 2 ply ½" gypsum wallboard one side, 1 ply on other, 3½" fiberglass core. — 33 44 52 60 63 53 **52**

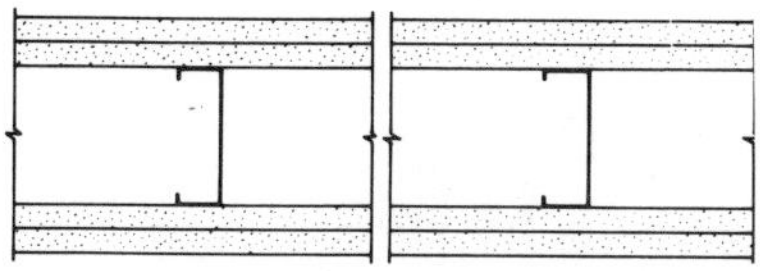 3⅝" steel studs, 24" o.c., 2 layers ½" gypsum wallboard on each side, (2 hr. rating - Type X)	30	40	49	55	58	52	**50**

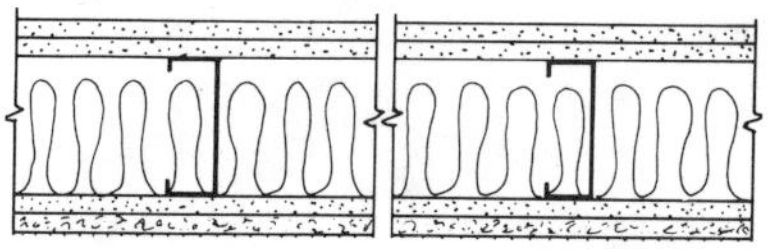 3⅝" steel studs, 24" o.c., 2 layers ½" gypsum wallboard on each side, (2 hr. rating - Type X) 3½" fiberglass core.	38	47	55	58	63	57	**56**

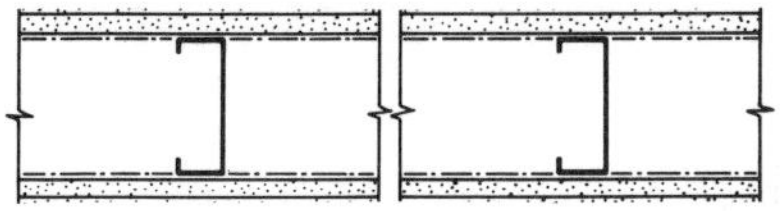 **14 ga. load bearing**, 3⅝" steel studs, 24" o.c., ⅝" gypsum wallboard each side[2].	25	36	41	47	38	42	**39**

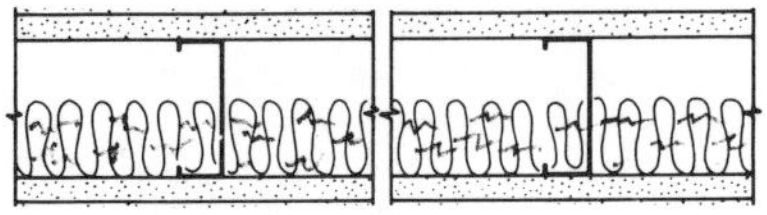 **14 ga. load bearing**, 3⅝" steel studs, 24" o.c., ⅝" gypsum board each side, 2" mineral wool insul. core[2].	31	32	44	48	38	40	**39**

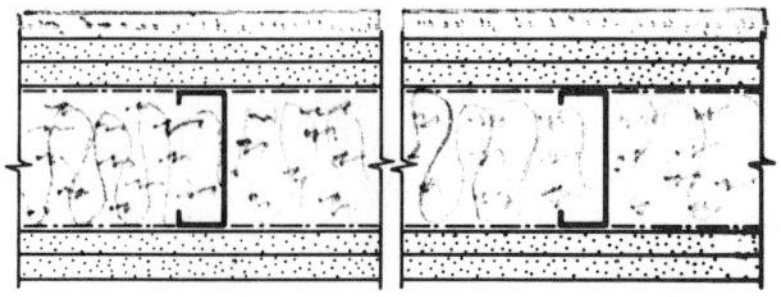 **20 ga.** 3⅝" steel studs 24" o.c., resilient furring channels 24" o.c. one side, 2 ply ½" gypsum wallboard over channels, 3 ply ½" gypsum board other side, 3" rock wool insulation core.	43	55	62	65	63	67	**62**

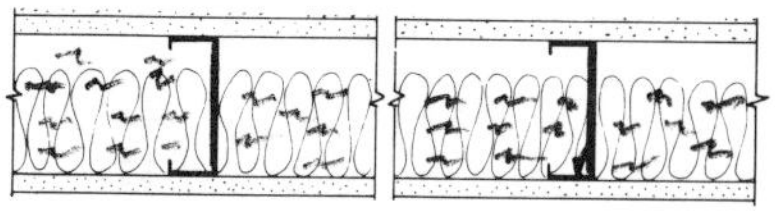 **14 ga. 6"** steel studs 24" o.c., ⅝" gypsum board both sides. 2" mineral fiber insul. core.	27	37	44	49	39	42	**40**

Sources:

1. Noise Control Manual, D. A. Harris, 1991, VNR publisher & D. A. Harris files.
2. United States Gypsum Co., 1995, Construction Selector & Design Data for Acousticians.
3. California Office of Noise Control, Sacramento, Calif., Catalog of STC and IIC Ratings for Wall & Floor Assemblies, 1981 & republished in 1987.

TABLE STL-3
FLOOR/CEILING ASSEMBLIES (STC & IIC)

2x8-10 wood floor joists 16" or 24" o.c.,
subfloor is plywood nailed and/or glued,
finish floors are vinyl, wood or carpet & pad
as noted and installed in standard manner.
Ceilings of gypsum wallboard applied perpendicular to
joists or furring channels, nails are spaced 8" o.c.,
screws are spaced 12" o.c., joints taped & spackled,
all perimeter is sealed airtight.
Test data obtained in ⅓ octave band frequencies.
Transmission Loss data presented in octave band
centers and STC. Impact Noise Rating (INC) is
given with and without carpet where available.
Data from California Office of Noise Control[1]
unless noted otherwise (#) at end of table.

OCTAVE BAND CENTER FREQ., HZ.

SKETCH	DESCRIPTION	125	250	500	1000	2000	4000	STC IIC
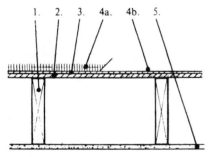	1. 2x8 wood joists 16" o.c. 2. ½" plywood nailed to joists. 3. ⅜" plywood nailed to joists. 4a. carpet & pad. 4b. no floor covering. 5. ½" Type X gypsum board nailed.	23	27	31	41	44	45	37 a. 66 b. 32
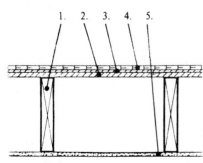	1. 2x10 joists, 16" o.c. 2. ⅝" plywood subfloor glued to joists & nailed 12" o.c. 3. ¼" particle board glued. 4. ½" parquet wood flooring glued. 5 ½" type X gypsum board with screws 12" o.c.	20	31	43	44	50	53	42 37

362

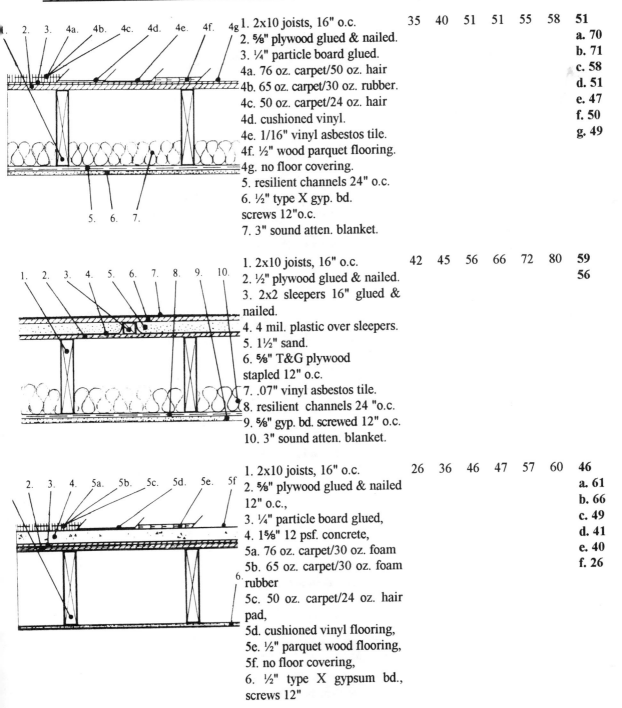

1. 2x10 joists, 16" o.c.   35 40 51 51 55 58 **51**
2. ⅝" plywood glued & nailed.   **a. 70**
3. ¼" particle board glued.   **b. 71**
4a. 76 oz. carpet/50 oz. hair   **c. 58**
4b. 65 oz. carpet/30 oz. rubber.   **d. 51**
4c. 50 oz. carpet/24 oz. hair   **e. 47**
4d. cushioned vinyl.   **f. 50**
4e. 1/16" vinyl asbestos tile.   **g. 49**
4f. ½" wood parquet flooring.
4g. no floor covering.
5. resilient channels 24" o.c.
6. ½" type X gyp. bd.
screws 12"o.c.
7. 3" sound atten. blanket.

1. 2x10 joists, 16" o.c.   42 45 56 66 72 80 **59**
2. ½" plywood glued & nailed.   **56**
3. 2x2 sleepers 16" glued & nailed.
4. 4 mil. plastic over sleepers.
5. 1½" sand.
6. ⅝" T&G plywood stapled 12" o.c.
7. .07" vinyl asbestos tile.
8. resilient channels 24 "o.c.
9. ⅝" gyp. bd. screwed 12" o.c.
10. 3" sound atten. blanket.

1. 2x10 joists, 16" o.c.   26 36 46 47 57 60 **46**
2. ⅝" plywood glued & nailed 12" o.c.,   **a. 61**
3. ¼" particle board glued,   **b. 66**
4. 1⅝" 12 psf. concrete,   **c. 49**
5a. 76 oz. carpet/30 oz. foam   **d. 41**
5b. 65 oz. carpet/30 oz. foam rubber   **e. 40**
5c. 50 oz. carpet/24 oz. hair pad,   **f. 26**
5d. cushioned vinyl flooring,
5e. ½" parquet wood flooring,
5f. no floor covering,
6. ½" type X gypsum bd., screws 12"

363

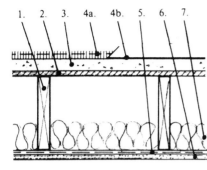

1. 2x10 joists, 16" o.c. | 40 | 51 | 57 | 66 | 73 | 84 | **61** |

2. ⅝" plywood subfloor nailed, **a. 79**
3. 1½" lightweight **b. 46**
concrete/15# felt
4a. 20 oz. carpet/40 oz. hair,
4b. 1/16" vinyl-asbestos tile,
5. resilient channels 24" o.c.,
6. ⅝" type X gypsum
bd./screws 12",
7. 3½" sound atten. blanket.

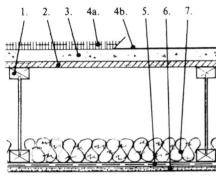

1. 12" plywood "I" beams, 24" | 43 | 50 | 53 | 58 | 67 | 73 | **58** |

o.c., **a. 77**
2. ¾" plywood subfloor nailed **b. 50**
6-10"
3. 1½" 15 psf. concrete,
4a. 44 oz. carpet/40 oz. hair,
4b. .07" vinyl floor tile,
5. resilient channels 24" o.c.,
6. ⅝" gyp. bd. screwed 12"o.c.,
7. 3" sound atten. blanket.

See sketch above

10" plywood web "I" beams | 37 | 44 | 56 | 65 | 64 | 69 | **57** |

24" o.c., **a. 64**
construction identical to **b. 40**
system described above except
concrete is replaced with:
**1" T&G damped plywood
nailed 8" o.c.[2]**
a. is carpet & pad,
b. is vinyl floor.

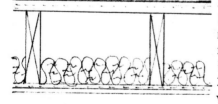

2x10 wood joists 16" o.c., | 27 | 41 | 55 | 64 | 62 | 67 | **51** |

**1" T&G damped plywood
nailed 8"** **34**
resilient channels 24" o.c.,
½" type X gyp. bd, screws 12"
3" mineral wool batt,
vinyl floor covering.
(note: footfall tests show
results equivalent to concrete[2])

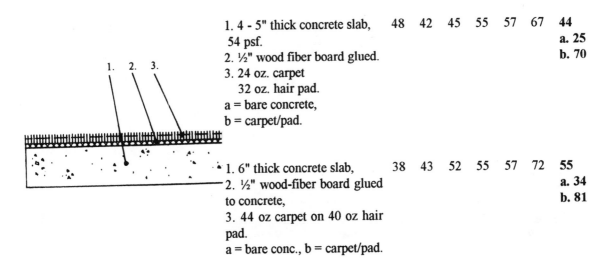

1. 4 - 5" thick concrete slab,   48   42   45   55   57   67   **44**
   54 psf.   **a. 25**
2. ½" wood fiber board glued.   **b. 70**
3. 24 oz. carpet
    32 oz. hair pad.
a = bare concrete,
b = carpet/pad.

1. 6" thick concrete slab,   38   43   52   55   57   72   **55**
2. ½" wood-fiber board glued   **a. 34**
to concrete,   **b. 81**
3. 44 oz carpet on 40 oz hair
pad.
a = bare conc., b = carpet/pad.

Sources:

1- California Office of Noise Control, Catalog of STC and NRC Ratings for Wall and Floor/Ceiling Assemblies, 1988
2- Greenwood Forest Products, Inc., Riverbank Acoustical Lab. tests in 1994.

TABLE STL-4
DOORS & DOOR ASSEMBLIES

Doors are operational and include frame
unless noted otherwise. Tests per ASTM E90
(i.e. door "shall be fully opened and closed
in a normal manner at least 5 times after
installation is completed and tested without
further adjustments"). The STC includes all
door system elements including the door, seals
& frame. Test source is noted by (#) & listed
at the end of table.

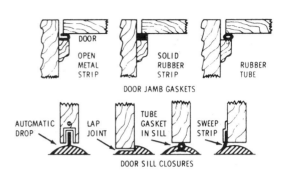

Door type & description	STC
1⅜" wood hollow core with flush faces designed for interior use[1].	
normal installation without seals	17
well sealed at edges and frame using weather strips, drop closure & caulked frame.	20
1¾" wood solid core, 3.9#/sq. ft.[1]	
normal installation without seals	20
well sealed with plastic or brass weather strips, drop closure & caulked frame.	28
1¾" acoustical wood door, 5 #/sq. ft.[2]	
manufacturers acoustical seals and center mounted drop seal	31
1¾" acoustical wood door, 5.2 #/sq. ft.[2]	
manufacturers acoustical seals and center mounted drop seal	36
1⅜" acoustical wood door, 5.8 #/sq. ft.[2]	
manufacturers acoustical seals and center mounted drop seal	38
1¾" acoustical wood door, 6.7 #/sq. ft.[2]	
manufacturers acoustical seals and center mounted drop seal	39
1¾" acoustical wood door, 7 #/sq. ft.[2]	
manufacturers acoustical seals and center mounted drop seal	45
2¼" acoustical wood door, 12.3 #/sq. ft.[2]	
manufacturers acoustical seals and center mounted drop seal	47
2¼" acoustical wood door, 15.8 #/sq. ft.[2]	
manufacturers acoustical seals and center mounted drop seal	47
(a rated/x ray shielding door)	

double doors composed of 2 1⅜" **hollow core** doors with 4" air space[1]
    normal installation without seals   22
    all edges sealed on both doors   26

double doors composed of 2 1⅜" **solid core** doors with 4" air space[1]
    normal installation without seals   28
    all edges sealed on both doors   40

double doors composed of 2 1⅜" solid core wood or steel doors,[1]
with 9" air space lined with sound absorbing material.
    normal installation without seals   42
    all edges sealed on both doors   50

2½" metal door with 16 ga. steel faces, cam hinges & magnetic seals[3]   47

Sources: 1. Noise Control Manual, D. A. Harris, VNR, 1991 and author files.
            2. Eggers Industries, Neenah, WI technical literature.
            3. Industrial Acoustics Co. Noise-Lock Door technical literature.

Table STL-5
WINDOWS AND WINDOW ASSEMBLIES

Windows are operational if listed in locked & unlocked columns.
Sealed STC means all potential sound leaks are blocked with a
permanently elastic acoustical sealant. Frame material is wood,
plastic wrapped wood or aluminum as indicated.
Window type is; standard double hung, storm sash, casement, sliding or fixed.
Glazing may be; single strength (ss), double strength (ds), or divided lights (dl).
Insulated or double glazing glass (ins) is shown with overall thickness.
Laminated glass (lam) is shown with overall thickness.

Frame	Type	Size	Glazing	SEALED STC	LOCKED STC	UNLOCKED STC
wood	double hung	3'x5'	ss	29		23
wood	double hung	3'x5'	ss/dl	29		
wood	double hung	3'x5'	ds	29		
wood	double hung	3'x5'	ds/dl	30		
wood	double hung	3'x5'	ins 7/16"	28	26	22
wood	fixed & sealed	6'x5'	ss/dl	28		
wood	fixed & sealed	6'x5'	ds	29		
wood	fixed & sealed	6'x5'	ins 1"	34		
wood/plastic	double hung	3'x5'	ss	29	26	26
wood/plastic	double hung	3'x5'	ins ⅜"	26	26	25
wood/plastic	storm sash	3'x5'	ds	30	27	
wood/plastic	storm sash	3'x5'	ins ⅜"	28	24	
wood/plastic	fixed casement	3'x5'	ds	31		
wood plastic	operable/casement	3'x5'	ds		30	22
wood/plastic	sliding		lam 3/16"	31	26	26

aluminum	sliding	ss	28	24	
aluminum	operable casement	ds	31	21	17
aluminum	single hung	ins 7/16"	30	27	25
test frame	single pane	ds 1/8"	30		
test frame	single pane	lam 1/4"	34		
test frame	single pane	lam ½"	38		
test frame	single pane	lam ¾"	40		
test frame	double glazed ⅛"	ins ½"	32		
test frame	double glazed ⅛"	ins 2"	38		
test frame	double glazed ⅛"	ins 4"	42		
test frame	double glazed ⅛"	ins 6"	44		
test frame	double glazed ¼"	ins 2"	42		
test frame	double glazed ¼"	ins 4¾"	46		
wood/plastic	sliding door	lam 3/16"	31	26	26
aluminum	sliding door	ss	28	24	--
aluminum	jalousie	lam ¼"	26	20	--

Source:
Noise Control Manual, D. A. Harris, VNR 1991 and author's files.

Table STL-6, EXTERIOR WALLS

Exterior walls are load bearing unless noted. Gypsum wallboard
nailed to wood studs 6"o.c. along edges and 8"o.c. in field.
Joints taped & spackled and perimeter sealed.
Octave band center data presented where available.

SKETCH	DESCRIPTION	OCTAVE BAND CENTER FREQ., HZ.						STC
		125	250	500	1000	2000	4000	
	2x4 wood studs 16" o.c. ½" wood fiber board sheathing, wood lap siding on exterior, ½" gypsum board interior.							37
	2x4 wood studs 16"o.c. ½" wood fiber board sheathing, wood lap siding on exterior, 3½" fiberglass (R-11) insulation, ½" gypsum board interior.							39
	2x4 wood studs 16"o.c. ½" wood fiber board sheathing, wood lap siding on exterior, reflective foil insulation, ½" gypsum board interior.							37
	2x4 wood studs 16"o.c. ½" wood fiber board sheathing, wood lap siding on exterior, 3" (R-11) rock wool insulation, ½" gypsum board interior.							38
	2x4 wood studs 16"o.c. ½" wood fiber board sheathing, wood lap siding on exterior, resilient furring channels 24"o.c., ½" gypsum board interior.							43
	2x4 wood studs 16"o.c. ½" wood fiber board sheathing, wood lap siding on exterior, resilient furring channels 24"o.c., 3½" fiberglass (R-11) insulation, ½" gypsum board interior.							47

2x4 wood studs 16"o.c.   **46**
1" cement/sand stucco on wire lath,
3½" fiberglass (R-11) insulation,
½" gypsum board interior.

2x4 wood studs 16"o.c.   **49**
1" cement/sand stucco on wire lath,
resilient furring channels 24" o.c.,
½" gypsum board interior.

2x4 wood studs 16"o.c.   **57**
1" cement/sand stucco on wire lath,
3½" (R-11) fiberglass insulation,
resilient furring channels 24" o.c.,
½" gypsum board interior.

2x4 wood studs 16"o.c.   **56**
brick veneer over sheathing,
3½" (R-11) fiberglass insulation,
½" gypsum board interior.

2x4 wood studs 16"o.c.   **54**
brick veneer over sheathing,
resilient furring channels 24" o.c.,
½" gypsum board interior.

2x4 wood studs 16"o.c.   **58**
brick veneer over sheathing,
3½" (R-11) fiberglass insulation,
resilient furring channels 24" o.c.,
½" gypsum board interior.

1. 4" face brick, mortared together[2]   32 34 40 47 55 51   **45**
2. 4" SCR brick mortared together   38 40 47 52 58 63   **51**

Table STL-6 continued

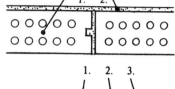

1. SCR brick, mortared together[2], 39 41 48 54 59 63 **53**
2. ½" gypsum/sand plaster on interior face.

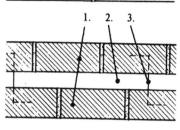

1. face brick, mortared together[2]. 37 37 47 55 62 66 **50**
2. 2" air space,
3. metal ties.

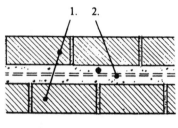

1. face brick, mortared together[2], 42 47 56 63 67 71 **59**
2. 2¼" cavity filled with concrete grout #6 bars vertically 48"o.c. & #5 bars horizontally 30" o.c.

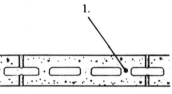

1. 4"x8"x16" 3-cell lightweight 25 28 38 41 48 51 **40**
concrete masonry units[2].
(17 #/block).

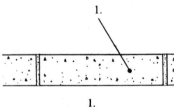

1. 4"x8"x16" solid lightweight 25 24 31 35 42 51 **35**
concrete masonry units.
(23# /block)[2].

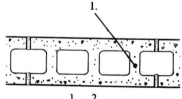

1. 6"x8"x16" 3-cell lightweight 37 34 41 46 54 57 **46**
concrete masonry units, (25#/block)[2],
2. paint both sides with primer-sealer coat and finish coat of latex.

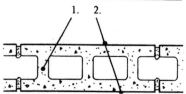

1. 6"x8"x16" 3-cell **dense** concrete 36 38 44 49 54 59 **48**
masonry units (36#/block)[2],
2. paint both sides with primer-sealer coat and finish coat of latex.

372

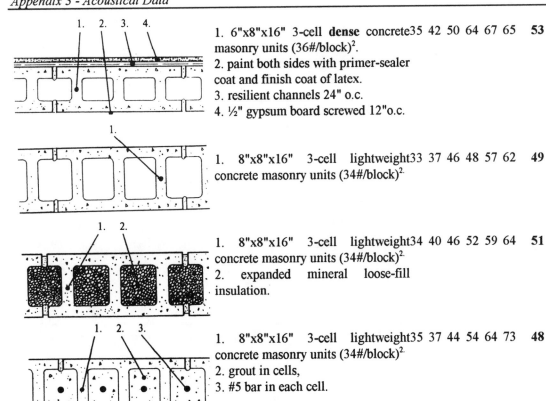

1. 6"x8"x16" 3-cell **dense** concrete 35 42 50 64 67 65 **53**
masonry units (36#/block)[2].
2. paint both sides with primer-sealer
coat and finish coat of latex.
3. resilient channels 24" o.c.
4. ½" gypsum board screwed 12"o.c.

1. 8"x8"x16" 3-cell lightweight 33 37 46 48 57 62 **49**
concrete masonry units (34#/block)[2.]

1. 8"x8"x16" 3-cell lightweight 34 40 46 52 59 64 **51**
concrete masonry units (34#/block)[2.]
2. expanded mineral loose-fill
insulation.

1. 8"x8"x16" 3-cell lightweight 35 37 44 54 64 73 **48**
concrete masonry units (34#/block)[2.]
2. grout in cells,
3. #5 bar in each cell.

1. 8"x8"x16" 3-cell lightweight 36 44 54 59 69 75 **55**
concrete masonry units (34#/block)[2.]
2. grout in cells,
3. #5 bar in each cell.
4. paint, 2 coats latex each side.

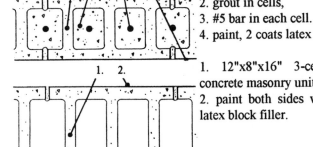

1. 12"x8"x16" 3-cell lightweight 34 42 44 51 58 60 **50**
concrete masonry units (43#/block)[2],
2. paint both sides with 3 coats of
latex block filler.

Sources:
1. Noise Control Manual, D. A. Harris, VNR, 1991
2. California Office of Noise Control, STC & IIC Ratings for Wall & Ceiling Assemblies, 1988.

Table STL-7 ROOF/CEILING ASSEMBLIES

The following test data was obtained by testing
three experimental homes constructed to be identical
in all respects except the type of insulation.
All 3 houses were basic single story, wood frame
construction - 3 bedrooms, 2 baths, living area & kitchen.
Roof/ceiling assembly - wood trusses 24" o.c.,
½" plywood sheathing & fiberglass shingles.
Ceiling - ½" gypsum board taped & spackled.
Fiberglass building insulation is R-11, R-19 & R-38
(1½", 3½", 6" thick).
Data obtained using steady state broadband noise source
located outside and via aircraft flyover test using
a Lear jet flyby at 250', 500' & 1000' altitude.
Steady state results are shown as STC.
Flyover results are shown as NIC.

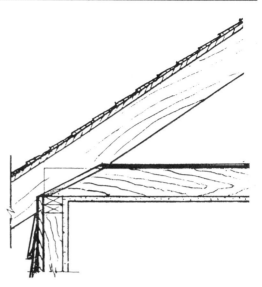

Location in House	SOUND TRANSMISSION CLASS (STC)/NOISE ISOLATION CLASS (NIC)		
	No insulation	R-19 insulation	R-38 insulation
Steady State Tests - STC			
living room	30	33	35
front bedroom	37	39	40
corner bedroom	32	36	37
rear bedroom	32	35	36
Aircraft Flyover Tests - NIC			
living room			
1000 feet altitude	34	34	36
500 ft.	34	35	37
250 ft.	34	35	37
rear bedroom			
1000 ft.	33	37	39
500 ft.	33	38	40
250 ft.	34	39	41
kitchen			
1000 ft.	30	33	--
500 ft.	29	32	--
250 ft.	31	33	--

(NOTE: For other types of roof ceiling assemblies see floor/ceiling assemblies with similar construction
details. Generally, the addition of a built up roof and roof insulation will improve the STC from 0 - 3 points.)

Source: Tests conducted by Owens Corning Fiberglas on thermal research homes at Granville Ohio and
reported in presentation to Acoustical Society of America by D. A. Harris and Barry Wyerman, 1982.

Section III - DUCT AND DUCT LINER MATERIALS

Sound attenuation down the duct (i.e. HVAC ducts) is based on sound pressure levels measured in a reverberation room after sound passes through a 10 foot specimen and enters the reverberation room. Attenuation data may be higher or lower depending on the number of turns, distribution of sound energy in various propagating duct modes, length of lined (or unlined) duct sections (which create discontinuities in the boundary conditions along the perimeter) and exit conditions at duct terminations.

Table III-1 provides sound attenuation per ASTM Method E 477 at an air velocity of 2000 feet per minute. Attenuation down the duct (i.e. sound reduction) is given in dB per lineal foot of duct.

Table III-2 provides radiated noise reduction through the duct wall in dB when tested per Air Diffusion Council (ADC) Flexible Air Duct Test Code FD72 pgr. 3.2.3. This value is essentially the single pass sound transmission loss through the duct wall composed of a 10 ft. length of 24 ga. sheet metal duct with duct liner.

TABLE III-1
DUCT ATTENUATION along the duct

Values are reduction (attenuation) in decibels for 1 foot long duct. Multiply by length of duct in use to obtain total attenuation.

PRODUCT OR TYPE DUCT	P/A* (duct size)	125	250	500	1000	2000	4000
1" thick fiberglass duct liner, 1.5# density.	3 (12"x 24")	0.5	0.5	1.5	2.8	4.0	2.7
	4 (12"x 12")	0.6	0.8	2.0	3.4	3.9	3.6
	5 (8"x 8")	0.5	1.2	2.1	3.4	5.1	3.8
	6 (6"x 12")	0.2	1.1	3.4	3.5	3.9	3.7
	7 (6"x 6")	0.4	1.7	3.1	4.0	4.8	4.4
2" thick fiberglass duct liner, 1.5 # density.	3 (12"x 24")	0.9	1.1	3.2	4.6	3.5	2.6
	4 (12"x 12")	0.9	1.5	3.0	4.1	3.9	3.8
	5 (8"x 8")	0.5	1.8	3.4	4.7	5.3	4.1
	6 (6"x 12")	0.4	1.3	3.1	3.6	3.9	3.5
	7 (6"x 6")	0.4	1.8	3.7	4.2	4.8	4.4
1" thick fiberglass duct liner, 2# density.	3 (12"x 24")	0.7	0.6	1.7	2.9	4.1	2.8
	4 (12"x 12")	0.6	0.7	2.0	3.4	4.1	3.7
	5 (8"x 8")	0.2	1.1	2.1	3.5	5.3	3.8
	6 (6"x 12")	0.3	1.0	2.4	3.4	3.8	3.4
	7 (6"x 6")	0.4	1.7	2.9	3.9	4.6	4.4
2" thick semi-rigid fine glass fiber board with black coating one side.	3 (12"x 12")	0.4	0.5	1.7	4.4	3.8	2.2
	6 (6"x 12")	0.3	0.9	2.7	4.7	5.2	4.1
1" thick semi-rigid fine glass fiber board with black coating one side.	3 (12"x 24")	0.6	1.0	3.8	4.7	3.6	2.3
	6 (6"x 12")	0.4	2.0	4.1	4.7	5.1	3.7

OCTAVE BAND FREQUENCIES (Hz)

* P/A: The inside perimeter of a lined duct in feet divided by the cross sectional free area of the duct in square feet. (P/A = 3 is based on a 12" x 24" duct, P/A = 4 is based on a 12" x 12" duct, P/A = 5 is based on a 8" x 8" duct, P/A = 6 is based on a 6" x 12" duct & P/A = 7 is based on a 6" x 6" duct.

Table III-2
RADIATED NOISE REDUCTION

Noise transmission through duct wall
(25 ga. sheet metal plus duct liner) in
dB reduction. (i.e. single pass sound
transmission loss through duct sidewall.)

PRODUCT OR TYPE DUCT	P/A* (duct size)	OCTAVE BAND FREQUENCIES (Hz)					
		125	250	500	1000	2000	4000
1" thick fiberglass duct liner,	3 (12"x 24")	10	14	20	25	31	32
1.5# density.	4 (12"x 12")	10	13	26	29	34	35
	5 (8"x 8")	5	13	22	26	30	31
	6 (6"x 12")	2	11	22	26	32	33
	7 (6"x 6")	2	10	14	20	24	29
2" thick fiberglass duct liner,	3 (12"x 24")	13	16	24	30	36	37
1.5 # density.	4 (12"x 12")	12	15	27	31	34	36
	5 (8"x 8")	7	17	24	26	30	33
	6 (6"x 12")	5	14	26	28	31	34
	7 (6"x 6")	3	11	16	20	25	28
1" thick fiberglass duct liner,	3 (12"x 24")	10	13	23	28	30	27
2# density.	4 (12"x 12")	10	13	26	30	34	36
	5 (8"x 8")	4	14	24	27	30	32
	6 (6"x 12")	2	14	24	27	30	32
	7 (6"x 6")	2	14	19	25	29	34
1" thick semi-rigid fine glass fiber	3 (12"x 12")	9	13	24	25	30	35
board with black coating one side.	6 (6"x 12")	4	16	26	30	33	36
1" thick semi-rigid fine glass fiber	3 (12"x 24")	12	17	28	30	37	42
board with black coating one side.	6 (6"x 12")	4	16	23	28	33	34

Source:
Noise Control Manual, D. A. Harris, VNR, 1991.

Appendix 4 - BIBLIOGRAPHY

ASTM Committee E33 *Annual Book of Standards* Volume 04.06 Thermal Insulation; Environmental Acoustics, PCN 01-04695-61, Philadelphia, PA, ASTM

Baron, R. A.   *The Tyranny of Noise*
                  New York & London: Harper & Row, 1970

Bell, L. H.   *Fundamentals of Industrial Noise Control.*
                  Trumbull, CT: Harmony, 1973.

Bell, L. H.   *Industrial Noise Control*
                  New York, NY: Marcel Dekker, 1982.

Beranek, L. L.   *Acoustic Measurements*
                  New York, NY: Wiley, 1949

Beranek, L. L.   *Acoustics*
                  New York, NY: McGraw-Hill, 1954

Beranek, L. L.   *Noise and Vibration Control*
                  New York, NY: McGraw-Hill, 1971

Brendt, R. D. and E. L. R. Corliss, E.L.R. *Quieting: A Practical Guide to Noise Control*
                  Washington, DC: National Bureau of Standards, 1976

Brendt, Winzer & Burroughs *A Guide to Airborne, Impact, and Structure Borne Noise-Control in Multifamily Dwellings* U. S. Dept. of Housing & Urban Development
                  Washington, DC: US Gov. Printing Office, for FHA, 1967

Broch, J. T.   *Mechanical Vibrations and Shock Measurements*
                  Marlborough, Massachusetts: Bruel and Kjaer

Crede, C. E.   *Vibration and Shock Isolation*
                  New York, NY: Wiley, 1951

Dupree, R.B.   *Catalog of STC and IIC Ratings for Wall and Floor Ceiling Assemblies*
                  Berkeley, CA: Office of Noise Control, Calif. Dept. of Health Services, 1987

Egan, D. M.   *Concepts in Architectural Acoustics*
                  New York, NY: McGraw-Hill, 1972

Faulkner, L.   *Handbook of Industrial Noise Control*
                  New York, NY: Industrial Press, 1975

Gypsum Assoc.   *Fire Resistance Design Manual* 12th edition
                  Washington, DC, Gypsum Association, 1988

Harris, C. M.   *Handbook of Noise Control*
                  New York, NY: McGraw-Hill, 1957

Harris, C. M. and C. E. Crede *Shock and Vibration Handbook*
                  New York, NY: McGraw-Hill, 1961

Harris, C. M.   *Handbook of Acoustical Measurements & Noise Control*, Third Edition
                  New York, NY: McGraw-Hill, 1991

Harris, D. A., et al *Planning and Designing the Office Environment*
                  New York, NY: Van Nostrand Reinhold, 1981 - 2nd edition (expanded) 1991

Harris, D. A. *Noise Control Manual*
                  New York, NY: Van Nostrand Reinhold, 1991

Harris, D. A.   *Sound Isolation Technology*
                 Fibreboard Corporation, San Francisco, CA, 1967
Jens, T. B.   *Acoustic Noise Measurements*
                 Marlborough, Massachusetts: Bruel and Kjaer, 1971
Irwin, J. D. and E. R. Graf *Industrial Noise and Vibration Control*
                 Englewood Cliffs, NJ: Prentice-Hall, 1979
Kinsler, L. E. and A. R. Frey *Fundamentals of Acoustics*
                 New York, NY: Wiley, 1950
Morse, P. M.   *Vibration and Sound, 1st ed.* 1936
Morse, P. M. and K. U. Ingard *Theoretical Acoustics*
                 New York, NY: McGraw-Hill, 1968
Owens Corning Fiberglas *Noise Control Manual* Pub. #5-BMG-8277-G
                 Toledo, OH: Owens Corning Fiberglas, 1986
Peterson, A.P. and E. E. Gross Jr. *E.E. Handbook of Noise Measurement*
                 Concord, MA,: GenRad, 1972
Purcell, W. E.   *Systems for Noise and Vibration Control*
                 Westlake, OH: Sound and Vibration Magazine, August 1982.
Purcell, W. E.   *Materials for Noise Control*
                 Westlake, OH: Sound and Vibration Magazine, July, 1982
U. S. Gypsum   *Design Data for Acousticians*
                 Chicago, IL: United States Gypsum Company Pub. #CS-139USG 10-84

Appendix 5 - Noise Control Check List

APPENDIX 5 - NOISE CONTROL CHECK LIST

COMPLAINT	PROBABLE CAUSE	POTENTIAL REMEDY	REFERENCE SOURCE
"This room is noisy." or "It is too difficult to concentrate in this room."	1. Noisy appliances, or other noise sources.	Eliminate or reduce noise source: Turn it off, buy quiet equipment, be sure moving parts are well balanced, damp vibrating panels, place on vibration isolation mount, surround with noise enclosure, install flexible connectors, reduce speeds.	Chapter 3 - Equipment noise. Appdx. 2 - Worksheets to calculate noise reduction effect. Appdx. 3. - noise reduction data.
	2. Room is excessively reverberant; there are echoes or an impact sound persists for several seconds.	Install sound absorbing material such as carpets, drapes, sound absorbing ceiling or wall treatment materials, soft furniture, etc.	Chapter 2 - Room acoustics & Sound Absorption. Appdx. 2 - Calculate reverberation time & noise reduction using worksheet. Appdx. 3 - absorption data.
	3. Outdoor noise intrusion; nearby freeway, railway, airport, factory, industry, pipeline, ball park, construction, etc.	Identify source/s. Contact local authorities to determine if noise ordinances apply and implement procedures to require compliance. If noise code compliance is not effective or additional noise quieting measures are necessary, upgrade sound attenuation performance of exterior envelope; caulk sound leaks, install gaskets on doors & windows, sound traps in ducts, upgrade STC of windows, doors, exterior walls and roofs.	Chapters 6 - Identify responsible entity, regulations & examples of compliance. Chapter 6.5 - design tools. Chapter 7 - case study examples. Appdx. 2 - worksheet to calculate transmission loss of a composite partition. Appdx. 3 - STC data on systems of construction.

Appendix 5 - Noise Control Check List

COMPLAINT	PROBABLE CAUSE	POTENTIAL REMEDY	REFERENCE SOURCE
"I hear my neighbors through the walls & ceiling in my condo." (or apartment)	1. Sound leaks via cracks at wall and/or ceiling connections, back to back utilities/ cabinets & air ducts serving both units.	Seal all cracks, penetrations, joints, etc. Eliminate common service access & back to back installation. Surface mount all cabinets. Relocate outlet boxes & stuff with insulation. Relocate common HVAC outlets, line ducts with F/G, install sound traps.	Chapter 2.10 - Airborne Flanking Paths. Chapter 4 - Regulations/Codes Chapter 5 - TL & caulking. Chapter 7.3 - Quiet design --.
I hear my neighbors through walls & ceiling in my condo." (or apartment)	2. Common wall or floor/ceiling separating living units has insufficient STC or has windows or doors that lower the effective STC.	Eliminate all windows and doors or install new units with STC appropriate to the need. Upgrade wall or floor/ceiling system: add a second wall or false ceiling with a sound absorbing core, add resilient furring channel, another layer of wallboard and sound absorbing core, upgrade system with better STC rating.	Chapter 2 - Sound isolation. Chapter 4 - Regulations/codes Chapter 5.2, 5.8 & 5.9 - Barriers Chapter 7.3 & 7.5 - Case Studies Appdx. 3 - Wall & floor STC's
"Footstep noises from the apartment above annoy me."	Impact sounds are prevalent due to poorly constructed floor/ceiling system.	Install carpet and pad in source area. If floor must be hard surface, tear up floor and install damped plywood subfloor and/or lightweight concrete before installing finish floor. Measures to improve airborne noise as described in 2 above will also improve impact noise. Upgrade system IIC.	Chapter 2.5 - Design techniques Chapter 4 - Regulations/codes Chapter 5.9 - Floor design Chapter 7.3 & 7.5 - Case Studies

380

Appendix 5 - Noise Control Check List

COMPLAINT	PROBABLE CAUSE	POTENTIAL REMEDY	REFERENCE SOURCE
"I hear whistling noise."	High frequency sounds that are constant due to air, water or gas passing a constricted area in a pipe or duct.	1. Air noise is usually generated at HVAC outlet grill. Open or remove grill to identify source. Replace with grill that has more open area or better closure seal. Remove pipe constriction/s, reduce air flow (i.e. lower fan speed), or line duct with sound absorbing material. 2. Water and gas noise occurs generally at a valve or faucet. Replace worn or defective washers and valve seals. Remove constrictions, reduce pressure and areas of potential flow turbulence.	Chapter 3.3, 3.4 & 3.5 - HVAC, Appliance & Plumbing Noise.
"Airplanes interrupt my conversation and sleep."	Flight patterns too near residential neighborhood.	If outside and inside noise level are excessive contact airport management, determine noise criteria, measure levels and implement compliance measures. If criteria is insufficient, upgrade STC of acoustical envelope. (i.e. Improve walls, roof, windows, doors, ducts, etc.)	Chapter 4.2 - Regulations Chapter 6 - Environment noise Chapter 7.8 - Residential sound insulating program.
"Let's design this community to be quiet."	Up front effort to control noise is highly commendable, has proven very effective and produces excellent dB/$.	Prepare a "Noise Element" for an Environmental Impact Study (EIS) and implement the findings. Adopt & enforce a Noise Ordinance for the community. Implement good noise design in the design stages of project and buildings.	Chapter 4 - Regulations, codes Chapter 6 - Environment noise Chapter 7.6 - Noise assessment case study.

381

Appendix 5 - Noise Control Check List

COMPLAINT	PROBABLE CAUSE	POTENTIAL REMEDY	REFERENCE SOURCE
"I hear rumbling, as if there is an earthquake"	Low frequency noise can be caused by trains, industry, aircraft or a nearby sound system. It is usually transmitted via the ground or structure.	Train and industrial noise can be reduced by mounting tracks or equipment on resilient mounts or by adding resilient pads to structural supports for building. Low frequency aircraft noise and sonic boom noise is typically left over after other efforts to reduce the sound transmission loss of the building envelope have been implemented. Best solution is to eliminate source. Other measures are best left to the acoustician.	Chapter 4 - Regulations/Codes Chapter 6 - Environment noise Chapter 7.7/7.8 - Case studies
"Street and freeway noises are lowering the value of my new development."	Busy street and/or freeways are too close to the residential area, but the land is cheap and desirable.	Relocate the project or live with the consequences. Some mitigation is possible by utilizing good noise control design practices in siting the buildings, using natural and constructed noise barriers, designing acoustical envelopes with superior STC and utilizing masking sound principals.	Chapter 1 - Noise control design considerations Chapter 4 - Regulations/Codes Chapter 6 - Environment noise Chapter 7.6,7.8 - Case Studies
"My neighbor's pool pump and air conditioner drive me nuts."	Equipment is improperly selected and located.	Buy quiet equipment, locate it where it will not impact neighbor's outdoor living area or bedroom window. If unit cannot be quieted by these means, utilize vibration damping techniques, mufflers or enclose it with a noise barrier.	Chapter 3 - Equipment noise Chapter 4 - Regulations/codes Chapter 5.3 - Source quieting Chapter 7.1 - Case study Appdx. 2 - Design Worksheets
"What is the difference between STC and NRC?"	Sound Transmission Class (STC), Noise Reduction Coef. (NRC)	Block sound by building a barrier with a good STC rating. Add sound absorbing material with high NRC to improve room acoustics & reduce echoes.	Chapter 1- Noise Control Principles.

382

INDEX

(Page number where found.   + means additional pages follow.)